FROM OBSERVATIONS TO SELF-CONSISTENT MODELLING OF THE ISM IN GALAXIES

FROM OBSERVATIONS TO SELF-CONSISTENT MODELLING OF THE ISM IN GALAXIES

A JENAM 2002 Workshop
Porto, Portugal
3–5 September 2002

Edited by:
MIGUEL A. de AVILLEZ
University of Évora,
Évora, Portugal

DIETER BREITSCHWERDT
Max-Planck-Institut für Extraterrestrische Physik
Garching, Germany

Reprinted from *Astrophysics and Space Science*
Volume 289, Nos. 3–4, 2004

KLUWER ACADEMIC PUBLISHERS
DORDRECHT / BOSTON / LONDON

Library of Congress Cataloging-in-Publication Data is available

ISBN 1-4020-1939-4 (alk. Paper)

2003055495

Published by Kluwer Academic Publishers,
P.O. Box 17, 3300 AA Dordrecht, The Netherlands.

Sold and distributed in the U.S.A. and Canada
by Kluwer Academic Publishers,
101 Philip Drive, Norwell, MA 02061, U.S.A.

In all other countries, sold and distributed
by Kluwer Academic Publishers,
P.O. Box 322, 3300 AH Dordrecht, The Netherlands.

Printed on acid-free paper

Printed in the Netherlands

TABLE OF CONTENTS

TURBULENCE, MAGNETIC FIELDS AND CRs

EXTRAPLANAR ISM AND IGM

SELF-CONSISTENT MODELLING OF THE ISM

List of Posters (CDROM)

BIRK / The Magnetic Buoyancy Instability In Partially Ionized Plasmas

BOOMSMA/OOSTERLOO/SANCISI/VAN DER HULST / The Effervescent Disk of NGC 6946

DAMIANI/BOCCHINO/SCIORTINO/MICELA/HARNDEN / Chandra X-Ray Observation of NGC 604

EHLE/DAHLEM / The Quest for Halo Emission in the Starburst Galaxy NGC 1511

ELWERT/DETTMAR / Stellar Input Spectra as a Significant Parameter in Simulations of the Diffuse Ionized Gas

GODDI/MOSCADELLI/ALEF / EVN Observations of H_2O Masers towards the High-Mass Young Stellar Object AFGL5142

GÓMEZ/COX / 3D MHD Modeling of the Gaseous Structure of the Galaxy

HAMIDOUCHE / Simulation of ESEs by Kolmogorov Turbulence in the Ionised ISM

KNUDE/FABRICIUS / Extinction and Distance from a Tycho-2 Calibration of Michigan dwarfs. LDN 1622 a Nearby Molecular Cloud?

KOSIŃSKI/HANASZ / Parker Instability under the Influence of Cooling and Heating Processes in the ISM

KOWAL/HANASZ/OTMIANOWSKA-MAZUR / Parker Instability, Resistivity and Magnetic Diffusion in Galactic Disks

MARCOLINI/BRIGHENTI/D'ERCOLE / 3D Simulations of Galactic Winds in Dwarf Galaxies

MONTERO-CASTAÑO/GÓMEZ/MORALES/BOWYER/EDELSTEIN/KORPELA / Extreme and Far UV Spectroscopic Observations of the ISM

NIETEN/NEININGER/GUÉLIN/LUCAS/UNGERECHTS/BECK/BERKHUIJSEN/WIELEBINSKI / High-Resolution CO(1-0) Survey of the Andromeda Galaxy

OTMIANOWSKA-MAZUR/HANASZ/KOWAL / Topological Evolution of Parker Unstable Galactic Magnetic Fields, Turbulent EMF Calculations

PYLSKAYA/WIEBE / Open Cluster Distance Scales and the Chemical Evolution of the Galaxy

WÜNSCH/PALOUŠ/EHLEROVÁ / Hydrodynamic Simulations of Expanding Shells – The Gravitational Instability

LIST OF PARTICIPANTS

Alves, J., jalves@eso.org
ESO, Garching, Germany
Ballesteros-Paredes, J., j.ballesteros@astrosmo.unam.mx
UNAM, Morelia, Mexico
Beck, R., p181bck@mpifr-bonn.mpg.de
MPI für Radioastronomie, Bonn, Germany
Berkhuijsen, E.M., eberkhuijsen@mpifr-bonn.mpg.de
MPI für Radioastronomie, Bonn, Germany
Birk, G., birk@usm.uni-muenchen.de
University of Munich, Germany
Boomsma, R., boomsma@astro.rug.nl
Kapteyn Astronomical Institute, The Netherlands
Bregman, J.N., jbregman@ra.astro.lsa.umich.edu
University of Michigan, USA
Breitschwerdt, D., breitsch@mpe.mpg.de
MPE, Garching, Germany
Brinchmann, J., jarle@MPA-Garching.MPG.DE
MPI für Astrophysik, Garching, Germany
Brüll, M., bruell@ph1.uni-koeln.de
University of Köln, Germany
Cardwell, A., cardwell@ll.iac.es
Instituto de Astrofísica de Canarias, Spain
Comerón, F., fcomeron@eso.org
ESO, Garching, Germany
Cox, D.P., cox@wisp.physics.wisc.edu
University of Wisconsin, Madison, USA
Dahlem, M., mdahlem@eso.org
ESO, Chile
Damiani, F., damiani@oapa23.astropa.unipa.it
Osservatorio Astronomico di Palermo, Italy
de Avillez, M.A., mavillez@galaxy.lca.uevora.pt
University of Evora, Portugal
Dettmar, R.-J., dettmar@astro.ruhr-uni-bochum.de
Rurh-University Bochum, Germany

Astrophysics and Space Science **289:** ix–xii, 2004.

Dib, S., dib@mpia-hd.mpg.de
MPI für Astronomie, Heidelberg, Germany
Dorfi, E., ead@astro.univie.ac.at
Institut für Astronomie, Wien, Austria
Ehle, M., mehle@xvsoc01.vilspa.esa.es
XMM-Newton Science Operations Centre, Spain
Ehlerova, S., sona@ig.cas.cz
Astro. Inst., Acad. of Sc., Czech Republic
Elwert, T., elwert@astro.ruhr-uni-bochum.de
Ruhr-Universität Bochum, Germany
Ferlet, R., ferlet@iap.fr
Institut d'Astrophysique de Paris, France
Fraternali, F., ffratern@astro.rug.nl
Osservatorio Astronomico di Bologna, Italy
Freyberg, M.J., mjf@xray.mpe.mpg.de
MPE, Garching, Germany
Gazol, A., a.gazol@astrosmo.unam.mx
Instituto de Astronomia, UNAM, Mexico
Goddi, C., cgoddi@ca.astro.it
Osservatorio Astronomico di Cagliari, Italy
Gomez-Reyes, G.C., gomez@wisp.physics.wisc.edu
University of Wisconsin, Madison, USA
Hamidouche, M., hamidouc@odyssee.cnrs-orleans.fr
LPCE-CNRS, Orleans, France
Hanasz, M., Michal.Hanasz@astri.uni.torun.pl
Nicholas Copernicus University, Poland
Harfst, S., supas122@astrophysik.uni-kiel.de
University of Kiel, Germany
Hebrard, G., hebrard@iap.fr
Institut d'Astrophysique de Paris, France
Heitsch, F., heitsch@solarz.colorado.edu
JILA, University of Colorado, USA
Hojaev, A., ash@astrin.uzsci.net
UBAI/UAS, Uzbekistan
Jenkins, E.B., ebj@astro.Princeton.EDU
Princeton University Observatory, USA
Jungwiert, B., bruno@ig.cas.cz
Astro. Inst., Acad. of Sc., Czech Republic
Kalberla, P.W., kalb@astro.uni-bonn.de
University of Bonn, Germany
Kendall, T., tkendall@oal.ul.pt
University of Lisbon, Portugal

Knude, J., indus@astro.ku.dk
Niels Bohr Institute, Denmark
Komossa, S., skomossa@xray.mpe.mpg.de
MPE, Garching, Germany
Konz, C., konz@usm.uni-muenchen.de
Ludwig-Maximilians-University, Germany
Korpi, M.J., mkorpi@sun3.oulu.fi
University of Oulu, Finland
Kosinski, R., Rafal.Kosinski@astri.uni.torun.pl
Nicholas Copernicus University, Poland
Kowal, G., kowal@oa.uj.edu.pl
Jagiellonian University, Poland
Lazarian, A., lazarian@astro.wisc.edu
University of Wisconsin, Madison, USA
Morales, C., morales@laeff.esa.es
INTA, Spain
Mac Low, M.-M., mordecai@amnh.org
AMNH, New York, USA
Marco, O., omarco@tmpdhcp217.sc.eso.org
ESO, Chile
Marcolini, A., andrea@giano.bo.astro.it
University of Bologna, Italy
Morganti, R., morganti@nfra.nl
ASTRON, Dwingeloo, The Netherlands
Oey, S., Sally.Oey@lowell.edu
Lowell Observatory, USA
Oosterloo, T., oosterlo@nfra.nl
ASTRON, Dwingeloo, The Netherlands
Otmianowska-Mazur, K., otmian@oa.uj.edu.pl
Jagiellonian University, Poland
Paladini, R., paladini@ap.sissa.it
S.I.S.S.A., Italy
Paolillo, M., paolillo@na.astro.it
Osservatorio di Capoedimonte, Italy
Plana, H., hplana@yahoo.com.br
Observatorio Nacional – MCT, Brazil
Rosen, A., rar@star.arm.ac.uk
Armagh Observatory, Northern Ireland
Rosolowsky, E., eros@astron.berkeley.edu
University of California at Berkeley, USA
Samland, M., samland@astro.unibas.ch
University of Basel, Switzerland

Sasaki, M., manami@xray.mpe.mpg.de
MPE, Garching, Germany
Schindler, S., Sabine.Schindler@uibk.ac.at
University of Innsbruck, Austria
Shadmehri, M., MShadmehri@science1.um.ac.ir
Ferdowsi University, Iran
Shukurov, A., anvar.shukurov@ncl.ac.uk
University of Newcastle, Newcastle, UK
Smette, A., alain.smette@ulg.ac.be
University of Liege, Liege, Belgium
Smith, M.D., mds@arm.ac.uk
Armagh Observatory, Northern Ireland
Tanaka, Y., ytanaka@mpe.mpg.de
MPE, Garching, Germany
Trinchieri, G., ginevra@brera.mi.astro.it
INAF-Osservatorio Astronomico di Brera, Italy
Xilouris, M., xilouris@discovery.saclay.cea.fr
CEA Saclay, France
Wakker, B., wakker@astro.wisc.edu
University of Wisconsin, Madison, USA
Wiebe, D., dwiebe@inasan.rssi.ru
Urals State University, Russia
Wong, T., Tony.Wong@atnf.csiro.au
CSIRO Australia Telescope National Facility, Australia
Wunsch, R., richard@ig.cas.cz
Astronomical Institute, Academy of Sciences, Czech Republic
York, D.G., don@oddjob.uchicago.edu
University of Chicago, Chicago, USA

FROM OBSERVATIONS TO SELF-CONSISTENT MODELING OF THE ISM IN GALAXIES

JENAM 2002

PREFACE

Although a classical field of Astronomy, research of the interstellar medium (ISM) has been advancing rapidly during the last 10 years, mainly due to immensely improved observational facilities and techniques in all wavelength ranges. We are now able to investigate the ISM in external galaxies and even the intergalactic and intracluster medium in great detail. Increased spatial and spectral resolution have provided us with a great deal of information on the interstellar gas in its various phases, the magnetic field and the cosmic rays, and of course, also the stellar component, which is the driving agent of the interstellar matter cycle.

In this respect, it is somewhat embarassing for us theoreticians (to which the organizers belong) that modelling of the *global ISM* has not progressed all that far from the original models of the 70's and 80's (e.g. three-phase ISM or chimney model). Of course we are aware and appreciate the excellent work of many of our colleagues who have contributed to the understanding of many *phenomena*, but the global picture is still vague. We believe that the reason lies in the inherent complexity of the ISM, in the non-linear feedback and the interaction of virtually all of its components. Therefore analytic modelling cannot be carried very far, and semi-analytic and full-blown numerical codes are required to gain more insight. Since many processes operate on vastly different time and length scales this poses a tremendous challenge, and *spatial resolution* amongst others is a necessary requirement. Only fairly recently, a sufficient amount of computing power has become available to individual scientists to tackle these problems with some prospect of obtaining a more realistic picture. Still many, in some cases drastic, approximations have to be made to run simulations at sufficient resolution over realistic time scales.

A major goal of this workshop was therefore to bring together observers and theoreticians, sufficiently close so that intense discussion about the necessities and desiderata of modelling could be stimulated. In retrospective we are very happy and pleased to have had such an enthusiastic audience who really challenged the excellent speakers in numerous discussions. As evidence for this we consider the well attended and two hours lasting panel discussion that we held on Friday evening, when usually many people prefer to be at home and with their families.

Quite naturally there is still some controversy about which path towards a self-consistent picture of the ISM is the most promising to go in future. Nevertheless, a few interesting suggestions emerged. In our own, and consequently not unbiased view, these were the inclusion and proper description of turbulence in the ISM,

Astrophysics and Space Science **289:** 177–178, 2004.

magnetic fields and cosmic rays, the ionization history of plasmas out of equilibrium and magnetic reconnection. Moreover, we still need to explain the volume filling factors of the ISM gas phases, the existence and maintainance of diffuse ionized extraplanar gas and magnetic fields, the X-ray and EUV emission of the Local Bubble and superbubbles and why they are not consistent with similarity solutions and the nature and dynamics of high velocity clouds, just to name a few. This embarassingly long list will certainly ensure an ongoing discussion, which we hope to have stimulated with this workshop. We look forward to similar meetings in the future.

We would like to thank all participants for their attendance, all the speakers and poster contributors for their effort and excellent presentations, the chairmen for their patience and rigour in keeping the speakers in time, the SOC for their valuable and competent advice in setting up an interesting program, and all the people from the JENAM 2002 LOC for their help and advice without which this meeting would not have been possible.

We thank the European Astronomical Society, the Portuguese Astronomical Society (PAS) and the Center of Astrophysics of the University of Porto (CAUP) for allowing us to organize this workshop under the auspices of the EAS and JENAM. We would like also to thank Drs. Mário Monteiro, the head of PAS, and Jorge Filipe Gameiro for all the support in the preparation of and during the meeting. We also thank to the systems managers of the CAUP for keeping everything running and solving all the problems related to notebook presentations. Finally, we are grateful for finnancial support from the Portuguese Science Foundation.

Guest Editors May 2003

Miguel A. de Avillez (Évora, Portugal)
and
Dieter Breitschwerdt (MPE Garching, Germany)

OPENING TALK

THE NEED FOR A SELF-CONSISTENT MODEL OF THE ISM

JOEL N. BREGMAN
University of Michigan; E-mail: jbregman@umich.edu

Abstract. The interstellar medium of the Milky Way possesses a remarkable range of components, which have individual lifetimes considerably less than a Hubble time. Consequently, there has been a major effort in understanding the properties of these components, their heating and cooling mechanisms, and the process by which one component is transformed into another. We review several of the issues in an effort to highlight controversies that are under study presently. One of the central debates involves star formation, where there are two general and conflicting theoretical models. Another issue deals with the efforts to understand the fractional amounts of gas in the various phases, which brings together star formation with global gas dynamical reactions. This interaction is responsible for heating of the gas by supernovae, causing superbubbles, and subsequently, either galactic winds or galactic fountains. The relative amounts of material in the hot and cold phases not only depends upon the supernova rate, it also depends upon details of interstellar models, such as whether thermal conduction is an important process. We review the progress that people have made on self-consistent modeling of the ISM. The constraints on such models come from a wide variety of observations of molecular, neutral atomic, warm ionized, and X-ray gas. Both observations and theory are represented at this meeting and we highlight areas in which contributions are likely to resolve the existing controversies.

Keywords: ISM, models

1. Introduction

Occasionally a friend of mine will remind me that part of the study of the interstellar medium is much like the study of weather, which he says in a derogatory tone. However, I remind my friend that one of his favorite television channels is *The Weather Channel*, and he defends his fondness by telling me that knowing the weather is essential to planning his day. It doesn't really affect how he will teach his class that day, and he really watches because it is interesting. There are some similarities to our work in that it is simply an interesting field with wide-ranging implications. Of course, one of the big differences is that we do not have a television channel dedicated to the interstellar medium in galaxies and I wonder why that has not happened yet. On *The Weather Channel*, there are detailed forecasts based on numerical hydrodynamic models, and this seems to be one of the big attractions. So how far away are we from doing something similar: self-consistent models of the interstellar medium. This is the general topic of this conference, which has several parts. I will begin with a general census of the interstellar content of the Milky Way, with emphasis on some of the more controversial issues. Then, I will discuss the progress in the area of interstellar modeling, which has been more rapid then I would have predicted and is a testament to the inventiveness of the investigators.

2. The Census of the Galactic ISM

I will move from the components that I think we know the most about to those that we know least about, trying to highlight some of the outstanding issues involved.

2.1. Neutral Atomic Gas

The component for which we have the most knowledge is probably HI, the cool atomic component of the ISM. Part of the reason for this is that 21 cm emission provides a linear measure of this gas component, so we have good masses, which is nearly a unique situation. For the other components, we measure either a surface brightness through an emission measure ($n^2 l$, where l is the lengthscale) or we measure an opaque surface (CO emission). But for HI, not only is the optical depth low, but the velocity of the gas relates to the location in the disk of the Galaxy, so we obtain distance information as well.

Whereas the early picture of the HI was a "raisin-pudding" model with cool dense clouds imbedded in a more dilute intercloud, the current picture is that the cold, dense HI (10^2 K, 30 cm^{-3}) is largely sheet-like (Kulkarni and Heiles, 1988). The HI between the clouds is warmer and of lower density (300-3,000 K; 1 cm^{-3}), but unlike the early picture, this does not occupy all of the volume. Some of the most remarkable images such as the the one shown in Figure 1, have been obtained with the Canadian Galactic Plane Survey, where images and references can be found at http://www.ras.ucalgary.ca/CGPS/gallery/.

The pressure is provided not only by the usual gas pressure but also by magnetic pressure and turbulent motion. The temperature of the gas is determined by the balance of photoelectric grain heating (a photon ejects an electron from a grain, which heats the gas) and radiative cooling from forbidden fine-structure lines (e.g., CI 157 μm) that are collisionally excited.

Despite the great strides that have been made, I have a few items on my 'wish list' that I hope will be resolved in the future. The first has to do with the filling factor of the HI, where my best guess is 20–50%, based upon the comparison of HI studies to models (Bregman, Kelson and Ashe, 1992). The other issue has to do with the High Velocity Clouds of HI, where three mechanisms have been proposed: galactic fountains; infall of gas from the LMC and SMC; and a population of gravitationally bound HI clouds that lie throughout the Local Group. There is good observational evidence that the first two mechanisms occur. For the infall from the LMC and SMC, there is little doubt that the Magellanic Stream has been torn away, and there is also good evidence that other HVC gas originated from the LMC based on its location and metallicity arguments (Wakker, Oosterloo and Putman,

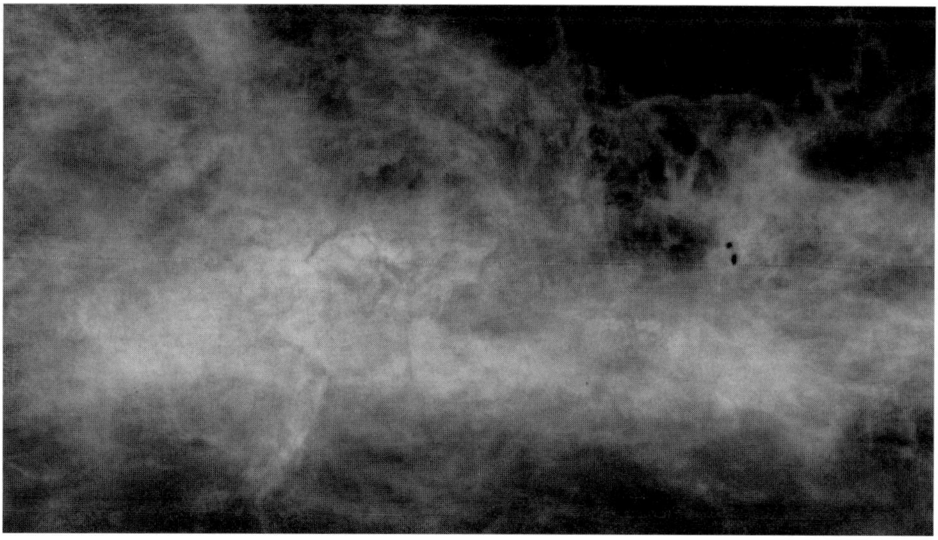

Figure 1. The HI total intensity map for the region of the Perseus arm in the Galactic plane (image size is 16° by 9°) from the Canadian Galactic Plane Survey, taken with the DRAO array. This shows shells, sheets, and dense regions (the two small holes toward the right side are star forming regions filled with molecular gas and are bright IRAS regions).

2002; Wakker, this meeting). For the galactic fountain picture, we detect the hot X-ray emitting halos that are needed (both from the Milky Way and around other galaxies). The third models, Local Group clouds, predict an ensemble of about 10^3 clouds of mass 5×10^7 $M_\odot$ with a typical distance of 500 kpc (Blitz et al. 1999). Some problems with this suggestion is that such clouds are not seen in other galaxy groups, and optical emission line studies place the clouds at a distance closer than 100 kpc and more likely 10-50 kpc (Weiner, Vogel and Williams, 2002; Tufte et al., 2002). However, the relative importance of these three mechanisms remains a hotly debated topic but one that can be resolved with further observations.

2.2. Warm Ionized Gas

The warm ionized gas is a component that constitutes a significant amount of volume and mass in the ISM yet the study of this component has been led but just a few people. It is a component where we have an extremely good measure of the mass along sightlines because the pulsar dispersion measures are linearly proportional to the column of electrons (Taylor and Cordes, 1993). With their model for these dispersion measures, they can determine distances to targets given a dispersion measure (see their updated model and tool at http://pulsar.princeton.edu/ pulsar/converters.shtml). These dispersion measures imply an electron column density that is about 1/3 of the HI column density, although with a scale height that is significantly higher, about 1.5 kpc (Reynolds, 1989). The volume filling factor can be estimated from the comparison of emission and the dispersion measure and it is

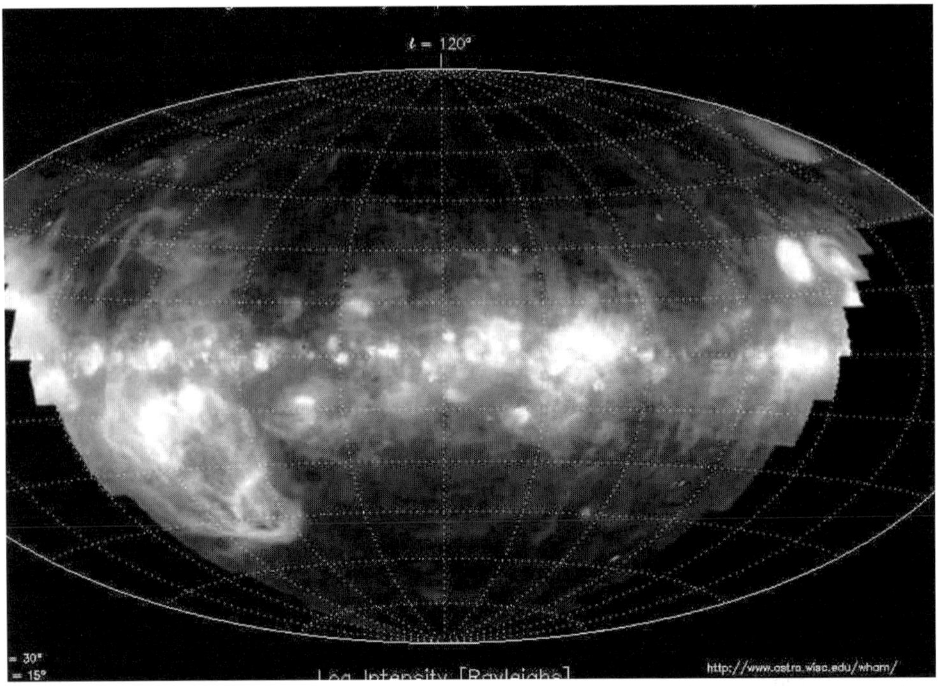

Figure 2. The Hα total intensity emission for part of the northern sky from the Wisconsin Hα Mapper (WHAM). In addition to small bright regions that are the classical HII regions, one see diffuse emission from all locations as well as large structures that denote shells and chimney walls.

found to be about 20% (Reynolds, 1991), or possibly greater. The average electron density of this HI gas is $n_e \approx 0.08$ cm^{-3}, and if the temperature is about 6×10^3 K, the pressure is p/k $\approx 10^3$ K cm^{-3}, which is below the typical gas pressure and is significantly below the effective pressure associated with the colder gas (when including magnetic fields and turbulence, which leads to p/k $\approx 10^{4.3}$ K cm^{-3}). Either these warm ionized regions are underpressure and are regions that are collapsing, or other forms of pressure (magnetic field, turbulence) plays a major role. A situation that could lead to underpressure warm ionized gas is the cooling of superbubbles from the hot phase (10^6 K), where they are in pressure equilibrium. Then, until these regions are compressed by their surroundings, they will have a pressure lower than the mean value for the ISM.

The WHAM survey (Reynolds et al., 1988) highlights lower density bubbles that are not the classical HII regions which one can see just from an examination of their survey data, shown in Figure 2 and updated on their home page http://www.astro.wisc.edu/wham/index.html. Some warm ionized regions are part of star-forming regions and photo-dissociation fronts (PDFs), which are among the most complex regions in the Galaxy, with the time-evolution of molecular, atomic, and ionized gas. Finally, we note that the heating of the warm ionized component is by photoionization while the cooling is largely from optical forbidden lines.

2.3. HOT IONIZED GAS

Gas in the 2×10^4-2×10^5 K range requires special conditions for gas to exist, whether it be in photoionization or collisional ionization equilibrium. The reason for this is that the cooling functions rise very quickly above 1×10^4 K and reaches a peak at about 2×10^5 K, so the heating must fall in only a narrow range for thermal equilibrium. However, the collisional cooling function decreases from 2×10^5 - 2×10^7 K, so a single value of the heating function can lead to two equilibrium points at different temperatures. Also, the gas in this temperature regime is subject to a thermal instability that will cause the gas either to heat up or to cool down. Gas at about 2×10^5 K has a relatively short cooling time so this situation leads to a relative paucity of gas in the temperature range 10^4-$10^{5.5}$ K.

Continuous heating, such as by photoionization, rarely produces gas above 10^4 K in the ISM, but shocks naturally produce hot gas in the $10^{5.5}$-$10^{7.5}$ K range. Supernovae, which dominate the energy input into the ISM, produce shocks with velocities of 100-1000 km/sec, and these shocks jump the gas temperature over the temperature regime of very rapid cooling and into the higher temperature regime.

The gas pressure of the shocked gas is greater than the surroundings, so it expands, as is seen in supernova remnants, superbubbles, and reheated low-density regions in the Galaxy. The all-sky mapping by *ROSAT* was essential for understanding the extent of the hot gas in the Milky Way. In addition to the Local hot bubble (100–200 pc), there are several other superbubbles of hot gas, possibly an extensive hot medium in the disk, along with a hot halo that is flattened (Snowden et al., 1998). The densities of this hot gas is typically 1-3×10^{-3} cm^{-3} and the typical temperature is 10^6 K. The volume filling factor of the hot component in the plane of the Galaxy remains a topic of debate and we will hear much about it at this meeting. Recent calculations place the value at about 20%, although the value is greater in the halo, where it may occupy most of the volume (de Avillez, 2000).

One of the new controversial topics relating to the hot component of Galactic gas is the extent of the halo and whether there is a hot group medium. Data has been accumulating for a few years, but the controversy really began last year when *Chandra* spectroscopic observations toward bright AGNs showed absorption lines of OVII and OVIII at zero redshift (Nicastro et al., 2002). Previously, OVI absorption was detected along many sight lines out of the Galaxy, which is detailed in a set of three comprehensive papers that describe the *FUSE* observations (Sembach et al., 2003; Wakker et al., 2003; Savage et al., 2003). Also, OVII was detected in emission in the X-ray band from a rocket flight (McCammon et al., 2002) and OVI was detected in emission in the UV band with *FUSE* (Shelton et al., 2000).

There are multiple ways of interpreting these data and one taken by Nicastro was to assume that the absorption by OVI, OVII, and OVIII were cospatial. This places several constraints on the nature of the ionization and they find that a pure collisional model will not fit the data, but they find that a low density (10^{-6} cm^{-3}) system in photoionization equilibrium can reproduce the data. Since the column densities for these high ionization oxygen species are well-determined (10^{16} cm^{-2}), this implies a size for the absorbing cloud of 4.9 /$[O/H]_{\odot}$ Mpc, so for a metallicity of 0.3, it implies a lengthscale of 15 Mpc, much larger than the local group. Their solution that has the highest density and smallest column permitted lead to a lengthscale of 3 Mpc. This corresponds to about 5×10^{13} $M_{\odot}$, more than an order of magnitude more mass than in all the known galaxies in the Local Group. This would have tremendous implications for the baryonic mass of the local universe if our Local Group is typical.

Alternatively, these three oxygen species may not be cospatial, as there is no compelling theoretical reason for such an arrangement. Instead, one might imagine that the OVII emission and absorption are cospatial, in which case, one obtains a minimum lengthscale of 50 kpc (Sanders et al., 2002). Then, the OVI absorption is produced in a lower temperature region and there are few constraints on the lengthscale; it could easily lie within 100 kpc. For completeness, the OVI emission is believed to originate from the lower Galactic halo, within a few kpc of the midplane.

2.4. The molecular component

Finally, we come to the molecular component of the ISM, where the primary molecule by mass is H_2. Unfortunately, it has no dipole moment, so it is a poor emitter and we are forced to find a proxy for it in emission. This substitute is usually CO, where the rotational lines in the millimeter region are strong and reasonably accessible. The problem with this situation is that the CO lines are generally optically thick (such as the 1–0 line), so this is a poor tracer of the total mass. For giant molecular clouds, the line strength is probably a good measure of the mass, but as we discuss below, this is clearly not true for less massive molecular clouds, which are more common. The typical density of the giant molecular clouds is 10^2 cm^{-3} and the temperature is around 10 K. Magnetic fields are detected in these cloud systems and are likely to contribute significantly to the total pressure of the system (as well as playing an important role in star formation).

Absorption has been used to study molecular gas, and here we can study H_2 directly through the many electronic transitions ($\sim$100) in the far ultraviolet region. Through this approach, molecular columns were determined in about 100 sight lines with *Copernicus* (Savage et al., 1977), and many more are being determined with the *Far Ultraviolet Explorer* (*FUSE*). The picture that has emerged for molecular gas is that it occupies an insignificant amount of the volume of the ISM, but it probably accounts for half of the mass.

One of the issues that must be addressed is the amount of molecular gas, which may be underestimated by a substantial amount. In the Milky Way, the *Copernicus* observatory showed that for $N_{HI} > 5 \times 10^{20}$ cm^{-2}, H_2 is very prevalent and a significant fraction of HI (25% of the total HI + H_2 column; Savage et al., 1977). In a completely independent study, where molecular gas is identified through the absorption strength of the HCO^+ millimeter line, HCO^+ absorption is seen in 18/22 lines of sight below 39 degrees (Lucas and Liszt, 1996), which corresponds to column densities above 5×10^{20} cm^{-2}; this is consistent with the *Copernicus* result. In another independent study in yet another waveband, X-ray absorption toward extragalactic sight lines indicates substantial absorption when $N_{HI} > 5 \times 10^{20}$ cm^{-2} (at latitudes below 40 degrees). The only possible ISM component that could be responsible for the extra absorption is H_2, which would account for about 40% of the total neutral gas column (Arabadjis and Bregman, 1998), consistent with the above results.

These three results paint a consistent picture of the H_2 content in the Solar circle, but it differs from the picture that one obtains from CO emission studies, which find that CO is not prevalent in the Solar neighborhood. Recently, there was a survey of CO at latitudes above 10 degreees (Hartmann, Magnani and Thaddeus, 1998), which detected little CO. They found a covering factor of 0.004–0.008 and using the usual CO to H_2 conversion ratio, a mass surface density of 0.015-0.035 M_o pc^{-2}. The inferred H_2 mass is about 50-130 times less than the H_2 mass determined from the studies above.

The difference between these two determinations of the H_2 mass can be understood bynoting that the interaction between molecules and the radiation field is different for CO and H_2 at these column densities. This interaction is discussed by van Dishoek and Black (1988; and others), who show that for the ambient radiation field in the disk, H_2 can shield itself against photodestruction when $A_V > 0.25$, which corresponds to N(HI) $> 5 \times 10^{20}$ cm^{-2}. However, CO is still easily destroyed under these conditions, which leads to the situation where CO is nearly absent while H_2 is still abundant. This can lead to values for the CO to H_2 conversion factor (X) that is orders of magnitude larger than the value derived for giant molecular clouds. Therefore, for the Galaxy, CO observations can lead to a significant underestimate of the molecular gas mass when using the typical value of X.

3. Progress Toward a Self-Consistent Model of the ISM

Our knowledge of the various interstellar components, although not perfect, is extensive and will allow us to build models and subsequently compare them to the data. The challenges to the modelers of the ISM are daunting both in scope and on the demands they place on codes and computers. To begin, the dynamic range of the density is enormous, from 10^{-4} cm^{-3} (hot gas) - 10^4 cm^{-3} (molecular gas), an eight order of magnitude range. It might be possible to reduce this range a couple of order of magnitude, which is still enormous. There is a similar range in the temperature of the gas, although the range in the pressure is far less, only a couple orders of magnitude. In addition, magnetic fields must be included, as it is an integral part of the pressure term. Also, it can lead to heating through reconnection, and it is possibly of dynamic importance as well. Turbulence is essential in a realistic treatment, and if one is to model this, it must be carried out in three-dimensions, as this is required if turbulence is to evolve. All this occurs in a rotating galaxy where differential rotation can be important in the evolution of large structures such as superbubbles or large sheets of gas.

For the gas components, the heating and cooling processes determine the nature of the different components. Supernovae provide the dominant input of energy through shocks and possibly through thermal conduction between cold and hot gas. The radiative cooling of the gas is relatively easy to model, provided that regions are optically thin. However, heating of gas in the range $10–10^4$ K is driven by ultraviolet photon, through photoionization and photoelectron ejection off of grains. The transfer of UV photons will depend upon the absorbing properties of the gas and the ability of molecular gas to shield itself. In addition, molecules are formed and destroyed, and as this is one of the important connections with observations, astrochemistry must be dealt with.

To finish this list of relevant processes, there is a transfer of gas from the ISM to stars through star formation, and the ejection of gas from aging stars (the red giant phase) as well as those at the ends of their lives (planetary nebulae and supernovae).

This is a tremendously daunting list of processes that one must keep track of and the task is formidable, but there have been a wide range of approaches taken that have shed light on this problem. We will hear about many of these at this meeting so I will only briefly review a few of the successes and approaches. Parts of the problem can be broken off and studied fruitfully. For example, turbulence is most important on moderately scales small compared to the Galaxy and there has been a lot of activity in studying energy transfer. A different example is the flow of hot gas in galactic halos, where the scales are kpc, so one can ignore the detailed structure of cold gas as being unimportant to the main aspects of the flow.

There has been a good deal of progress understanding the role of turbulence in theprocess of energy transfer. This addresses the issue of how the mechanical energy of a magnetized fluid is transformed into thermal energy. Along these lines, an important theory is the Alfvén Wave Decay Model of Goldreich and Sridhar (1995) in which they study incompressible magnetized fluids. Previous 'traditional' theories posited the isotropy of the energy cascade in Fourier space, but as Goldreich and Sridhar point out, this cannot be true in the ISM because B fields are more easily bent on large than on small scales. Therefore, the energy cascade should be anisotripic in Fourier space, so in response to this, they developed a theory of strong turbulence of shear Alfvén wave. In their theory, there is a critically balanced cascade between linear wave periods and nonlinear turnover timescales, which results in a Kolmogorov-type spectrum. This has been further developed by these and other authors and numerical simulations support this picture (Cho, Lazarian and Vishniac, 2002; Lazarian, this meeting).

An important extension of these physical conditions occurs when the fluid becomes compressible. These conditions have been investigated numerically and although they find that the turbulence is also Kolmogorov-like, there are phenomena not possible in the incompressible case: shocks and density fluctuations. MacLow and coworkers (e.g., this meeting) suggest that the density fluctuations are important in that they are the regions that ultimately undergo gravitational collapse in the star formation process. In addition, they show that shocks can be of greater importance than Alfvén waves in transferring energy on these scales.

The large-scale numerical efforts to simulate that ISM have taught us a great deal and have the potential of tapping into the wealth of observational data available. These numerical simulations divide space into a large number of cells, each of which has a single value for the various fluid properties, such as density, temperature, velocities, and magnetic field strengths. These calculations are carried out on the fastest computers available and a limiting constraint is usually the number of cells used. If one wants to investigate a region large enough to contain superbubbles and halo gas (a scale of several kpc), while still resolving relevant detail in the molecular component (1 pc), the number of cells required becomes very great and the calculation timestep becomes unacceptably small; such calculations are unfeasible. This difficulty is now resolved by using larger and fewer cells in most of the calculation, but adding cells to regions that require a greater degree of resolution. This subgridding scheme is dynamic in that it number and location of the extra cells changes with time and location in the grid.

The results of these calculations in the past few years (e.g., de Avillez, 2000 and this meeting) show results that are common with older calculations (e.g., Rosen and Bregman, 1995), but with several new surprises. Gas is generally pushed into sheets of material with lower density gas separating the sheets. The sheets can be quite dense, especially where they intersect with other sheets and these are regions that are the denser HI clouds or H_2. Supernovae tend to work collectively, creating large hot low density regions, which can be well-defined as superbubbles, or rather irregular regions. This hot gas will naturally rise, but it does not have a clear path out of the galaxy, since it interacts with (and becomes) the gaseous halo above the disk that is a consequence of the supernova heating.

For the Galactic value of the heat input from supernovae, the various scale heights for the different temperature components can be reproduced, but in addition to that, there are other important structural features that emerge. For example, these calculations produce hot superbubbles of approximately the size that is observed in the X-ray observations. Also, one finds the formation of chimneys in the HI component as well as the formation of HI 'mushrooms' that are observed (de Avillez, 2001a,b).

The simulations that focus on molecular clouds yield an appearance that, to visual inspection, is generically similar to CO maps found by observers. Also, these simulations appear to reproduce the observed size-linewidth relationship. These calculations necessarily include magnetic fields and turbulence (Ostriker, Stone and Gammie, 2001; Pavlovski et al., 2002) and there are significant morphological differences between the results of these investigations that we can identify whether or not these clouds lie in the strong or the weak turbulent limits. I am impressed with these calculations since I thought that numerical studies with this level of realism were still years away.

One of the long-standing disagreements in the development of a global model of the ISM has been the role of thermal conduction. Thermal conduction was introduced as a crucial component of these models by McKee and Ostriker (1977), probably the most often quoted paper on the subject. They argued that the undisturbed interstellar medium is naturally a two-density medium at nearly constant pressure: there are many dense cold HI clouds imbedded in a lower density warmer medium. A supernova shock races through the low density medium, heating it rapidly, but the shocks do not heat the dense clouds effectively, so there are dense clouds within a hot, high-pressure medium. The juxtaposition of hot and cold gas leads naturally to conductive heat transfer, provided that tangled magnetic fields do not reduce the effectiveness of the process. Heat conduction would lead to the transfer of the cold clouds to the hot medium. This process, in addition to providing an elegant mechanism by which one component is transformed into another, can act as a regulation mechanism. In addition, it seemed to solve an apparent problem with the size-mass relationship for SNRs, where they appeared to increase their mass more rapidly than the increase in their volume.

This theory seemed to solve many problems at once and was consistent with much of the data. However, the issue with the size-mass relationship for SNRs went away when Hamilton, Chevalier and Sarazin (1983) showed that the relationship could be reproduced if one included time-dependent ionization effects. In the discussion of the effectiveness of conduction, the central issue became the predicted 'intermediate temperature gas' that lies between the temperature of the hot gas and of the cold gas. Many of the important ions in this temperature range (10^5-10^7 K) have their resonance transitions at energies above 13.6 eV and cannot be studied, but OVI, representative of gas near $10^{5.5}$ K, is the most common heavy element and has strong transitions in the far ultraviolet region, a doublet centered at 1035 Å. While OVI absorption is detected, it is far weaker than predicted from the McKee-Ostriker picture (Shelton and Cox, 1994) and unless this deficit of OVI can be explained, the conflict with the theory is great enough to rule out the theory. We might add that if one were to replace thermal conduction with a mixing layer model (Slavin, Shull and Begelman, 1993), it would suffer the same problem since this process produces a significant amount of gas in the 10^5-10^6 K range. This is good new for modelers in that it appears that they can ignore thermal conduction or mixing layers in their models.

Some of the other assumptions of the McKee-Ostriker model may not be present, such as the mixture of many cold clouds within a lower density medium. While there are undoubtedly small clouds of HI, such as the one that the Sun is located within (although it is not high density), observations and theory suggest that most of the cold gas will lie in sheets that are separated from the lower density gas, rather than being intertwined.

4. Confronting Theory with Observations

The confrontation between observations and theory is at the center of the advancement in our knowledge. In the past, most of the comparisons have been with a few scalars that were extracted from the data, such as a density, temperature, pressure, or scale height. However, both observations and theory contain far more information that cannot adequately be summarized by a few scalars. For example, the Canadian Galactic Plane HI Survey has a vast amount of detailed structural information (and as a function of velocity), yet this richness is not being exploited in the comparison with models. Similar situations exist for molecular gas studies, where there are many excellent CO maps, and for the warm ionized component, where WHAM is producing beautiful Hα maps. This problem has faced other fields at some point in their history and there have been some interesting solutions to this problem. For example, in cosmology, a geometric method was developed (the 'genus' method; Gott, Melott and Kickinson, 1986) that could quantify the topology of the distribution of galaxies. The quantity that is fit in this case is a curve rather than a scalar, which is necessary to summarize the data. There have been a number of other methods as well but the most successful ones begin from physical principles and craft a statistical test around the most important physical quantities. An example of the opposite approach is to take some mathematical form that may or may not have physical significance, such as fractals, and perform an analysis based on it. In most cases, I have not learned much from such studies.

In an effort to progress toward these more complex analysis, Adams and Wiseman (1994) suggest an interesting approach that can be applied to molecular gas, but has broader applications. They discuss the analysis of maps (although their formalism is more general) whereby one would compare an observed map to a theoretical one. The difference between two maps is measured as a distance in map space, with the goal of minimizing this distance. They give an example of this method applied to *IRAS* maps of molecular clouds in star-forming regions (Wiseman and Adams, 1994). Whereas there are other possible approaches even to this problem (e.g., a 'tree' algorithm for molecular cloud maps; Houlahan and Scalo, 1992), it points out that there are more sophisticated techniques that need to be employed if we are to compare the new sets of theoretical models to the wonderfully detailed maps of data.

References

Adams, F.C. and Wiseman, J.J.: 1994, *ApJ* **435**, 693.
Arabadjis, J.S. and Bregman, J.N.: 1998, *ApJ* **510**, 806.
de Avillez, M.A.: 2000, *MNRAS* **315**, 479.
de Avillez, M.A. and Berry, D.L.: 2001, *MNRAS* **328**, 708.
de Avillez, M.A. and MacLow, M.-M.: 2001, *ApJ* **551**, L57.
Blitz, L., Spergel, D.N., Teuben, P.J., Hartmann, D. and Burton, W.B.: 1999, *ApJ* **514**, 818.

Bregman, J.N., Kelson, D.D. and Ashe, G.A.: 1992, *ApJ* **409**, 682.
Cho, J., Lazarian, A. and Vishniac, E.T.: 2002, *ApJ* **564**, 291.
Cox, D.P.: 1981, *ApJ* **245**, 534.
Van Dishoeck, E.F. and Black, J.H.: 1988, *ApJ* **334**, 771.
Goldreich, P. and Sridhar, S.: 1995, *ApJ* **438**, 763.
Gott, III, J.R., Melott, A.L. and Kickinson, M.: 1986, *ApJ* **306**, 341.
Hamilton, A.J.S., Chevalier, R.A. and Sarazin, C.L.: 1983, *ApJS* **51**, 115.
Hartmann, D., Magnani, L. and Thaddeus, P.: 1998, *ApJ* **492**, 205.
Houlahan, P. and Scalo, J.M.: 1992, *ApJ* **393**, 172.
Kulkarni, S.R. and Heiles, C.: 1988, in: G.L. Verschuur and K.I. Kellermann (eds.), *Galactic and Extragalactic Radio Astronomy*, 2nd edn.; Springer, Berlin, p. 95.
Lucas, R. and Liszt, H.: 1996, *A&A* **307**, 237.
McKee, C.F. and Ostriker, J.P.: 1977, *ApJ* **218**, 148.
Ostriker, E.C., Stone, J.M. and Gammie, C.F.: 2001, *ApJ* **546**, 980.
Pavlovski, G., Smith, M.D., MacLow, M-M. and Rosen, A.: 2002, *MNRAS* **337**, 477.
Reynolds, R.J.: 1989, *ApJ* **339**, L29.
Reynolds, R.J.: 1991, *ApJ* **372**, L17.
Reynolds, R., Tufte, S.L., Haffner, L.M., Jaehnig, K. and Percival, J.W.: 1998, *Pub. Astron. Soc. Aust.* **15**, 14.
Rosen, A. and Bregman, J.N.: 1995, *ApJ* **440**, 634.
Sanders, W.T. et al.: 2002, *AAS*, 201.4705.
Savage, B.D., Bohlin, R.C., Drake, J.F. and Budich, W.: 1977, *ApJ* **216**, 291.
Savage, B.D. et al.: 2003, *ApJS*, in press.
Sembach, K.R.: 2003, *ApJS*, in press.
Shelton, R.L. et al.: 2001, *ApJ* **560**, 730.
Shelton, R.L.: 2002, *ApJ* **569**, 758.
Shelton, R.L. and Cox, D.P.: 1994, *ApJ* **434**, 599.
Slavin, J.D., Shull, J.M. and Begelman, M.C.: 1993, *ApJ* **407**, 83.
Snowden, S.L., Egger, R., Finkbeiner, D.P., Freyberg, M.J. and Plucinsky, P.P.: 1998, *ApJ* **493**, 715.
Taylor, J.H. and Cordes, J.M.: 1993, *ApJ* **411**, 674.
Tufte, S.L., Wilson, J.D., Madsen, G.J., Haffner, L.M. and Reynolds, R.J.: 2002, *ApJ* **572**, L153.
Wakker, B.P. et al.: 2003, *ApJS*, in press.
Wakker, B.P., Oosterloo, T.A. and Putman, M.E.: 2002, *AJ* **123**, 1953.
Weiner, B.J., Vogel, S.N. and Williams, T.B.: 2002, in: J.S. Mulchaey and J. Stocke (eds.), *Extragalactic Gas at Low Redshift*, *ASP Conference Proceedings* Vol. **254**, ASP, San Francisco, p. 256.

INTERSTELLAR MATTER IN GALAXIES

AN OBSERVER'S OVERVIEW OF THE GALACTIC INTERSTELLAR MEDIUM

DONALD G. YORK
Department of Astronomy and Astrophysics, and Enrico Fermi Institute, The University of Chicago, 5640 South Ellis Avenue, Chicago, IL 60637, USA

Abstract. A number of problems related to understanding diffuse interstellar clouds are the focus of this contribution. Observations are discussed relating to the high velocity clouds, to the morphology of diffuse interstellar clouds and to the time variability of features in the diffuse clouds. Abundances and ionization issues are dealt with as needed. In some of these areas, progress may be substantial over the next few years, through observation and modeling, but in others, it may be that a new generation of space satellites will be required to make substantial progress.

Keywords: ISM: general, galaxies: halos, galaxies: starburst, X-rays: galaxies

1. Introduction

Observations from UV satellites probe the interstellar diffuse clouds. In the forty years over which such observations have been possible, there have been significant changes in our understanding of the diffuse material in space. However, there are several dilemmas, some of which are addressed in this brief summary, that challenge our understanding of the diffuse interstellar medium. The dilemmas I choose to focus on here are the distances of the high velocity clouds (HVCs) in the Galactic halo, the physical properties of the diffuse clouds in the disk and the time variability of interstellar lines.

2. The High Velocity Clouds

2.1. Distances

The system of high velocity clouds (HVCs) around the disk of the Milky Way was discovered in the 1960's. Various possible origins for this system of gas were recognized early. Some are discussed in the volume, in the contribution by B. Wakker. See also, Schwarz et al. (1995), Wakker (2001).

Determining the distances to the HVCs is among the most important of interstellar problems because the clouds are appealed to so often to explain end results of other problems. In the 1970s, interpretations involving infall of primordial gas were popular, to provide enrichment of the clouds in the just-discovered Galactic

Astrophysics and Space Science **289:** 195–201, 2004.

deuterium abundance (Rogerson and York, 1973). In the 1980s, the HVCs were popularly interpreted as returning gas from the Galactic fountain (Shapiro and Field, 1976). At the end of the century, the realization that theories of galaxy formation predicted the existence of more dwarf galaxies than can be seen today helped the interpretation that the HVCs are distant members of the Local Group (Blitz et al., 1999) that never formed stars (Simon and Blitz, 2002; Willman et al., 2002). The wide range of interpretations would not be possible if we actually knew the distances to the clouds.

The existence of new surveys for RR Lyr stars and the availability of a new generation of large telescopes make this an ideal time to mount a new attack on this fundamental problem. RR Lyrae stars are ideal for the determination of the distances to the HVCs. By looking for interstellar absorption lines in the spectra of stars that lie within the contours of a known HVC with known 21 cm velocity (typically, $v_{lsr} > 100$ km/sec), and by picking stars at successively greater and greater distances until Na I and/or Ca II absorption are detected at or near the 21 cm velocity, one can achieve rough estimates of the distances. Using similar techniques and a finer grid of stars, the estimates can be refined. RR Lyr stars are often found to be metal poor, so they have relatively smooth continua. The use of one type of star means that the few stellar blends with HVC interstellar absorption will be well understood. The stars have a wide range of velocities, so the star chosen for a given HVC can be chosen to avoid having their strong stellar Ca II and Na I lines overlapping the target interstellar lines. Periods must be determined to plan observations at the peak of the 3900A brightness to maximize the signal-to-noise ratio at the interstellar Ca II lines $\lambda\lambda 3933,\ 3969$).

Several of us are starting such a program, including Jenkins, Wakker and myself, to name individuals at this conference, along with Jon Barentine, Zeljko Ivezik, Ben McCall, Tim McKay and Jeff Pier. RR Lyr stars will be taken from the Sloan Digital Sky Survey (York et al., 2000; Ivezic et al., 2000), the ROTSE III survey (Akerlof et al., 2000) and the LONEOS survey. Between the surveys, we expect to sift between an initial sample of 0.5 RR Lyr per square degrees to find an appropriate distance sample for at least 20 HVCs. The sample should reach to 50 kpc, and to 80 kpc for some cases. Very large telescopes will be needed for the final observations, detection of optical interstellar lines of strength $\sim$10 milliAngstroms.

The HVCs probably consist of several types of clouds, with different distances. Each cloud will have to be separately considered, without prejudice. However, the RR Lyr stars can now be easily found and the large telescopes are adequate to complete the study. It may be hoped that over a timescale of 5–10 years, significant progress can be expected.

2.2. IONIZATION

There are other issues associated with the physics of the clouds that will inform the studies of global interstellar models being discussed at this conference. One of these is the ionization state of the HVC's. A new UV spectral atlas of halo sightlines to AGNs, using FUSE (Wakker et al., 2003) shows clearly, for instance, that a given cloud may or may not have H I 21cm emission when O VI is present, and vice versa. For instance, refer to the Wakker et al atlas: the case of NGC 588 shows O VI, H I and C II in absorption out to as high as –300 km/sec, yet PG0052+251 shows strong OVI, a little C II, but no H I out to the same velocity. Mrk 116 shows detectable C II and 21cm out to –200 km/sec, but no O VI is present. It is unlikely that star samples dense enough to address, observationally, this problem of different ionization states of halo gas will be available (RR Lyr stars have inadequate flux at 1030 A to be useful in this regard). Sparse sampling of cloud regions coupled with detailed modeling (discussed in other papers of this volume; Konz et al., 2002), will be required.

An alternative approach is to study halos of other galaxies, which presumably have systems of HVCs as well, using background QSOs for the sources. Sources are starting to be available in great numbers, from the Sloan Digital Sky Survey (Schneider et al., 2001) However, to get a sample of multiple sightlines through halos of galaxies of given types, one has to rely on high-resolution observations of 19th magnitude QSOs with space satellites: ionization studies require ions only available in the ultraviolet region of the spectrum. These studies are beyond what is possible with the Hubble Space Telescope, for instance. I will return to the need for a new UV spectroscopy instrument later.

A second issue is the abundance of elements in the clouds. It is obvious, for instance, that Galactic fountain gas and primordial gas with no stars yet formed should have drastically different abundances. I cannot deal with this topic here, but will return to it a the end of this paper, in a more general discussion of abundances of interstellar clouds.

3. Densities in Normal Diffuse Interstellar Clouds

The physical properties of diffuse clouds are derived from UV observations of modest resolving power, compared to what is needed, supplemented by very high resolution, optical observations (>500,000 resolving power) of K I, CH and other species. The optical profiles are then used to define the profiles for the UV lines. One is usually limited by the process of modeling profiles because regions of different temperatures and densities overlap in velocity space and the warmer regions have no unique lines that can be used to identify such regions in the optical part of the spectrum. Ca II and Ti II could be used for separating cool and warm regions, but these have a large range of depletion, not suitable when analyses of volatile species such as Zn II and even Si II and Mg II are involved. Numerous assumptions have to be made.

A particular dilemma, now well defined, before us, is the derivation of densities of interstellar clouds, or, to be more precise, regions within interstellar clouds.

Namely, the densities derived using simple radiative recombination of ions with electrons leads to contradictions with other data. For instance, in a thorough examination of the sight line to 23 Ori, Welty et al. (1999) found a wide range of values of n_e from ratios such as N(Mg I)/N (Mg II), N(Fe I)/N(Fe II), N(Ca I)/N(Ca II), etc. Here, N is the column density of the species given in parenthesis. Not only do the values of n_e so derived vary widely, but they are all higher than those inferred from the normal expectation that the electrons derive from carbon and that n_e/n_H is the abundance ratio of C to H, since all C is ionized in diffuse clouds. In other words, partial ionization of H is required. Lepp et al. (1988) noted a discrepancy between the electron density inferred from collisional population of the fine-structure levels of C+ and from recombination of C+ with an electron. They argued that large molecules might be involved in neutralizing ions, leading to an overestimate of the electron density if the molecular process is ignored. Weingartner and Draine (2002) have examined this process in detail and found that it may reduce the electron densities by a factor of two in some cases, adequate to resolve some problems in the densest diffuse clouds but not problems of the extent found in the 23 Ori study, noted above. See also the survey of Ca I by Welty et al. (2003). Lizst (2003) shows that grain neutralization of ions leads to similar effects and resolves some old and some new dilemmas if adopted: in particular, the ionization rate of hydrogen by cosmic rays can be up to 40 times higher than has been assumed in the last few years, helping to resolve a dilemma associated with the presence of H3+ in diffuse clouds (McCall et al., 2003, and references therein). However, the problem in 23 Ori remains and the problem has shown up in other cases, referenced in Welty et al. (2003). Interestingly, a recent analysis of the sight line to HD185418 by Sonnentrucker et al. (2003) gets more consistent results from the C+ fine-structure analysis and the neutral/ion ratios, assuming only radiative recombination, but the fractional ionization remains high ($\sim$0.01, not 0.0001).

This all-too-brief summary makes no attempt to resolve the physical misinterpretations we may have. I rather want to raise questions about the analysis techniques we are forced to use. For analysis of densities or of abundances (for diffuse clouds or HVCs), if one were trying to understand diffuse clouds with known ranges in temperature between 10 and 8000 K and hydrogen densities between 1 and 5000 per cubic centimeter, it is clear that one would build instruments with resolving powers adequate to resolve the thermal widths of the lines, which are as low as 0.1 km/sec, corresponding to a resolving power of 3,000,000. It is also clear that one would want signal-to-noise ratios in excess of 100 from 920 A to 3000 A, to determine the point spread functions accurately and to allow detection of the weakest lines, which are most free of saturation effects. Of course, such an instrument would involve a UV telescope in orbit with enormous aperture (20 meters?), reason enough to adapt to the instruments and modeling techniques we have. But, there is no guarantee that we are now getting the correct answers and the existence of dilemmas as great as that noted above at least gives room for doubt. The arguments by Jenkins based on C I pressures (see later in this volume) that challenges the application of simple geometries or time stability to interstellar clouds suggests that the most simple assumptions may be incorrect.

My own major concern is that we are not properly discerning the distribution of the main ions between warm and cool gas for lack of adequate resolving power to separate the two. Imagine, for instance, a simple, neutral region surrounded by a photo-ionized shell, caused by ambient starlight in the interstellar medium. The coolest part of that entity would have narrow, low equivalent width lines, while the outer shell could produce much stronger lines, by virtue of the velocity dispersion being over 10 times larger. Yet, both regions would have the same space velocity and would be completely blurred together at even the best UV resolving power now available for UV spectroscopy, about 1.5 km/sec. A combination of very high resolution ground based profiles of K I and Ca II might hint at this situation and allow it to be modeled, but the uncertain abundances of K II and Ca III, as an example, would leave such models uncertain. In fact, those two quantities are examples of exactly what we would like to derive at the end of the modeling process, so one cannot assume the values for the analysis.

Assuming, for the moment, that the values for N and n_e we are deriving are not woefully incorrect, the conclusion of all of this discussion is that simple radiative recombination analyses are not adequate. The description of neutralization of ions in diffuse cloud may resemble a network of chemical reactions with large molecules, small grains and free electrons involved in the process. With regard to the large molecules, the ever-expanding data and curiosity of the diffuse interstellar bands (Thorburn et al., 2003) reflects the fact that, never mind speculation that we need more detailed models, there is a an entire set of entities in the diffuse interstellar medium of which we have no knowledge based on firm identifications. Perhaps the dilemma of the identity of the diffuse interstellar bands is somehow related to the other puzzles of diffuse clouds noted here.

For lack of an instrument capable of doing the job needed, we can hope that more sophisticated modeling of the interstellar medium itself, as opposed to just the line profiles needed for interpretation, might become an effective tool of investigation. Several papers in this volume point the way to such models.

4. Time Variability of Absorption Lines in Diffuse Interstellar Clouds

A third problem I want to bring up is the problem of variable 'interstellar': absorption lines, words that sound to most of us as a non sequitur. It has been found by a number of authors that various types of interstellar lines are variable: medium velocity clouds (Welty, 2001); low velocity, low density, diffuse clouds with molecular fractions, f (H2) below 10^{-1} $\{f={}_2N(H_2)/[N(HI) + {}_2N(H_2)]\}$ (Blades et al., 2000; Lauroesch et al., 2000); and high density clouds with f>0.1 (Rollinde et al., 2003). Lines of CH, CH+, K I and Zn II have been found to vary. As might be expected, lines that are optically thin show more variation than do saturated lines. The time scales are 2 months to years and the implied scale sizes, assuming the variations are due of proper motions of the clouds with respect to the stars, are 10 AU and above. Variations on scales of 1000 AU are confirmed by variations in the same velocity component in spectra of close binaries (Lauroesch and Meyer, 1999). (There are several older cases of variability and many new ones, announced almost monthly: I could not give a complete review here.)

Discussions so far focus on geometrical interpretations, a structured cloud with a proper motion compared to a background star, so that a series of observations finds the line of sight to the star sampling slightly different parts of the cloud. If this is true, there is a wealth of information about cloud structure, and perhaps dissipation of that structure, in a well thought out, systematic set of time series observations of a large set of stars. While a program such as this can be designed based on what we know, it may turn out that the reason for the variability is much more interesting than we can imagine. Of course, the phenomenon was not discovered until recently, in UV and optical absorption lines, because the equipment was not good enough, the question was not well posed and telescope time was not readily available for such 'fishing' expeditions. Now, we do have adequate instruments and analysis techniques, the question is well posed and there is an abundance of time on telescopes adequate to this task.

5. Models and New Instruments

The problems noted here: the distances and physics of the interstellar clouds; the neutralization of ions in diffuse clouds; and the serendipitously- discovered time variability of absorption lines in a number clouds of many types suggests to me that we are on the verge of a new age of discovery concerning diffuse matter in space. We cannot be satisfied with using the same, inadequate instruments; with relying on numerous, simplifying assumptions that are gradually being undermined; and with getting by without detailed models.

I am advocating the unthinkable: a 20 meter, UV/optical telescope in space with spectroscopic capability adequate to resolve the thermal widths of any absorption line from a diffuse cloud. Such an instrument would need to be assembled in space, with the optics coated in space and with a mirror built in some segmented fashion. A diffraction limited, UV telescope would certainly have many other uses, but the resulting, small image is essential for the interstellar science, to keep the size of the spectrograph(s) within reason.

Until such a dream is realized, and certainly should it ever come to pass, there is a great need for models and for a completely fresh effort to compare models with the sparse data we have on diffuse clouds. The comments above imply that there are structures in diffuse clouds much smaller that the resolution of the models presented at this conference; that these have dynamical aspects; that the chemistry of diffuse clouds is very complex and cannot be left out of models; and that the ionization structure of clouds at high latitude represents a prime dilemma not directly addressable by observation (because of the sparse sampling of suitable sources). All of the dilemmas require detailed models to gain any insight from the observations, as our current equipment cannot reach the required resolving powers, density of space sampling, or range of ions and completeness of chemical products required.

Acknowledgements

In preparing my remarks and this manuscript, I was greatly aided by, and am grateful for, discussions with Miguel de Avillez, Ed Jenkins, Lew Hobbs, Ben McCall, Warren Moos, Tak Oka, Bart Wakker and Dan Welty.

References

Akerlof, C. et al.: 2000, *ApJ* **119**, 1901.
Blades, J.C., Sahu, M.S., He, L., Crawford, I.A., Barlow, M.J. and Diego, F.: 1997, *ApJ* **478**, 648.
Blitz, L., Spergel, D.N., Teuben, P.J., Hartmann, D. and Burton, W.B.: 1999, *ApJ* **514**, 818.
Ivezic, Z. et al.: 2000, *AJ* **120**, 963.
Konz, C., Bruns, C. and Birk, G.T.: 2002, *A&A* **391**, 713.
Lauroesch, J.T., Meyer, D.M. and Blades, J.C.: 2000, *ApJ* **543**, L43.
Lauroesch, J.T. and Meyer, D.M.: 1999, *ApJ* **519**, L181.
Lepp, S., Dalgarno, A., van Dishoeck, E.F. and Black, J.H.: 1988, *ApJ* **329**, 418.
Lizst,, H.S.: 2003, *A&A* **398**, 621.
McCall, B.J. et al.: 2003, *Nature* **422**, 500.
Rogerson, J. and York, D.G.: 1973, *ApJ* **186**, L96.
Rollinde, E., Boisse, P., Federman, S.R. and Pan, K.: 2003, *A&A* **401**, 215.
Schneider, D.P. et al.: 2001, *AJ* **123**, 567.
Schwarz, U.J., Wakker, B.P. and van Woerden, H.: 1995, *A&A* **302**, 364.
Shapiro, P.R. and Field, G.B.: 1976, *ApJ* **205**, 762.
Simon, J.D. and Blitz, L.: 2002, *ApJ* **574**, 726.
Sonnentrucker, P. et al.: 2003, *ApJ* **596**, 350.
Thorburn, J. et al.: 2003, *ApJ* **584**, 339.
Wakker, B.: 2001, *ApJS* **136**, 463.
Wakker, B. et al.: 2003, *ApJ*, in press.
Weingartner, J.C. and Draine, B.: 2002, *ApJ* **563**, 842.
Welty, D.E., Hobbs, L.M., Lauroesch, J.T., Morton, D.C., Spitzer, L. and York, D.G.: 1999, *ApJS* **512**, 686.
Welty, D.E. and Fitzpatrick, E.L.: 2001, *ApJ* **551**, L175.
Welty, D.E., Hobbs, L.M. and Morton, D.C.: 2003, *ApJS* **147**, 61.
Willman, B., Delcanton, J., Ivezic, Z., Schneider, P. and York, D.G.: 2002, *AJ* **124**, 2600.
York, D.G. et al.: 2000, *AJ* **120**, 1579.

PROBING THE IRC +10° 216 CIRCUMSTELLAR ENVELOPE USING SPECTROSCOPIC OBSERVATIONS OF BACKGROUND STARS *

TIM KENDALL
Centro de Astronomia e Astrofísica da Universidade de Lisboa, Observatório Astronómico de Lisboa, Tapada da Ajuda, 1349-018 Lisboa, Portugal; E-mail: tkendall@oal.ul.pt

NICOLAS MAURON
Groupe d'Astrophysique, CNRS & Université de Montpellier, Case 072, Place Bataillion, 34095 Montpellier Cedex 05, France; E-mail: mauron@graal.univ-montp2.fr

JUNE MCCOMBIE and PETER SARRE
School of Chemistry, The University of Nottingham, University Park, Nottingham, NG7 2RD, UK

Abstract. A unique and novel set of observations has been undertaken to study the circumstellar envelope (CSE) of the nearby (130 pc) carbon star IRC +10° 216 using optical absorption spectroscopy towards two *background stars* lying beyond the envelope. The primary aim of the observations is to search for diffuse band (DIB) carriers in the CSE. The *circumstellar* $H+2H_2$ column density expected along the line of sight towards the targets is large compared to that derived from the small interstellar reddening in this region, $E_{B-V} < 0.03$ mag. We summarise our main findings, already reported in Kendall et al. (2002): The 6284 Å DIB is detected in the VLT/UVES spectrum of one target, but clearly arises in the foreground ISM. No other DIB is seen. Hence the DIB carriers, if present in the CSE, have a low abundance relative to H in the C-rich envelope of IRC +10° 216, in comparison with this ratio in the ISM. An important new result, previously unreported, concerns the detection of circumstellar C_2 *via* absorption of the Phillips band near 8760Å.

Keywords: IRC +10° 216, carbon stars, circumstellar matter, ISM: molecules

1. Introduction

Diffuse interstellar bands (DIBs) have been observed in the visible spectra of reddened OB stars for nearly eighty years; over 200 are now known, but no carrier molecules have yet been definitively assigned. Their identification remains a major challenge in molecular astrophysics.

We have addressed the key issue of the site of the origin of DIB carriers by undertaking a search for DIB absorptions, arising in the CSE of the prototypical extreme carbon star IRC +10° 216, against the continua of background stars (Kendall et al., 2002). The dense, dusty outflows of such asymptotic giant branch (AGB) stars inject into the ISM a large amount of carbon-rich material. Carbon chains (polyynes, C_3, C_5), to which the DIB carriers may be related, are known to be

* Based on observations made with the ESO-VLT, WHT, and OHP 1.2 m telescopes.

Astrophysics and Space Science **289:** 203–206, 2004.

TABLE I

Parameters for the observed stars and circumstellar quantities. The distances d are best estimates derived from our photometric and spectroscopic data; the angle Θ ($''$) is the angular offset from IRC +10° 216; $N_{\rm H}^{tang}$ is the expected line-of-sight circumstellar column density; $A_{\rm V}$ is the *circumstellar* extinction derived from $N_{\rm H}^{tang}$ when *assuming* a $N_{\rm H}/A_{\rm V}$ ratio typical of diffuse interstellar matter. USNO identifiers are prefixed with 0975-0633-

USNO	*	V	Spec.	d/pc	Θ	r/cm	$N_{\rm H}^{tang}/{\rm cm}^{-2}$	$A_{\rm V}$/mag
6975	6	16.00	G2:	~1400	37	$6.8\ 10^{16}$	$1.8\ 10^{21}$	0.9
6812	C	12.4	F8:	~500	153	$3.0\ 10^{17}$	$4.1\ 10^{20}$	0.2

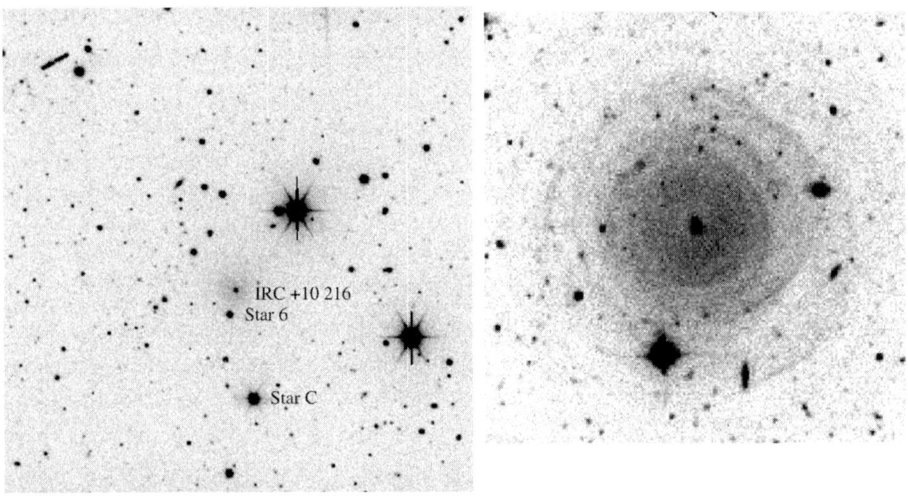

Figure 1. Left panel: The 12.4 × 12.4′ field around IRC +10° 216, imaged in the V-band at the 1.2m Haute-Provence telescope. Stars 6 and C are labelled. *Right panel*: Deep 2 × 2′ B-band CFHT image showing the IRC +10° 216 envelope in dust scattered ambient galactic light. Star 6 is the bright star located 37″ below and slightly to the left of IRC +10° 216.

present in the CSE of IRC +10° 216, the nearest dusty, molecule-rich carbon star with a high mass-loss rate.

2. The Field of IRC +10° 216

Figure 1 shows potential background targets close to IRC +10° 216 (left panel). The CSE itself, extending out to $\sim 3'$, is revealed in dust-scattered, ambient galactic light (right panel). Using a smooth r^{-2} law to model the envelope, an estimate of the total hydrogen column density at a given radius can be found using well-constrained parameters for IRC +10° 216. Further circumstellar quantities so de-

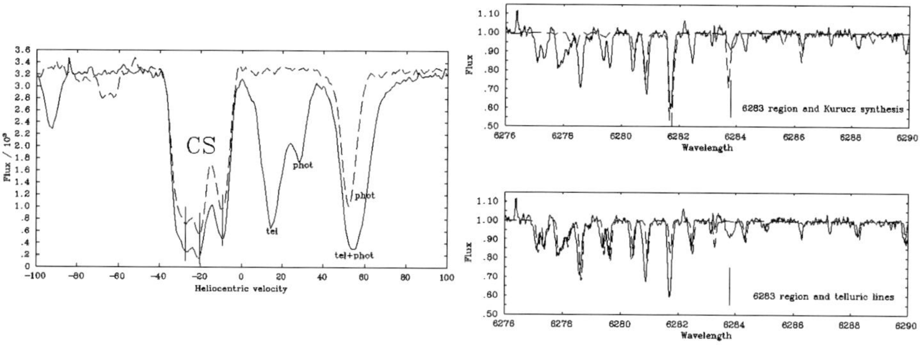

Figure 2. Left panel: K I lines in the VLT spectrum of Star 6, shown in heliocentric velocity space. The solid line is the K I 7665 Å profile, the dashed line K I 7699 Å. The three vertical lines trace circumstellar components in K I, centred very closely to the heliocentric velocity of IRC +10° 216 at –19.3 km s^{-1}. *Right panel*: The region of the 6284 DIB in the Star 6 spectrum. The interstellar DIB feature is indicated by the vertical line. All spectra are at the heliocentric velocity. The observational data (solid line) is compared to a Kurucz synthesis (top panel) and a pure telluric spectrum (bottom panel).

rived are given in Table I; details can be found in Kendall et al. (2002). The extinction A_V towards stars 6 and C compares favourably with that toward stars with intervening diffuse clouds, in the spectra of which DIBs are readily detected. Methods for the derivation of spectral type, and hence distance, (also given in Table I) are fully detailed in Kendall et al. (2002) and will not be repeated here.

3. Atomic Lines and DIBs in the UVES Star 6 Spectrum

Figure 2 summarises our findings, as reported in Kendall et al. (2002). We detect the K I resonance lines with equivalent width ~600mÅ, far too large to be attributed to the foreground diffuse ISM. The K I absorption is centred at the heliocentric velocity of IRC +10° 216, with FWHM close to twice the outflow velocity of the CSE. It clearly arises in the CSE.

The 6284Å DIB has equivalent width ~27mÅ and its wavelength is far from that expected if it arose in the carbon star CSE. As described in Kendall et al. (2002), we can definitively show its origin in the foreground ISM. 3σ upper limits on the strengths of other, strong DIBs are ~ 5 times smaller than would be expected from the A_V value for star 6 given in Table I. This suggests that DIB carriers have a low abundance in the IRC +10° 216 CSE, and we speculate that some other processing not occurring in the CSE, perhaps strong UV irradiation, is required to fabricate DIB carriers from C-rich precursor material.

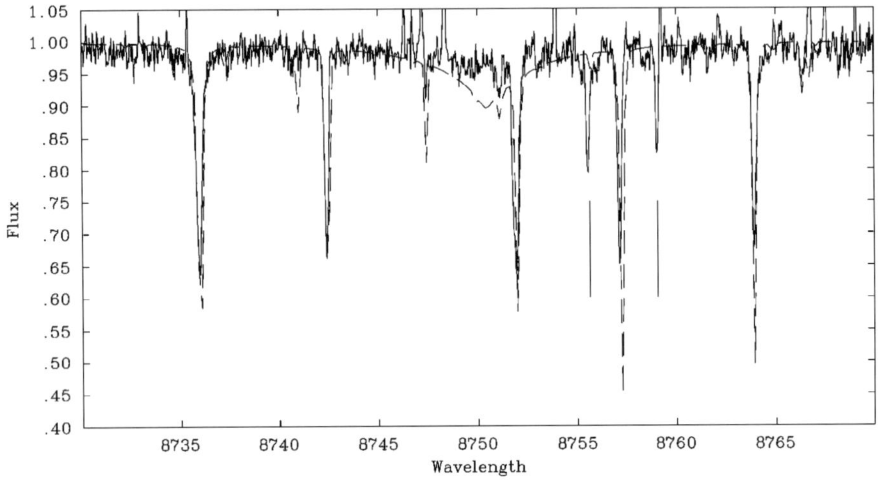

Figure 3. Detection of circumstellar C_2 absorption in the VLT/UVES spectrum of background target 6, indicated by vertical bars. The dashed line is a Kurucz spectral synthesis, which matches the star 6 photospheric features accurately.

4. Detection of Circumstellar C_2

Our spectra show the first detection of circumstellar C_2 in the outer layers of the CSE, at $\Theta = 37''$. In Figure 3, the two lines marked can be assigned to the R(0) and Q(2) lines of the Phillips (2,0) band of C_2, with heliocentric velocities respectively –18.7 and –19.9 $\mathrm{km\,s^{-1}}$, consistent with an origin in the IRC +10° 216 envelope. While we do not derive a rotational temperature from just these two lines, the detection supports the evidence for a low temperature CSE given by Le Bertre and Gérard (2001). This detection is also qualitatively consistent with C_2 seen towards the central star by Bakker et al. (1997), and a quantitative analysis will be made elsewhere. This result underlines the effectiveness of our new method to probe the cool, molecular layers of the circumstellar envelope of IRC +10° 216.

References

Kendall, T.R., Mauron, N., McCombie, J. and Sarre, P.J.: 2002, *A&A* **387**, 624.
Bakker, E.J., van Dishoeck, E.F., Waters, L.B.F.M. and Schoenmaker, T.: 1997, *A&A* **323**, 469.
Le Bertre, T. and Gérard, E.: 2001, *A&A* **378**, L29.

VOLUME FILLING FACTORS OF THE DIG IN M 31

ELLY M. BERKHUIJSEN
Max-Planck-Institut für Radioastronomie, Auf dem Hügel 69, 53121 Bonn, Germany

Abstract. The volume filling factor f_V of the diffuse ionized gas in the bright emission ring of the galaxy M 31 is derived from radio continuum observations. The dependence of f_V on the local mean electron density n_e is a power law, $f_V(n_e) = an_e^{-b}$ with $a = 0.011 \pm 0.003$ and $b = -1.2 \pm 0.3$, where n_e is in cm^{-3}. The same power law was recently found for the DIG in the solar neighbourhood from pulsar data.

Keywords: DIG, filling factor, H II: diffuse, ISM, M 31

1. Introduction

At least 20% of the total Hα emission from M 31 (corrected for extinction) is coming from the diffuse ionized gas (DIG), which is widespread in and around the spiral arms (Walterbos, 2000). In order to estimate the true volume density of the electrons and the mass of the DIG, the volume filling factor, f_v, is needed. Berkhuijsen (1998) showed that for the Milky Way f_v is nearly inversely proportional to the local mean density of the electrons, n_e, which is indicative of fractal structure of the DIG (Elmegreen, 1998). In this paper the first estimate of $f_v(n_e)$ for an external galaxy is made.

The volume filling factor of spherical clouds of free electrons can be derived from

$$f_v(n_e) = \frac{\langle n_e \rangle}{n_e} = \frac{\langle n_e \rangle^2}{\langle n_e^2 \rangle}, \qquad (1)$$

where n_e is the mean volume density in a cloud and brackets indicate averages along the line of sight in the same direction. In this work the variables in (1) are found from the emission measure EM and the rotation measure RM:

$$\mathrm{EM} = \langle n_e^2 \rangle L_2 \ \mathrm{cm^{-6}\,pc}\,, \qquad (2)$$

$$\mathrm{RM} = 0.81 \langle n_e \rangle \langle B_{\mathrm{reg}\parallel} \rangle L_1 \ \ \mathrm{rad\,m^{-2}}\,, \qquad (3)$$

where L_1 and L_2 are the effective lines of sight in pc and $B_{\mathrm{reg}\parallel}$ the line-of-sight component of the regular magnetic field in μG. EM is obtained from the thermal radio emission and both RM and $\langle B_{\mathrm{reg}\parallel} \rangle$ from the polarized radio emission.

Astrophysics and Space Science **289:** 207–210, 2004.

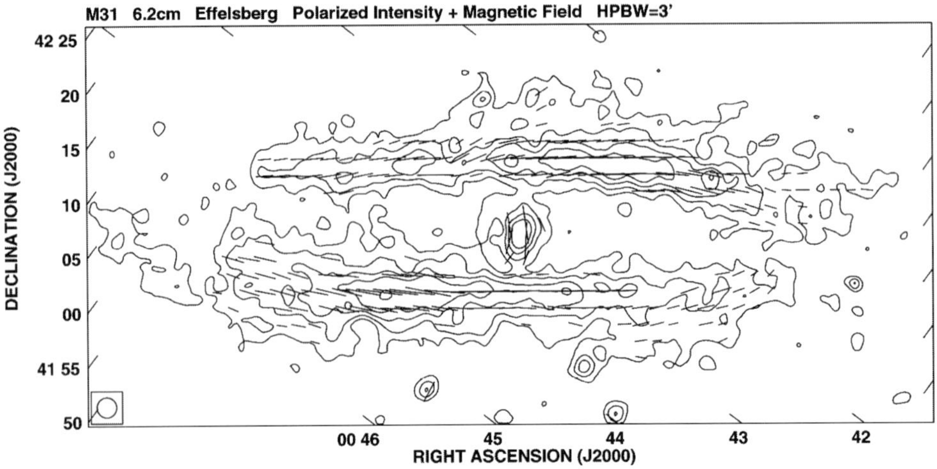

Figure 1. Linearly polarized continuum emission from M 31 at λ6.2 cm and magnetic field vectors, observed with the Effelsberg 100-m dish and smoothed to an angular resolution of 3′. Contour levels are 0.5, 1.0, 1.5, 2 and 3 mJy/beam area; the r.m.s. noise is 0.25 mJy/b.a. North is to the upper left, east to the bottom left. Note the general alignment of the magnetic field with the bright ring of emission.

2. M 31: Data and Analysis

Three high-resolution radio continuum surveys of M 31 are available in total and polarized emission: at λ20.5 cm (VLA + Effelsberg, 45″ resolution, Beck et al., 1998), λ11.1 cm (Effelsberg, 5′ resolution, Beck et al., 1980) and λ6.2 cm (Effelsberg, 3′ resolution, Berkhuijsen et al., 2003). The data at λ20.5 cm and λ6.2 cm yielded a spectral index map of the total intensity at a resolution of 3′, which was used to separate thermal and nonthermal emission from M 31 assuming a nonthermal spectral index $\alpha_n = -1.0$.

The intensities of nonthermal and polarized emission give the components perpendicular to the line of sight of the total and regular magnetic field, respectively, assuming energy equipartition between cosmic ray particles and total magnetic field. The line-of-sight component of the regular field can then be found on the assumption that B_{reg} is parallel to the disk.

Furthermore, the polarization angles of the polarized emission at λ11.1 cm and λ6.2 cm yield the rotation measures RM(11, 6) used to correct for Faraday rotation. Figure 1 shows the distribution of the polarized emission at λ6.2 cm with the orientation of $B_{reg\perp}$ superimposed (see Berkhuijsen et al. (2003) for details).

For further analysis the data were averaged in 20°-wide sectors in the plane of M 31 in the radial range R = 8–12 kpc centred on the bright emission, adopting a distance to M 31 of 690 kpc and an inclination angle of 78°. A comparison of the variations of RM(11, 6) and the intensity of the thermal emission at λ6.2 cm with azimuthal angle shows that RM(11, 6) is not increased in sectors with strong thermal emission. As the thermal intensity is dominated by high-density

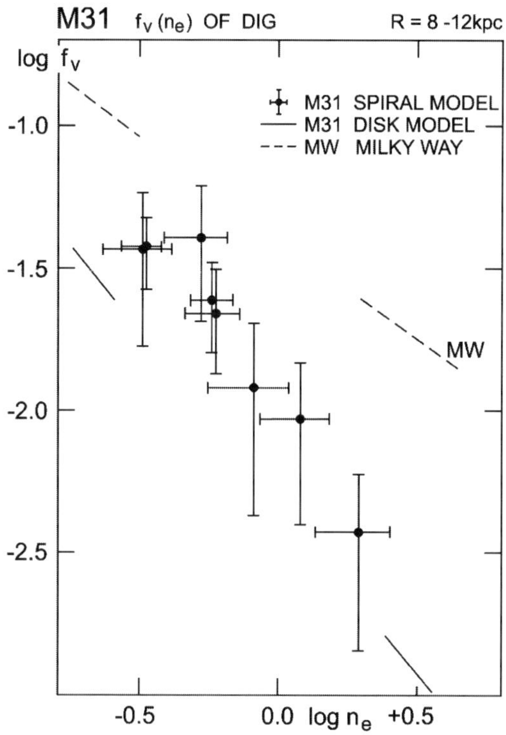

Figure 2. Volume filling factor f_V of the DIG in M 31 as a function of electron density (cm^{-3}) for 20°-wide sectors in the ring R = 8–12 kpc. The DIG was assumed to be concentrated in 2 spiral arms. The full lines indicate the mean result if the DIG were distributed in a homogeneous disk. The dashed lines indicate the result of Berkhuijsen (1998) for the Milky Way.

H II regions of small filling factors, this indicates that RM(11, 6) intrinsic to M 31 originates in the DIG.

The thermal emission from the DIG at R = 8–12 kpc mainly comes from two spiral arms seen along the same line of sight (Braun, 1991; Fig. 4 in Berkhuijsen et al., 1993). Assuming that these arms are tori of equal brightness and diameter (= 2× exponential scale height h), the effective lines of sight adopted are: $L_2 = 1$ kpc for the emission measure, and $L_1 = 1.4$ kpc for the rotation measure intrinsic to M 31 (Berkhuijsen, in prep.). The volume filling factors for the sectors can then be found from equations (2), (3) and (1).

3. Results and Discussion

The relationship between f_V and n_e for M 31 is presented in Figure 2. The errors are largely determined by the uncertainty in RM which was too large to get meaningful results for sectors near the major axis. The points represent a power law of the form:

$$f_{\rm v}(n_{\rm e}) = (0.011 \pm 0.003) n_{\rm e}^{-1.2\pm0.3} \quad \text{for } 0.3 \lesssim n_{\rm e} \lesssim 3\,\text{cm}^{-3} \,,$$

which is somewhat lower and steeper than that found for the Milky Way by Berkhuijsen (1998) for a much larger range of $n_{\rm e}$.

Recently Müller, Mitra and Berkhuijsen (in prep.) derived $f_{\rm v}(n_{\rm e})$ of the DIG in the solar neighbourhood (SN) using the dispersion measures of 13 pulsars at $|b| \gtrsim 5^\circ$ with parallactic distances and EM from the WHAM survey (Haffner et al., 2001). They obtained the same power law as found for M 31, i.e. $f_{\rm v}(n_{\rm e}) = (0.0111 \pm 0.0003) n_{\rm e}^{-1.15\pm0.02}$, which holds for the same range in $n_{\rm e}$ as in Figure 2. If this agreement is not accidental, it is surprising because M 31 is a less bright galaxy than the MW: it has a 5–10 times smaller star-formation rate (Walterbos, 2000) and the atomic gas density is 3 times smaller (Walterbos and Braun, 1994). On the other hand, the ring at R = 8–12 kpc in M 31 contains arm as well as interarm regions and the SN is located on the border of a weak local arm. Thus the properties of these DIG regions may indeed be comparable.

An inverse proportionality between $f_{\rm v}$ and $n_{\rm e}$ is expected for a fractal ISM (Fleck, 1996; Elmegreen, 1998). The ISM in galaxies of different type shows similar fractal structure. The identical power laws of $f_{\rm v}(n_{\rm e})$ in M 31 and the SN may be another sign of the universal fractal structure of the ISM in galaxies.

References

Beck, R., Berkhuijsen, E.M. and Wielebinski, R.: 1980, *Nature* **283**, 272–275.
Beck, R., Berkhuijsen, E.M. and Hoernes, P.: 1998, *Astron. Astrophys. Suppl.* **129**, 329–336.
Berkhuijsen, E.M.: 1998, in: D. Breitschwerdt, M.J. Freyberg and J. Trümper (eds.), *The Local Bubble and Beyond*, Lecture Notes in Physics, Springer, pp. 301–304.
Berkhuijsen, E.M., Bajaja, E. and Beck, R.: 1993, *Astron. Astrophys.* **279**, 359–375.
Berkhuijsen, E.M., Beck, R. and Hoernes, P.: 2003, *Astron. Astrophys.*, (in press).
Braun, R.: 1991, *Astrophys. J.* **273**, 54–66.
Elmegreen, B.G.: 1998, *Pub. Astron. Soc. Australia* **15**, 74–78.
Fleck, R.C.: 1996, *Astrophys. J.* **458**, 739–741.
Haffner, L.M., Reynolds, R.J., Madsen, G.J. et al.: 2001, *AAS* **199**, 58.01 (WHAM).
Walterbos, R.A.M.: 2000, in: E.M. Berkhuijsen, R. Beck and R.A.M. Walterbos (eds.), *The Interstellar Medium in M31 and M33*, Shaker, Aachen, 99–106.
Walterbos, R.A.M. and Braun, R.: 1994, *Astrophys. J.* **431**, 156–171.

COMPARING THE CO AND HI DISTRIBUTIONS IN NEARBY SPIRAL GALAXIES AT HIGH RESOLUTION

TONY WONG
CSIRO Australia Telescope National Facility; E-mail: Tony.Wong@csiro.au

LEO BLITZ
Astronomy Department and Radio Astronomy Laboratory, UC Berkeley, USA

J. M. VAN DER HULST*
Kapteyn Astronomical Institute, Groningen, The Netherlands

Abstract. With wide-field CO maps of spiral galaxies at $\sim 6''$ resolution provided by the BIMA Survey of Nearby Galaxies (BIMA SONG), it is now feasible to compare the CO and HI distributions in disks on scales of ~1 kpc or less. The results of such a comparison suggest that, when azimuthally averaged, the ratio of CO to HI emission, and by implication the molecular-to-atomic gas ratio, is determined by the midplane interstellar pressure, with the ISM metallicity as a secondary factor. A variety of observed phenomena, including the inhibition of molecular cloud and star formation in the outer disks of galaxies, can be understood in terms of this dependence. We present recent data on the edge-on galaxy NGC 891 that allow a comparison of the vertical distribution of CO and HI, and will therefore provide an independent test of these ideas.

Keywords: galaxies: ISM

1. Introduction: A Chicken and Egg Problem?

Do atomic clouds form from molecular clouds, or vice versa? Through direct observation one can find examples of locations where H I appears to be the result of H_2 photodissociation (e.g. the spiral arms of M51, Vogel et al., 1988), as well as locations where H_2 appears to have formed by compression of H I clouds (e.g. the Perseus arm of our Galaxy, Heyer and Terebey, 1998). The difference is that the former occur in regions where the dominant phase is molecular, and the latter in regions where the dominant phase is atomic. Depending on which phase is dominant, the appearance of the other phase is likely to be attributable to a localized event, although the underlying processes governing the phase transition remain the same.

The principal formation and destruction processes for molecular gas are expected to be as follows (Elmegreen, 1993): (1) formation of H_2 on dust grains, giving a metallicity and gas volume density dependence; (2) destruction of H_2 by far-ultraviolet (FUV) photons, giving a gas column density and FUV intensity

* Distinguished Visitor to CSIRO ATNF

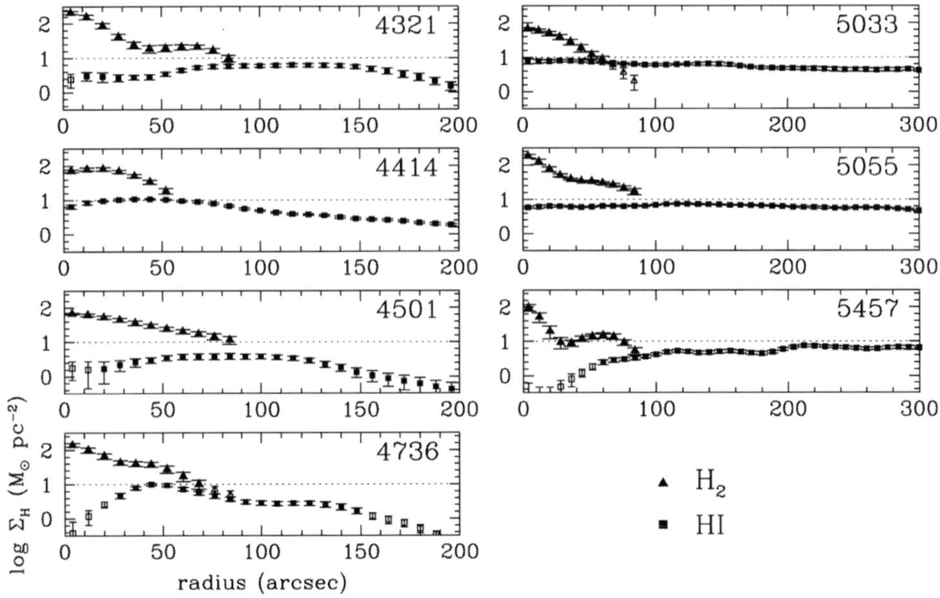

Figure 1. CO and H I radial profiles for seven galaxies studied by Wong and Blitz (2002). A Galactic value for the CO-to-H_2 conversion factor has been used.

dependence. If the volume density of a cloud scales with the ambient pressure, then the molecular fraction at a certain column density into the cloud will depend on the pressure P, metallicity Z, and photodissociative flux ϕ.

Observations of the nearest star-forming galaxies, such as the LMC and M33, are needed to probe the timescales for molecular cloud formation and destruction, and relate these to the star formation timescale. For more distant galaxies, one can examine the azimuthally averaged radial profiles of CO and H I to estimate how the overall H I/H_2 ratio varies with galactocentric radius. These radial profiles reflect a time average of the physical processes discussed above. Comparing them with the profiles of P, Z, and ϕ can provide insight into which of these factors dominates the H I-to-H_2 transition.

2. BIMA SONG Results

Using high-resolution ($\sim$6″) CO mapping with BIMA, mostly obtained as part of the BIMA Survey of Nearby Galaxies (BIMA SONG, Regan et al., 2001), and archival ($\sim$15″) H I mapping with the VLA, we have compared the CO and H I distributions in seven nearby (d=4–20 Mpc) spirals (Wong and Blitz, 2002). All of the CO maps have single-dish data from the NRAO 12m Telescope incorporated in them, and most extend out to a radius of $\sim$100″. Many galaxies show a central H I depression, whereas CO continues to increase in brightness towards the centre.

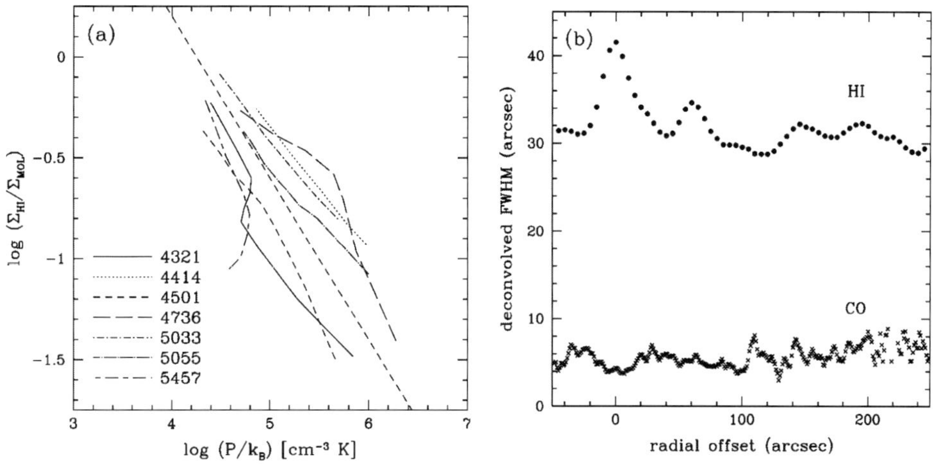

Figure 2. (a) Ratio of H I to H_2 radial profiles for seven galaxies plotted against the hydrostatic disk pressure at the midplane, as given by Elmegreen (1993). The straight dashed line represents the mean slope of −0.8. (b) Deconvolved thickness (FWHM) of the CO and H I layers in NGC 891 as a function of radius. The CO and H I beam sizes are taken to be $5''.5$ and $14''$ respectively ($1'' \approx 45$ pc).

Comparison of the radial CO and H I profiles confirms that CO is always more centrally concentrated than H I (Figure 1).

We also find that the H I/CO ratio varies systematically with radius as $kR^{1.5}$, where k is a constant that varies by a factor of ~10 among the sample galaxies. Although almost any quantity will vary with radius, we believe that the pressure P is the dominant contributor to this trend [Figure 2(a)]. Not only does this substantially reduce the scatter among galaxies, but the galaxies which still tend to have higher H I/CO ratios are those with lower metallicity, as expected from theory. We conclude that in massive spirals, variation in the interstellar pressure is primarily responsible for the H I-to-H_2 transition. The FUV field, which would produce more H I in the inner disk, is relatively less important.

3. The Edge-On Galaxy NGC 891

Recent BIMA observations of CO emission in the southern half of the edge-on galaxy NGC 891 (Wong, Howk and van der Hulst, in preparation) cover the necessary spatial dynamic range to detect both compact and extended molecular gas structures. We find that the CO layer is remarkably thin compared to the H I layer observed by Swaters et al. (1997) [Figure 2(b)], probably reflecting a steep vertical pressure profile. The CO midplane is also very well defined, apart from a strong 'kink' near the southern edge. A more quantitative analysis of both CO and H I datasets is in progress.

4. Implications

The pressure-dependent formation of molecular gas has implications for large-scale star formation in galaxies. In particular, it suggests that the lack of star formation in the outer disks of galaxies (and in low surface brightness galaxies) is due to insufficient pressure to support a cool H I phase (Elmegreen and Parravano, 1994), rather than to a lack of gravitational instability in the disk (Kennicutt, 1989). This may mitigate some of the observational problems with thresholds based on gravitational instability (Wong and Blitz, 2002). A closer comparison of the cool H I phase with the molecular gas is still needed.

We note that our analysis implicitly assumes an equilibrium pressure in an ISM that is intrinsically dynamic. Indeed, the ISM is continually being pounded by impulsive events such as supernovae, and one observes small-scale variations in the ISM pressure (E.B. Jenkins, these proceedings). However, it is nonetheless plausible that the galactic disk will be in hydrostatic equilibrium on large scales. In fact a large portion of the hydrostatic pressure may derive from turbulent motions.

Acknowledgements

We thank our BIMA SONG collaborators for making this study possible, and Jess O'Brien for a timely reduction of archival VLA data for NGC 891. A. Bosma, R. Braun, J. Knapen, C. Mundell, A. Thean, and M. Thornley kindly provided reduced H I cubes. T.W. was able to attend thanks to support from a Bolton Fellowship at the ATNF.

References

Elmegreen, B.G.: 1993, *ApJ* **411**, 170.
Elmegreen, B.G. and Parravano, A.: 1994, *ApJ* **435**, L121.
Heyer, M.H. and Terebey, S.: 1998, *ApJ* **502**, 265.
Kennicutt, R.C.: 1989, *ApJ* **344**, 685.
Regan, M.W., Thornley, M.D., Helfer, T.T., Sheth, K., Wong, T., Vogel, S.N., Blitz, L. and Bock, D.C.-J.: 2001, *ApJ* **561**, 218.
Swaters, R.A., Sancisi, R. and van der Hulst, J.M.: 1997, *ApJ* **491**, 140.
Vogel, S.N., Kulkarni, S.R. and Scoville, N.Z.: 1988, *Nature* **334**, 402.
Wong, T. and Blitz, L.: 2002, *ApJ* **569**, 157.

EVIDENCE AND IMPLICATIONS OF PRESSURE FLUCTUATIONS IN THE ISM

EDWARD B. JENKINS
Princeton University Observatory, Princeton, NJ 08544-1001, USA;
E-mail: ebj@astro.princeton.edu

Abstract. A recent survey of the fine-structure excitation of neutral carbon reveals that the interstellar medium in the Galactic plane exhibits a thermal pressure, nT/k, that ranges from about 10^3 to $10^4 \mathrm{cm}^{-3}$K from one location to the next, with occasional excursions in excess of about $10^5 \mathrm{cm}^{-3}$K. The large excitations for small amounts of gas indicate that some regions are either subjected to shocks or must be pressurized within time scales much shorter than the time needed to reach thermal equilibrium. These rapid fluctuations probably arise from the cascade of macroscopic mechanical energy to small scales through a turbulent cascade. One consequence of this effect is that changes in gas temperature can arise from near adiabatic compressions and expansions, and this may explain why investigations of 21-cm emission and absorption reveal the presence of hydrogen at temperatures well below the expected values derived from the balance of various known heating and cooling processes.

Keywords: dynamics, thermal pressure, gas phases, turbulence, structure

Abbreviations: ISM – Interstellar Medium

1. Introduction

While gases within the Milky Way show enormous ranges in density and temperature, the simplest, most general picture is that variations in the product of these two quantities, the thermal pressure, are very much smaller by comparison. This may be true at the most superficial level, and indeed a convincing theory (at the time) of the ISM (Field, Goldsmith and Habing, 1969) was built upon the premise that the observable phases of the ISM arise simply from a thermal instability (Field, 1965) with uniform heating at a nearly constant pressure. In this picture, the cool phase is reasonably static and confined by a uniform external pressure from a surrounding warm medium. It is now more apparent that this tranquil picture of the ISM is probably unjustified since we have known for some time that disturbances over macroscopic scales arising from supernova explosions, newly formed H II regions, stellar mass loss, bipolar flows, infalling gas from the halo and Galactic density wave shocks can all inject significant mechanical energy in various locations and disrupt the medium.

A major refinement in the theory of the ISM was created by McKee and Ostriker (1977), who recognized some important global consequences of energetic

Astrophysics and Space Science **289:** 215–223, 2004.

phenomena, which can lead to the creation of large filling factors for material at high temperatures and low densities [see also (Cox and Smith, 1974; Cox, 1979, 1981)]. An often overlooked property of the McKee and Ostriker model [e.g., (Cox, 1995)] is that the pressures should vary by significant amounts from one location to the next (Jenkins, Jura and Loewenstein, 1983). Ultimately, we should expect that large-scale enhancements and rarefactions in the pressures should become more widespread and chaotic as they propagate to different locations and migrate to smaller scales through the creation of a turbulent cascade (Mac Low et al., 2001).

Thermal pressures are a minor component of the total ISM pressure, which also includes magnetic, dynamical (i.e., turbulent) and cosmic ray components. Boulares and Cox (1990) estimated that the weight of material on either side of the Galactic plane should balanced by a total pressure $p/k = 2.5 \times 10^4\ \mathrm{cm}^{-3}$ K. Inasmuch as typical thermal pressures are about a factor 10 lower and thus seemingly of little importance dynamically, one might wonder why we are interested in their average value and distribution over different volumes of space. The answer arises from several considerations: first, thermal pressure values are molded by processes arising from the other forms of pressure and thus convey important information, second, they reflect the physical state of the gas at the atomic scale and thus have an influence on various reaction rates and equilibria, such as those pertaining to ionization, heating, cooling and chemistry. Finally, thermal pressures are easy to measure.

There are various ways to determine thermal pressures, but probably the most straightforward one is by measuring the relative populations of atoms in the excited fine-structure levels of their ground electronic states. This can be done by observing absorption lines of the atomic multiplets, seen in the uv spectra of hot background stars. The fine-structure levels are populated by collisions with other atoms, ions, molecules, and electrons. The excitations (and de-excitations) occur at rates proportional to the local densities of these collision partners, and a balance between these processes and the spontaneous radiative decay of the upper levels establishes an equilibrium (with a very short time constant) determined by the local density and temperature of the gas.

2. Thermal Pressures from C I Fine-structure Excitation

Of the various species that have ground electronic states with two or more fine-structure levels, in most circumstances neutral carbon seems to be the most useful one for indicating local conditions. While one is not able to determine unique values for either temperature or density, Jenkins and Shaya (1979) showed that under the most common conditions in the ISM their product was reasonably well defined from the two population ratios of the upper levels with respect to the total in all three states. Moreover, since the problem is over-determined by having more than one ratio, it is possible to differentiate between a uniform, medium pressure

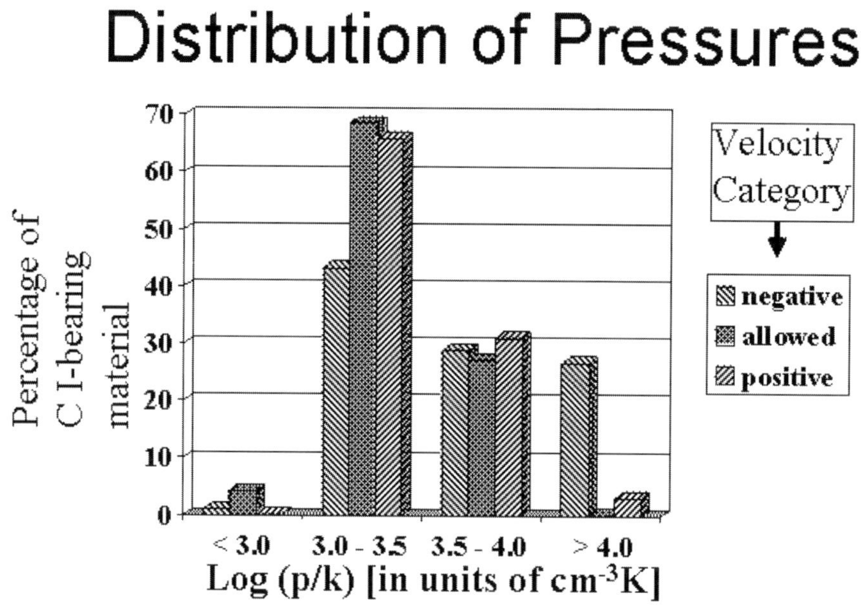

Figure 1. A histogram indicating the relative amounts of C I in different logarithmic intervals found by Jenkins and Tripp (2001), with the gas segregated according to whether the radial velocity is below, within, or above the allowed range of velocities between the observer and the star, as determined from estimates for the effects of differential Galactic rotation along the respective sight lines.

and an admixture of two or more regions having low and high pressures, even if they have the same radial velocity. The latter property was exploited by Jenkins and Tripp (2001) (hereafter JT01) to show that there was pervasive evidence for a small amount of gas at $p/k > 10^5\,\mathrm{cm}^{-3}\,\mathrm{K}$, some two orders of magnitude above the general average pressure.

Figure 1 shows the amounts of C I within different pressure intervals that were found in a survey of 21 stars by JT01, who used the highest resolution mode (E140H with the narrowest entrance slit, yielding $R = 200{,}000$) of the *Space Telescope Imaging Spectrograph* on *HST*. The pressure distributions show a clear link to the kinematical properties of the gas: disturbed material outside the range of normally allowed radial velocities are dominant at high pressures, while the lowest extremes in pressure appear to be quiescent. From this it seems clear that the thermal pressures are responding to dynamical processes, as we would expect. The preponderance of negative velocity gas at high pressures probably arises from material compressed by mass flows from the target stars or their neighbors in the early-type stellar associations.

As we consider the evidence from C I, we must acknowledge that the distribution of pressures reflects material weighted by the respective abundances of C I, which in turn are governed by the ionization equilibrium of C I with the more common form of carbon, C II. This effect exaggerates the amount of high-pressure

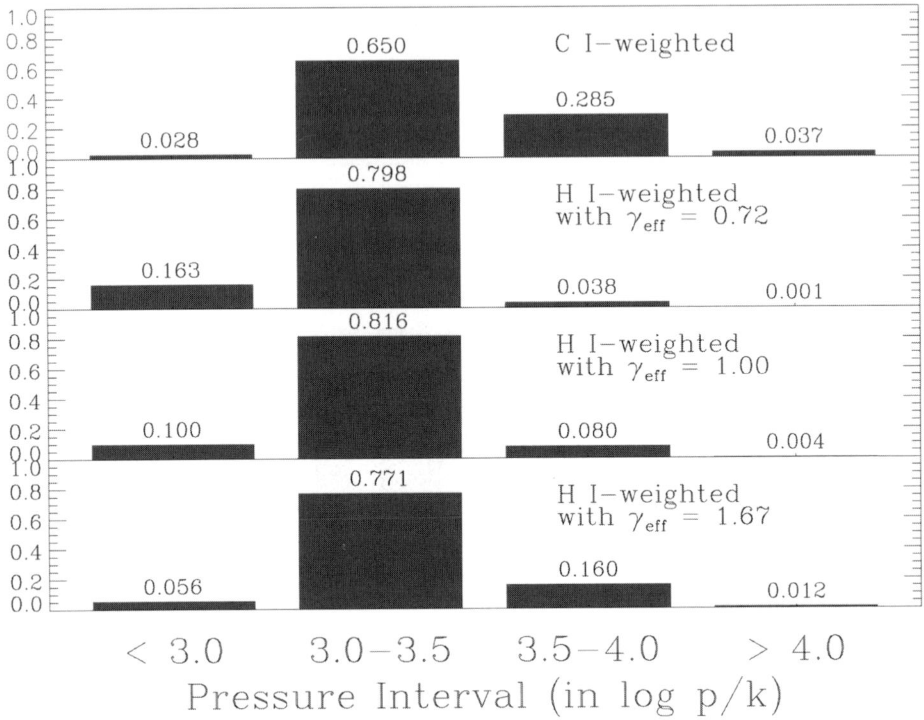

Figure 2. Transformations of the original C I pressure distribution (top panel: sum of all velocity intervals depicted in Figure 1) to the total amounts of gas for 3 values of the barytropic index $\gamma_{\rm eff}$ (lower 3 panels), derived by applying Eq. 1 to each pressure interval. The three values of $\gamma_{\rm eff}$ represent (1) 0.72: approximate slope of the thermal equilibrium relation, 1.00: isothermal gas, 1.67; purely adiabatic behavior.

material because there is a shift toward greater C I/C II at high densities. If the gas can be characterized with an equation of state (see §3) $p \propto n^{\gamma_{\rm eff}}$, where $\gamma_{\rm eff}$ is the barytropic index,

$$n({\rm H\ I}) \propto p^{0.324-1.622/\gamma_{\rm eff}} n({\rm C\ I})\ . \tag{1}$$

Eq. 1 shows the correction[1] that must be applied to the C I pressure distribution function to obtain an even-handed representation for all of the neutral gas. The transformations of the distributions for three different values of $\gamma_{\rm eff}$ are shown in Figure 2.

[1] The coefficients in the exponent of p are derived from the calculations of carbon ionization equilibria by Jenkins (2002) which include not only recombinations of ions with free electrons but also their neutralization by very small, negatively-charged dust grains (Weingartner and Draine, 2001).

3. The Effects of Changes in Pressure

We now imagine what happens when the ISM is in a chaotic state where dynamical processes force the pressures to vary with position and time on either side of some mean value. In this picture, we may characterize the pressure fluctuations as random. Even though temperatures and densities may vary, we must still insist that a balance between heating and cooling is maintained over a time interval much longer than the characteristic thermal time scales. Of course, our assumption about randomness is an idealization: in large part the real data will reflect this condition, but from time to time coherent effects (pressurizations) from specific sources are bound to manifest themselves and distort the results.

Wolfire et al. (1995) have calculated thermal equilibria of the ISM using recent refinements in the estimates for various processes that contribute to the heating and cooling. Two of their equilibrium curves are shown in the three panels showing $\log(p/k)$ vs. $\log n$ of Figure 3. If we propose that pressure changes are very slow compared to the thermal equilibration time, then the gas will adjust its temperature to agree with the equilibrium line as the pressure changes from one extreme to another (solid curve). This condition is shown in the top panel of the figure. However, the results of JT01 indicate that such slow changes for the more extreme excursions are not realistic, since large positive pressures would cause the temperature to drop to the point that the C I excitations become very small, contrary to what is observed. JT01 found that $\gamma_{\rm eff} > 0.90$, i.e., measurably larger than $\gamma_{\rm eff} \approx 0.72$ for the equilibrium curve.

At the opposite extreme, very rapid fluctuations in pressure will cause the gas to move along an adiabat, where $\gamma_{\rm eff}$ becomes the real value of γ, $c_p/c_v = 5/3$, as is illustrated in the middle panel of Fig. 3. Also, if the gas is subjected to supersonic turbulence, shocks will form and cause momentary excursions of the temperatures up to about 1000 K for Mach numbers $M \approx 10$.

Finally, we may consider a contrived but pedagogically useful situation where the pressure driving function is approximately a square wave. Here, a cycle may be established where rapid adiabatic expansions and compressions have interludes where the gas may approach its thermal equilibrium for the given pressure. In this last case, somewhat larger extremes in temperature may be achieved. The significance of this possiblity is discussed in the next section.

4. The Puzzle of Cold H I

Recent surveys of 21-cm emission and absorption are providing convincing evidence that temperatures of neutral hydrogen in the ISM in many cases deviate substantially from the values expected from thermal equilibrium (Gibson et al., 2000; Heiles 2001; Knee and Brunt, 2001; Dickey et al., 2003; Heiles and Troland, 2003). [The part of the equilibrium curve with a negative slope in Figure 3

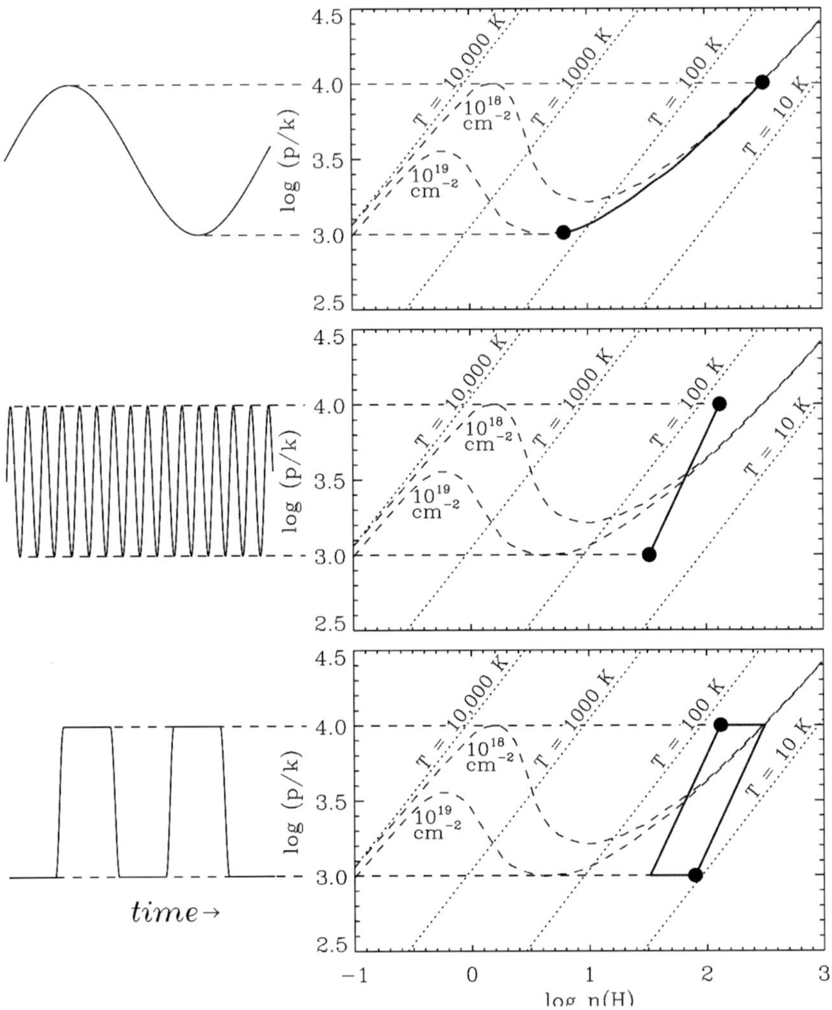

Figure 3. Three schematic illustrations of how the logarithms of the ISM pressures (ordinates) and densities (absicssae) could change for different hypothetical driving functions of pressure as a function of time (three periodic waves on the left). The dashed curves inside the panels show the thermal equilibrium curves of Wolfire et al. (1995) for column densities $N(\mathrm{HI}) = 10^{19}\,\mathrm{cm}^{-2}$ (lower curve) and $10^{18}\,\mathrm{cm}^{-2}$ (upper curve) for shielding against external EUV radiation. *Top panel:* Very slow changes in pressure (shown to the left of the panel) result in the gas conditions moving up and down the thermal equilibrium curve with a slope in the $\log p - \log n$ plane of $\gamma_{\mathrm{eff}} \approx 0.72$. The temperature extremes (heavy dots) for pressures ranging from 10^3 to $10^4\,\mathrm{cm}^{-3}$ K are 166 and 30 K. *Middle panel:* Very rapid changes in pressure will cause the gas to move on an adiabatic track ($\gamma_{\mathrm{eff}} = 5/3$) with temperature extremes of 30 and 76 K for the same pressure differences. *Bottom panel:* A square-wave driving function can achieve larger temperature extremes of 12 and 76 K, as the gas is forced to move on the parallelogram track in a clockwise fashion.

is unstable to bifurcation into separate warmer and cooler phases (Field, 1965; Shull, 1987; Begelman, 1990).] In a chaotic medium, one can understand why temperatures that are intermediate between the two principal stable phases may be present. It is easy to imagine that mixing of warm and cool phases can operate over timescales shorter than the time needed to segregate the gas into stable high and low temperature branches of the equilibrium curve (Vázquez-Semadeni, Gazol and Scalo, 2000; Gazol et al., 2001). However, H I temperatures *below* a value of about 30 K are more difficult to accept. Occasionally, temperatures as low as 10 K have been recorded.

One initially might propose that the very cold temperatures imply simply a diminution of the heating rate or enhancement of the cooling rate. For instance, perhaps there is a deficit of small grains that are responsible for photoelectric heating. While this may sound like an attractive solution, Wolfire et al. (1995) show that the dust-to-gas ratio must be lower than one-tenth the normal value to have temperatures as low as 15 K, a value that is still above that of some of the 21-cm absorbing clouds. Many of the cold clouds seem not to be identified with regions containing molecules which could add to the cooling rate (Gibson et al., 2000).

A possible solution to the cold H I problem could arise from the existence of mechanical cooling due to rapid pressure fluctuations. A single cycle of pressure changes, perhaps one that somewhat approximates the square-wave idealization shown in the bottom panel of Figure 3, could achieve a temperature as low as 12 K soon after a rapid expansion from a thermally stabilized condition starting at $p/k = 10^4$ K. However, in order for decompressions to drive the gas along an adiabat, they must occur over a time scale[2] that is much shorter than the heating time,

$$t_{\text{heat}} = (5/2)\frac{kT}{\Gamma} \tag{2}$$

≈ 5500 yr at $T = 30$ K if the heating rate $\Gamma = 6 \times 10^{-26}$ erg s^{-1}atom^{-1} that is typical for the cold, neutral medium (Wolfire et al., 1995). A cloud that is suddenly relieved of an external confining pressure will expand only at its sound speed,

$$c_s = (\gamma p/\rho)^{1/2} \tag{3}$$

which equals 0.54 km s^{-1} for $T = 30$ K. If the cloud's shape is a sheet, then the factor of 4 drop in n(H) shown in Fig. 3 could be accomplished in less than 5500 yr if the thickness were less than about 10^{-3} pc (cylindrical or spherical clouds could satisfy this condition if they had diameters larger by factors of $3^{1/2}$ and $3^{2/3}$, respectively). However, in a medium where there is supersonic turbulence,

[2] The arguments presented here apply to gas that is mostly in atomic form. See Ballesteros-Paredes, Vázquez-Semadeni and Scalo (1999) for a similar discussion that applies to dense, molecular clouds.

random inertial forces in the fluid can create large positive mass divergences and drag a dense region apart at an expansion rate comparable to the random velocities present, which may exceed the sound speed by a large factor. Thus, roughly speaking, if the average Mach number of the turbulence is 10, we may expect an increase of the limiting size for nearly adiabatic cooling to $\sim$0.01 pc.

Over recent years, new findings from optical absorption lines (Meyer and Lauroesch, 1999; Lauroesch, Meyer and Blades, 2000; Andrews, Meyer and Lauroesch, 2001) and 21-cm line absorption (Dieter, Welch and Romney, 1976; Diamond et al., 1989; Frail et al., 1994; Davis, Diamond and Goss 1996; Faison et al., 1998; Faison and Goss, 2001) have shown unmistakable evidence that in almost any viewing direction the ISM exhibits large contrasts in density[2] over scales ranging from 2×10^{-5} to 0.03 pc. These observations provide strong support for the pervasive character of small structures, a conclusion that is consistent with their being shaped by the action of turbulence.

5. Final Remarks

A major theme advanced in this paper is that the pressures and hence structure of the ISM change rapidly and are probably molded by chaotic velocity fields that converge and diverge. This is a significant departure from the picture that high densities arise simply from static 'clouds' that are stabilized by a uniform external pressure. The notion that such clouds may be an oversimplification is not new: the nondiscrete character of ISM phases was imagined by Chandrasekhar and Münch (1952) when they stated, '... the distribution of density is considered to be continuous but exhibiting fluctuations of a statistical character.' This outlook was formed long before interstellar turbulence was a fashionable topic.

An important consideration that arises from the dominant role of interstellar turbulence is that under many circumstances the density condensations are ephemeral (unless gravitational binding or the coherent, large-scale dynamical processes listed in §1 become important). As a consequence, pressure enhancements of very short duration may allow some the more rapid atomic and chemical processes to take place, but other reactions may not achieve an equilibrium state. For instance, the time needed to adjust the rotational temperatures of H_2 range from 7000 yr for moderate J levels to change by 2 as they collide with H, to about 90,000 yr for transitions between ortho and para hydrogen caused by collisions with protons. The equilbrium between molecular and atomic hydrogen in the general ISM has a time constant that ranges from about 600 yr for unshielded regions to over 10^7 yr for shielded regions. In view of the strong indications that pressures can change over time scales intermediate between these extremes, we must view with caution

[2] In interpreting raw information from either the visible absorption line or 21-cm absorption results, one must acknowledge that they have temperature dependences $T^{-0.62}$ and $T^{-1.00}$, respectively.

any interpretations that assume that these phenomena have advanced uniformly to an equilibrium state.

References

Andrews, S.M., Meyer, D.M. and Lauroesch, J.T.: 2001, *ApJ* **552**, L73–L76.

Ballesteros-Paredes, J., Vázquez-Semadeni, E. and Scalo, J.: 1999, *ApJ* **515**, 286–303.

Begelman, M.C.: 1990, in: H. Thronson and M. Shull (eds.), *The Interstellar Medium in Galaxies*, Kluwer, Dordrecht, pp. 287–304.

Boulares, A. and Cox, D.P.: 1990, *ApJ* **365**, 544–558.

Chandrasekhar, S. and Münch, G.: 1952, *ApJ* **115**, 103–123.

Cox, D.P.: 1979, *ApJ* **234**, 863–875.

Cox, D.P.: 1981, *ApJ* **245**, 534–551.

Cox, D.: 1995, *Nature* **375**, 185–186.

Cox, D.P. and Smith, B.W.: 1974, *ApJ* (Letters) **189**, L105–L108.

Davis, R.J., Diamond, P.J. and Goss, W.M.: 1996, *MNRAS* **283**, 1105–1114.

Diamond, P.J., Goss, W.M., Romney, J.D., Booth, R.S., Kalberla, P.M.W. and Mebold, U.: 1989, *ApJ* **347**, 302–306.

Dickey, J.M., McClure-Griffiths, N.M., Gaensler, B.M. and Green, A.J.: 2003, *ApJ* **585**, 801.

Dieter, N.H., Welch, W.J. and Romney, J.D.: 1976, *ApJ* (Letters) **206**, L113–L115.

Faison, M.D. and Goss, W.M.: 2001, *AJ* **121**, 2706–2722.

Faison, M.D., Goss, W.M., Diamond, P.J. and Taylor, G.B.: 1998, *AJ* **116**, 2916–2928.

Field, G.B.: 1965, *ApJ* **142**, 531–567.

Field, G.B., Goldsmith, D.W. and Habing, H.J.: 1969, *ApJ* (Letters) **155**, L149–L154.

Frail, D.A., Weisberg, J.M., Cordes, J.M. and Mathers, C.: 1994, *ApJ* **436**, 144–151.

Gazol, A., Vázquez-Semadeni, E., Sánchez-Salcedo, F.J. and Scalo, J.: 2001, *ApJ* **557**, L121–L124.

Gibson, S.J., Taylor, A.R., Higgs, L.A. and Dewdney, P.E.: 2000, *ApJ* **540**, 851–862.

Heiles, C.: 2001, *ApJ* (Letters) **551**, L105–L108.

Heiles, C. and Troland, T.H.: 2003, *ApJ* **586**, 1067.

Jenkins, E.B.: 2002, *astro-ph*, 0207579.

Jenkins, E.B., Jura, M. and Loewenstein, M.: 1983, *ApJ* **270**, 88–104.

Jenkins, E.B. and Shaya, E.J.: 1979, *ApJ* **231**, 55–72.

Jenkins, E.B. and Tripp, T.M.: 2001, *ApJ* (Suppl.) **137**, 297–340.

Knee, L.B.G. and Brunt, C.M.: 2001, *Nature* **412**, 308–310.

Lauroesch, J.T., Meyer, D.M. and Blades, J.C.: 2000, *ApJ* **543**, L43–L47.

MacLow, M.-M., Balsara, D., Avillez, M.A. and Kim, J.: 2001, *astro-ph*, 0106509.

McKee, C.F. and Ostriker, J.P.: 1977, *ApJ* **218**, 148–169.

Meyer, D.M. and Lauroesch, J.T.: 1999, *ApJ* **520**, L103–L106.

Shull, J.M.: 1987, in: D.J. Hollenbach and H.A. Thronson (eds.), *Interstellar Processes*, Reidel, Dordrecht, pp. 225–244.

Vázquez-Semadeni, E., Gazol, A. and Scalo, J.: 2000, *ApJ* **540**, 271–285.

Weingartner, J.C. and Draine, B.T.: 2001, *ApJ* **563**, 842–852.

Wolfire, M.G., Hollenbach, D., McKee, C.F., Tielens, A.G.G.M. and Bakes, E.L.O.: 1995, *ApJ* **443**, 152–168.

D/O AND D/H RATIOS IN THE LOCAL INTERSTELLAR MEDIUM FROM FUSE OBSERVATIONS *

GUILLAUME HÉBRARD
Institut d'Astrophysique de Paris, CNRS, 98 bis, boulevard Arago, F-75014 Paris, France

Abstract. Since HI, OI, and DI have nearly the same ionization potential, the deuterium-to-oxygen ratio (D/O) is an important tracer of the D/H ratio and of its putative spatial variations. D/O is indeed very sensitive to astration, both because of deuterium destruction and oxygen production. Moreover, D/O measurements are less subject to systematic errors than D/H.

A survey of D/O in the interstellar medium is performed by the Deuterium Working Group of the FUSE PI Team. The first results show that D/O is a better D/H proxy than D/N, and that D/O is homogeneous in the local interstellar medium, at least within about 100pc. The stability of D/O appears as a strong argument which supports both D/H and O/H spatial stability in the local interstellar medium.

Keywords: ISM: abundances, ISM: clouds, ultraviolet: ISM

1. Introduction

Deuterium is believed to be produced in appreciable quantities only in primordial Big Bang nucleosynthesis and destroyed in stellar interiors; it is thus a key element in cosmology and in galactic chemical evolution. Three samples of D/H measurements may be performed, each representative of a given epoch, at look-back times of $\sim$14 Gyrs (in primordial intergalactic clouds), 4.5 Gyrs (protosolar), and 0.0 Gyrs (interstellar medium, $(D/H)_{ISM}$). Although the evolution of the deuterium abundance seems to be qualitatively understood, measurements of D/H at similar redshifts show some dispersion and indicate that additional processes may be important in shaping the abundance of deuterium (see Lemoine et al., 1999, for a review).

The measurement of $(D/H)_{ISM}$ through Lyman absorption-line observations are among the main scientific objectives of *FUSE*, the Far Ultraviolet Spectroscopic Explorer (Moos et al., 2000). I present here its first results related to that subject.

* Based on observations made with the NASA-CNES-CSA Far Ultraviolet Spectroscopic Explorer. FUSE is operated for NASA by the Johns Hopkins University under NASA contract NAS5-32985.

Astrophysics and Space Science **289:** 225–228, 2004.

2. Previous $(D/H)_{ISM}$ Measurements

The first measurement of $(D/H)_{ISM}$ was reported by Rogerson and York (1973) toward β Cen, using *Copernicus*: $(D/H)_{ISM} = 1.4 \pm 0.2 \times 10^{-5}$. Since then, many other $(D/H)_{ISM}$ measurements have been performed using different instruments (*Copernicus*, IUE, *IMAPS*, HST) and the derived values show significant dispersion around the β Cen one.

For example, an average value $(D/H)_{ISM} = 1.50(\pm 0.10) \times 10^{-5}$ $(1\,\sigma)$ has been derived for the Local Interstellar Cloud by Linsky (1998) from the comparison of 12 nearby sight lines, but several other lines of sight show values outside this range (Lemoine et al., 1999). This dispersion may result from spatial variations due to some unknown physical processes or underestimation of systematic errors. There is still considerable debate on this issue which final resolution may have implications for understanding the physics of the ISM, as well as the chemical evolution of the Galaxy and the baryonic density of the Universe inferred from D/H measurements.

3. D/H with *FUSE*

Several tens of targets were observed by *FUSE* within the local ISM. D I absorptions lines are clearly detected in a large part of these observations, together with other species (O I, N I, Fe II...). Results for 7 lines of sight (white dwarfs and subdwarfs) were already published (Friedman et al., 2001; Hébrard et al., 2002; Kruk et al., 2002; Lehner et al., 2002; Lemoine et al., 2002; Sonneborn et al., 2002; Wood et al., 2002), with an overview by Moos et al. (2002).

The weighted mean of the D I/H I ratio for the five *FUSE* targets with reliable H I column densities is $D/H = 1.52\,(\pm 0.08) \times 10^{-5}$ $(1\,\sigma)$, without significant variation from one sight line to the other. Thus, the *FUSE* studies support the idea that the D/H ratio is constant out to a distance of about 100 pc.

However, H I column density measurements from Lyman absorption lines fits have been proved to be subject to systematic errors due to the possible presence of broad, weak extra absorbing components (Linsky and Wood, 1996; Lemoine et al., 2002; Vidal-Madjar and Ferlet, 2002), presumably arising from the heliosphere. Indeed, one of the challenges of D/H measurements is to evaluate simultaneously H I and D I column densities, which differ by about 5 orders of magnitudes. This could lead to systematic errors due to the fact that all lines from the same species may lie on the non-linear part of the curve of growth, or that there may be clouds seen in H I but without D I detectable counterpart.

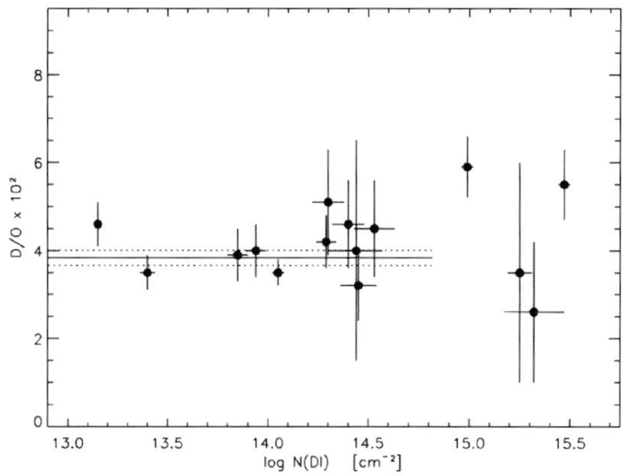

Figure 1. **D/O vs. N(D I).** The weighted mean of the 11 most local values is $(\mathrm{D/O})_{\mathrm{LISM}} = 3.84\,(\pm 0.17) \times 10^{-2}$. All the plotted error bars are 1 σ.

4. D/O with *FUSE*

An alternative may be to use the D I/O I ratio as a proxy to D I/H I. Many of the difficulties inherent to D/H measurements could be avoided by measuring D/O (e.g. Timmes et al., 1997): indeed, the average D/O ratio in the ISM is of the order of a few percent, instead of a few 10^{-5} for D/H, and many O I absorption lines with different oscillator strengths are present in the *FUSE* bandpass. Furthermore, O I is believed to be a good tracer of H I in the Galactic disk (Meyer, 2001) since both species have nearly the same ionization potential (the ionization of H I, D I, and O I are locked together by high charge transfer; Jenkins et al., 2000); neutral forms of these species dominate over ionized states in the diffuse ISM. Moreover, D/O is sensitive to astration, both from D destruction and O production.

For targets at distances lower than 100 pc, within the Local Bubble, *FUSE* results show a constancy in the D/O ratio (Moos et al., 2002). A *FUSE* survey of D/O in the local ISM that includes additional sight lines also reaches the same conclusion (see Figure 1), contrary to the behaviour of the D I/N I ratio. The discrepancy probably arises from O I being a better tracer of H I than N I, because of ionization effects (Jenkins et al., 2000). The weighted mean of D/O and the standard deviation of the mean are $(\mathrm{D/O})_{\mathrm{LISM}} = 3.84\,(\pm 0.17) \times 10^{-2}$.

The D/O constancy in the LISM agrees with that of $(\mathrm{D/H})_{\mathrm{LISM}}$. Indeed, the only possibility to have both D/O stability and D/H variations would be that D/H and O/H vary precisely in the same way for D/O to remain constant. That is unlikely since (i) O/H is uniform in the ISM over paths of several hundreds parsecs (e.g. Meyer, 2001), and (ii) astration should lead to an anti-correlation of D I and O I abundances. The stability of D/O in the LISM appears rather as a strong argument which supports both D/H and O/H spatial stability in the LISM. We note however that D I and O I anti-correlation is not detected over limited metallicity range sampled by these *FUSE* sight lines. By assuming the value O/H =

$3.43(\pm 0.15) \times 10^{-4}$ found by Meyer (2001), the *FUSE* D/O result implies D/H $= 1.32(\pm 0.08) \times 10^{-5}$ ($1\,\sigma$). This value is slightly lower than the average *FUSE*-value reported above.

First *FUSE* results obtained for $\log N(\mathrm{D\,I}) \geq 15$ and $d \geq 100\,\mathrm{pc}$ suggest that D/O variations may occur at larger column densities (larger pathlengths). Obviously, such studies should be extended to more distant lines of sight.

5. HD/H_2 with *FUSE*

These distant sight lines will be observed through D I Lyman absorption lines; however, some of them could probe dense molecular clouds in which H_2 and HD molecules will be observed through multiple absorption lines in the *FUSE* band pass. Indeed, the high sensitivity of *FUSE* allows denser clouds to be observed (Ferlet et al., 2000), in which most of the deuterium and hydrogen are expected to be in their molecular form. Thus, D/H can be obtained from the HD/H_2 ratio without strong model-dependent uncertainties. Up to now, HD is detected in more than 100 lines of sight with *FUSE*.

References

Ferlet, R., André, M., Hébrard, G., Lecavelier, A. et al.: 2000, *ApJ* **438**, L69.
Friedman, S.D., Howk, J.C., Chayer, P. et al.: 2002, *ApJS* **140**, 37.
Hébrard, G., Lemoine, M., Vidal-Madjar, A. et al.: 2002, *ApJS* **140**, 103.
Jenkins, E.B., Oegerle, W.R., Gry, C. et al.: 2000, *ApJ* **538**, L81.
Kruk, J.W., Howk, J.C., André, M., Moos, H.W. et al.: 2002, *ApJS* **140**, 19.
Lehner, N., Gry, C., Sembach, K.R., Hébrard, G. et al.: 2002, *ApJS* **140**, 81.
Lemoine, M., Audouze, J., Ben Jaffel, L. et al.: 1999, *NewA* **4**, 231.
Lemoine, M., Vidal-Madjar, A., Hébrard, G. et al.: 2002, *ApJS* **140**, 67.
Linsky, J.L. and Wood, B.E.: 1996, *ApJ* **463**, 254.
Linsky, J.L.: 1998, *SSRv* **84**, 285.
Meyer, D.M.: 2001, in: R. Ferlet et al. (eds.), *XVIIth IAP Colloquium*, Paris, p. 135.
Moos, H.W., Cash, W.C., Cowie, L.L., Davidsen, A.F. et al.: 2000, *ApJ* **538**, L1.
Moos, H.W., Sembach, K.R., Vidal-Madjar, A., et al.: 2002, *ApJS* **140**, 3.
Rogerson, J. and York, D.: 1973, *ApJ* **186**, L95.
Sonneborn, G., André, M., Oliveira, C., Hébrard, G. et al.: 2002, *ApJS* **140**, 51.
Timmes, F.X., Truran, J.W., Lauroesch, J.T. and York, D.G.: 1997, *ApJ* **476**, 464.
Vidal-Madjar, A. and Ferlet, R.: 2002, *ApJ* **571**, L169.
Wood, B.E., Linsky, J.L, Hébrard, G., Vidal-Madjar, A. et al.: 2002, *ApJS* **140**, 91.

X-RAYS FROM THE LOCAL BUBBLE

MICHAEL J. FREYBERG
Max-Planck-Institut für extraterrestrische Physik, Giessenbachstraße 1, 85748 Garching, Germany

Abstract. Thermal plasma emission in the soft X-ray band (0.1 – 2.0 keV) is believed to be responsible for the bulk of the X-ray intensity seen from the Local Bubble, a low-density cavity extending over $\sim$ 70 – 200 pc around the Sun. The state of the hot plasma is still a matter of discussion as previous instrumentation like aboard ROSAT was not able to unambiguously distinguish between equilibrium and non-equilibrium emission models and thus to pin-point the origin of the Local Bubble. Recent missions like DXS, XQC, and XMM-Newton have shed more light on this subject and observations indicate that collisional ionization equilibrium with solar abundances cannot explain the data: lines appear at positions and with intensities in contradiction to standard models. Analysis of EPIC-pn data of X-ray shadowing observations (MBM 12, Ophiuchus molecular cloud) suggest a component with higher temperature ($kT \sim 0.14$ keV) besides the standard $kT \sim 0.09$ keV plasma.

Keywords: local bubble, local interstellar medium, galactic halo, diffuse X-ray emission, XMM-Newton mission

1. Introduction

Since the discovery of the soft X-ray (0.1 – 2.0 keV) diffuse background (SXRB) its origin has been controversially discussed (see e.g. Snowden et al. 1998; Breitschwerdt 2001, and references therein). The general observed anti-correlation of X-ray intensity with galactic neutral hydrogen column density favoured an 'absorption model' which placed the emitting regions behind the absorbing material. This could not explain a non-vanishing flux in the galactic plane where the large absorbing column density would not allow distant flux to be observable. 'Interspersed' models mixed the emitting and absorbing material, however, had to clump the absorbing gas by more than allowed from 21 cm observations (Burrows, 1989). A 'displacement model' suggests that the emitting X-ray flux originates in front of the absorbing material in a 'Local Bubble' (LB) and that larger intensities are related to longer lines-of-sight through the emitting plasma (Snowden et al., 1990). This model could best explain the obervations in the pre-ROSAT era. Not much soft X-ray flux from distant regions was expected.

After the launch of ROSAT the picture changed significantly. The ROSAT PSPC All-Sky Survey (RASS) provided a new large-scale data base in the soft X-ray band with higher spatial and spectral resolution than available before, accompanied by an extensive follow-up programme of pointed observations. This survey, however, suffered from the only moderate spectral resolution of proportional counters. A detailed spectroscopy was therefore still not possible because of ambiguities in the spectral modelling. Various approaches were used to disentangle local and distant plasma components.

Looking toward opaque absorbers (e.g. dense molecular clouds) gives the foreground emission while lines-of-sight off-cloud sample both local and distant emission components. Such an X-ray shadow (not predicted from a simple displacement model) was first detected in the *Draco* cloud complex at a distance of $\geq$ 300 pc (Snowden et al., 1991; Burrows and Mendenhall, 1991) which had proven that a significant fraction ($\sim 1/2$) of the soft X-ray emission originates beyond of the bulk of the absorbing material and that the distribution of material responsible for the SXRB may be more complicated.

The moderate spectral resolution of the ROSAT PSPC allowed only about 6 independent energy bands, i.e. R1 $\sim$ 0.08 – 0.28 keV, R2 $\sim$ 0.14 – 0.28 keV, R4 $\sim$ 0.44 – 1.01 keV, R5 $\sim$ 0.56 – 1.21 keV, R6 $\sim$ 0.73 – 1.56 keV, and R7 $\sim$ 1.05 – 2.04 keV (Snowden et al., 1997). Colour information can be obtained by means of band ratios, e.g. the R2/R1 (Snowden et al., 1998) and R4/R1 (Freyberg and Schmitt, 1998). While this can give information on global spectral properties it does not try to fit the model spectra in the other bands. Information on lines inside a single band is almost completely hidden unless the intensity in this band is dominated by a single spectral line or if the presence of a line (complex) is strongly correlated to lines in other bands. Important X-ray emission lines for a million K thermal plasma are C VI (368 eV), O VII (triplet 561 – 574 eV), O VIII (653 eV), Fe XVII ($\sim$ 740 eV and 810 eV). Many more lines contribute to the X-ray spectrum, depending on the actual state and history of the plasma (see e.g. Breitschwerdt and Schmutzler, 1999).

Due to the strong energy dependence of the absorption cross sections the corresponding absorbing column density for optical depth $\tau = 1$ for a broad band depends on the incident spectrum as it is gradually hardened while passing through the interstellar medium. Typical values for $\tau = 1$ for temperatures of 1 to several million K for the abovementioned bands are 0.8 – 1.0, 1.0 – 1.5, 6.0 – 12, 10 – 15, 18 – 27, and 30 – 60 $\times$ 10^{20} cm^{-2}, respectively. The interstellar absorption for 1 keV X-rays is comparable to the infrared *J* band.

Recently, with the launch of Chandra and XMM-Newton with their superior spatial resolution and large collecting area, respectively, somewhat complementary observatories were placed into orbit. The spectral resolution of the imaging CCD cameras is better as compared to proportional counters, but these are still not able to resolve important line complexes like the O VII triplet. Spectrometers like the LETG or RGS aboard these satellites, however, do not have the necessary sensitivity (i.e. effective area) for studies of low surface brightness diffuse objects and lose much of its resolving power due to the window function and a missing entrance slit which convolves spatial and spectral information into one CCD pixel. Therefore still an ambiguity in spectral deconvolution remains which could only be

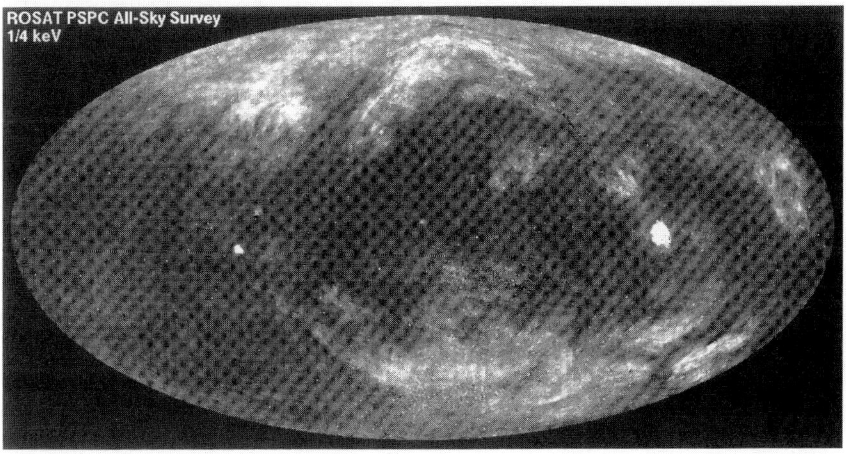

Figure 1. ROSAT All-Sky Survey in the 1/4 keV (R12) band in galactic coordinates

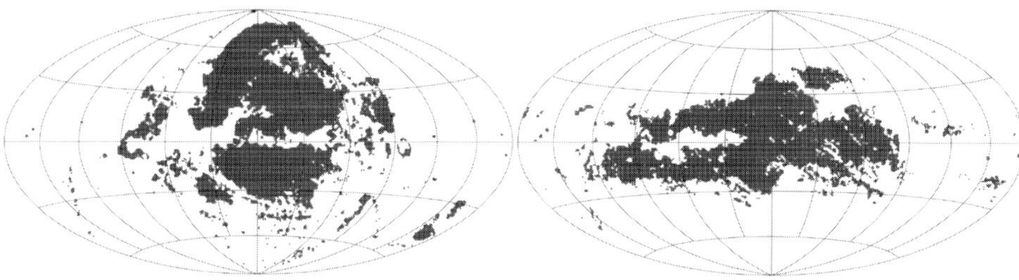

Figure 2. ROSAT PSPC All-Sky Survey: analysis of the R4/R1 band ratio. Left: sight-lines with low R1 band and high R4 band rates (R1 $<$ 100, R4 $>$ 80): features Cygnus (and apparent outflow to halo), Eridanus, North Polar Spur, Loop I. Right: Local Bubble emission model subtracted, (R1 $<$ 200, R4 $>$ 50): residual structure indicates deviation from a million degree plasma, part of it can be assigned to a galactic bulge emission, 3$^{\rm rd}$ quadrant ($l \sim 240°$) shows significantly lower intensity (Freyberg and Schmitt, 1998).

solved by dedicated missions that combine both high spectral and spatial resolution or by a new generation of multi-purpose observatories.

2. Spatial Structures

2.1. The Local Bubble in X-rays

The RASS was the first soft X-ray survey with imaging optics and unveiled a wealth of new details of the X-ray sky: a large number ($\sim 10^5$) of new sources (Voges et al., 2000) as well as detailed imaging of large-scale diffuse emission (Snowden et al., 1995, 1997). Figure 1 shows the 1/4 keV map in galactic coordinates as derived after completion of the RASS in 1997 (Freyberg and Egger, 1999),

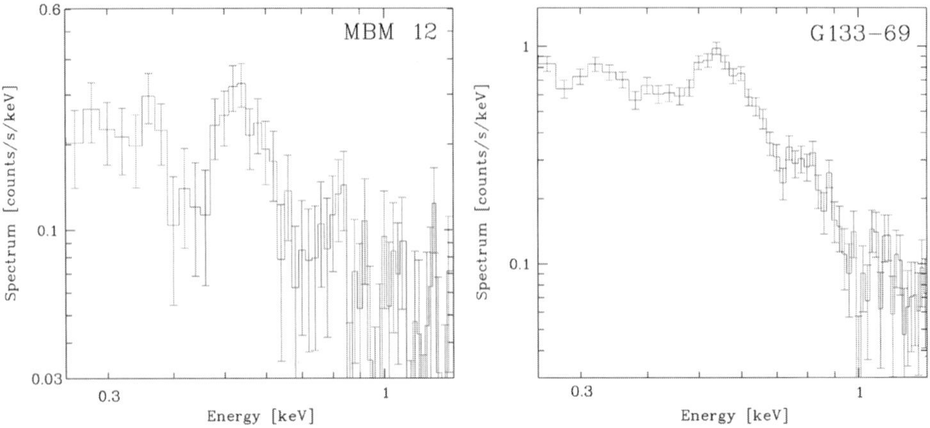

Figure 3. XMM-Newton EPIC pn spectra of MBM 12 (left) and G133-69 (right) as representative of molecular cloud shadow and of galactic halo spectra, respectively. For details see text (Freyberg and Breitschwerdt, 2003a).

with almost constant intensity in the galactic plane and about a factor of 2 to 3 higher intensity toward the galactic poles.

Snowden et al. (1998) have analysed the R1 and R2 band intensities and correlated these with the column densities of neutral hydrogen to derive the foreground intensities. Assuming constant emission measure along the line-of-sight this can directly be converted to the radius of the LB. It was found to be about $50 - 80\,\mathrm{pc}$ in the galactic plane and $70 - 120\,\mathrm{pc}$ toward higher latitudes. The distribution of R2/R1 band ratios was well peaked at 1.12 ± 0.10 away from localized regions (Loop I, Eridanus, Monogem). From this ratio spectral parameters for local and distant emission were derived and found to be almost equal ($T \sim 10^{6.07}$ and $\sim 10^{6.02}\,\mathrm{K}$, respectively). The pressure and electron density were found to be $P/k \sim 15000\,\mathrm{cm}^{-3}\,\mathrm{K}$ and $n_\mathrm{e} \sim 0.007\,\mathrm{cm}^{-3}$.

2.2. The Local Bubble in Na I Lines

Besides mapping the local 'hot' environment (LHB) in the soft X-ray band it is also possible to use stellar absorption lines to correlate distances to these stars with the absorbing column density and thereby to derive the extent of the neutral 'cold' bubble. Such an analysis (Sfeir et al., 1999) with Na I D lines (equivalent line width 20 mÅ corresponds to log(NaI)= $11.0\,\mathrm{cm}^{-2}$ or log(HI)= $19.3\,\mathrm{cm}^{-2}$) has revealed that the LB is irregularly shaped with extent of $\sim 70\,\mathrm{pc}$ in the galactic plane (with a large extension to $\sim 200\,\mathrm{pc}$ in the direction to β CMa) and $\geq 200\,\mathrm{pc}$ toward the poles. Since polar extent is rather uncertain due to paucity of stars in these directions, the absorption line data cannot answer the question whether the LB is closed or open at the poles and whether hot gas enters the lower galactic halo.

2.3. Comparison

The extents of the LB derived from R2/R1 band ratios (Snowden et al., 1998) and from Na I D lines (Sfeir et al., 1999) are different. The X-ray size seems to be more homogeneous and in particular no significant enhancements of 1/4 keV band emission are seen toward β CMa. Recently, Berghöfer and Breitschwerdt (2002) suggested, that multiple supernovae caused by a moving cluster, or with stars now found in the Sco-Cen association (Maíz-Apellániz, 2001), may be responsible for the current shape of the hot bubble while the boundaries of the neutral bubble were produced by much older events. This would not only explain the difference but also an irregular shape and a possible open polar cap.

2.4. Large-scale structures

Freyberg and Schmitt (1998) have extended the analysis of Snowden et al. (1998) to the R4 band. The R4/R1 band ratio is a steeper function of plasma temperature and thus more sensitive to temperature variations. Figure 2 shows the distribution of low R1 band and high R4 band intensities, e.g. likely regions of emission originating from higher temperature. Clearly the Loop I emission in the centre is seen, caused by an active superbubble with $T \sim 4 \times 10^6$ K. Moreover, parts of the Eridanus superbubble at $(l, b) \sim (195°, -50° \ldots -30°)$ and of the Cygnus region $(l, b) \sim (80°, 0° \ldots 30°)$ (*Cygnus-Draco Plume*, c.f. Snowden et al. (1997)) appear. Quantitative analysis of the R4/R1 band ratio away from these excess regions indicated that there is a component with a equivalent plasma temperature of $kT \sim 0.14 - 0.16$ keV. Note, that the R4/R1 ratio is *not* very sensitive to changes at low temperatures, i.e. some structures in the R2/R1 band ratio are insignificant here. In the third quadrant there are a couple of regions with very low R2/R1 ratio, e.g. at $(l, b) \sim (225°, -9°)$ and $(l, b) \sim (248°, -13°)$, caused by increased R1 band and below-average R2 band intensities. These enhancements are in the general direction of the extension of the neutral cavity.

3. XMM-Newton EPIC pn Data

3.1. MBM 12

One key observation of the ROSAT mission was the X-ray shadow in *MBM 12* (Snowden et al., 1993). This molecular cloud is believed to be located in or at least very close to the edge of the LB. The foreground spectrum toward the cloud should then be a good measure of the LB spectrum. Unfortunately, very high scattered solar radiation deteriorated the data quality especially in the 3/4 keV (R45) band and no strong constraints could be put: $< 25\%$ of the 3/4 keV band intensity is due to foreground emission. This result was obtained by modelling various components as R12 and R45 *band intensities* separately. An alternative approach (Freyberg et

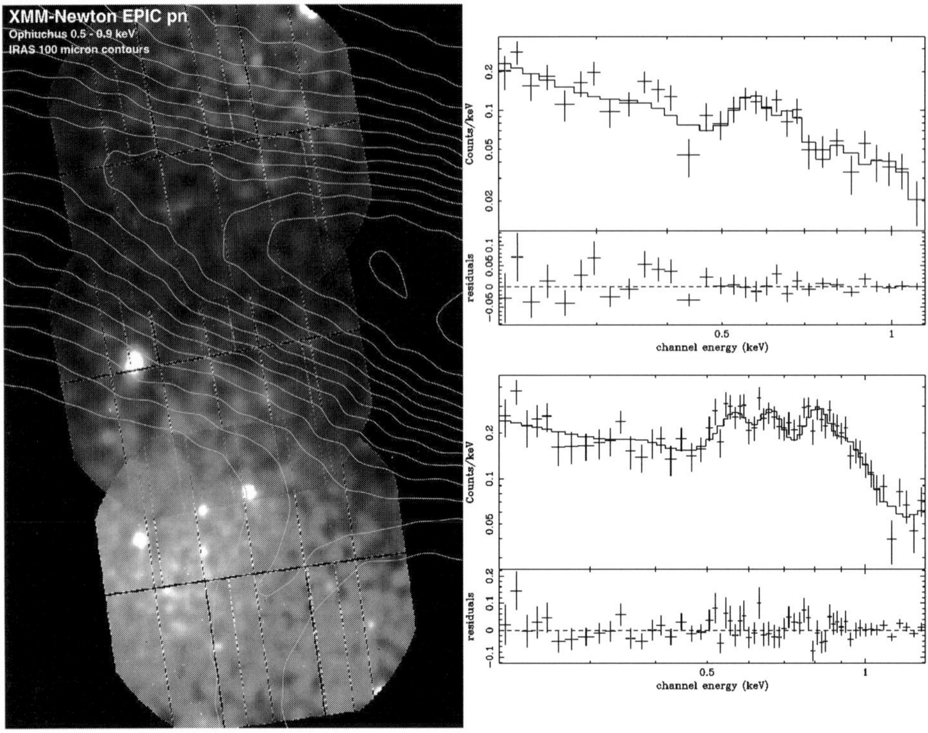

Figure 4. Left: merged image of three XMM-Newton EPIC-pn observations of the *Ophiuchus* molecular cloud in the $0.5 - 0.9$ keV band with IRAS 100 μm contours, the cloud casts a strong shadow onto the X-ray background. Right: on-cloud (top) and off-cloud (bottom) spectra (data, folded model, and residuals) extracted from the central pointing, note the large difference at ~ 0.8 keV, probably due to iron lines from Loop I emission (Freyberg and Breitschwerdt, 2003a).

al., in preparation) modelled all *component spectra* and yielded slightly different results, a higher local temperature ($kT \sim 0.13 \pm 0.03$ keV), or higher oxygen and iron line intensities (by adding extra Gaussian lines) than predicted by a million K thermal plasma in collisional ionization equlibrium (CIE). Like the ROSAT observation also the XMM-Newton observation was affected by high background and thus derived spectral parameters remain uncertain, however, indicate a temperature of $kT \sim 0.14 \pm 0.02$ keV in addition to the standard 0.09 keV component. Figure 3 shows the *MBM 12* (foreground) spectrum (left) and for comparison an observation with significant contribution by galactic halo emission (right), in both cases strong O VII line emission is visible. No strong X-ray shadow is seen at *MBM 12*.

3.2. Ophiuchus

Three overlapping pointings on the *Ophiuchus* molecular cloud were performed to determine foreground and background spectra by X-ray shadowing in the direction of Loop I (Figure 4, left). The EPIC pn spectra (Figure 4, right) on-cloud (top) show strong oxygen lines, the off-cloud spectrum (bottom) also strong iron lines from Loop I. A multi-component spectral model always required an unabsorbed component similar to *MBM 12* of $kT \sim 0.14 \pm 0.02$ keV (Freyberg et al., in preparation). As the *Ophiuchus* molecular cloud is behind the LB wall inside Loop I a contribution by hotter gas from the Sco-Cen superbubble is still present in the foreground spectral fits, however, the weakness of the on-cloud Fe lines indicates that this contribution is small.

4. Other Instruments

4.1. FUSE, DXS, AND XQC

Using FUSE (*Far Ultraviolet Spectroscopic Explorer*) Shelton et al. (2001) detected the first O VI lines ($\lambda\lambda$1032, 1038 Å) from the general interstellar medium. The measured intensities are too large to be solely due to Local Bubble emission from a hot plasma with $T \sim 3.210^5$ K in CIE. Contributions by a galactic thick disk and from the lower halo are necessary to explain the data. Shelton et al. suggest that the Local Bubble is not homogeneously filled with X-ray emitting material. The *Diffuse X-ray Spectrometer* (DXS) was flown on the STS-54 Shuttle flight in January 1993 (Sanders et al., 2001). The Bragg crystal spectrometers have an energy resolution of about 4 – 14 eV, with a spatial resolution of $\sim 15°$. The quite narrow energy band pass of 148 – 284 eV (44 – 84 Å) contains a large number of spectral lines for Fe IX-XVI, Mg VI-X, Si VI-XII, S VII-XII, and Ne VIII. During the mission a region of $l = 225° \pm 85°$, $b = +5° \pm 10°$ was observed, i.e. Crux, Vela, Puppis, Monogem, Auriga. The spectra show several prominent lines (e.g. at 149, 161, 168, 185, 197, 225 eV), however, line positions and intensities are not explained by standard plasma models (but the underlying atomic physics may be inaccurate). Interestingly, the general galactic background was found to be equal to the spectrum of the old *Monogem* SNR but different from the younger *Vela* SNR, which could be interpreted as the LHB being the result of one or more old SNR explosions. More recently, an *X-ray Quantum Calorimeter* (XQC) was flown on a sounding rocket (McCammon et al., 2002). The wide energy range of 60 – 1000 eV (12.4 – 207 Å) can be resolved with $\sim 5 - 12$ eV, but practically without spatial resolution (the field-of-view is about 1 sr). Since the observed region ($l = 90° \pm 30°$, $b = +60° \pm 30°$) samples quite high galactic latitudes and thus low absorbing galactic column densities distant emission from the halo and beyond contributes to the spectrum. It shows a number of prominent lines (e.g. at 368, 561 – 574, 653 eV), but also a significant deficiency in the iron lines (Fe M at ~ 70 eV) and a lower continuum than expected. McCammon et al. estimate that less than 20% in the R4 band is due to truly diffuse extragalactic emission.

4.2. CHIPS

The *Cosmic Hot Interstellar Plasma Spectrometer* (CHIPS) will operate in 2003 to perform an all-sky survey in the 90 – 260 Å range with a resolution of $\lambda/\Delta\lambda \sim 10^2$ and a slit-limited field-of-view of $5^\circ \times 26.7^\circ$ (Dixon et al., 1998). With the diagnostics of medium ionized Ne, Fe, O, Ni, S, and Si lines it will be possible to distinguish between equilibrium and non-equilibrium models for the local hot plasma.

4.3. ROSITA AND XEUS

ROSITA (*Röntgen Survey with an Imaging Telescope Array*) shall perform an all-sky survey in a higher energy band (0.5 – 12 keV) than ROSAT (Predehl et al., 2003). The telescope design is similar to ABRIXAS (Trümper et al., 1998), instead of a satellite payload it will be mounted on the International Space Station (ISS) and expected to become operational in 2009. The viewing geometry on the ISS is more difficult, however, the resulting sky exposure will be more homogeneous (i.e. less pronounced at the ecliptic poles but higher in the ecliptic plane). Currently under investigation is the *X-ray Evolving Universe Spectroscopy* (XEUS) mission, a formation flight of a detector and a telescope satellite with focal length of 50 m and effective area of 30 m^2 at 1 keV (Parmar, 2003). Since it is not specially designed for studies of large-scale low surface brightness objects, its major weakness in this respect will be the small field-of-view. The prime mission goal is to study the high-redshift universe and therefore good spectral and spatial resolution at low energies (0.1 – 1.0 keV) is required for an analysis of redshifted Fe and O lines and absorbing material. This could also be utilized to resolve galactic thermal emission (FWHM of NFI $\sim$ 2 eV) related to arcmin-scale objects (Freyberg and Breitschwerdt, 2003b).

5. Conclusions

Analysis of current XMM-Newton data of X-ray shadowing observations indicates that the temperature of the LB is higher than in previous 'CIE works', or more than one local temperature is required ('non-equilibrium'). In the case of *MBM 12* the distance to the cloud is still uncertain, which could place it outside the LB and behind the bulk of the emitting material without strong illumination from behind or it is really close but most of the emission is still in front of it, in both cases no strong shadow were expected. Moreover, the weak absorption gradient in the field-of-view also reduces the contrast. The Loop I spectrum is completely different from the LB spectrum while the galactic halo spectrum is similar to the LB spectrum. For a definite determination of the local X-ray spectrum in the 0.15 – 1.5 keV range one would need absorbing column densities of $N_H \sim 10^{23}\,cm^{-2}$ like in dense globules (e.g. Barnard 68, Breitschwerdt and Freyberg, in preparation). Future missions with higher spectral resolution will diminish modelling ambiguities.

Acknowledgements

The organizers of the workshop are thanked for all the assistance before, during, and after the meeting. The XMM-Newton project is an ESA Science Mission with instruments and contributions directly funded by ESA Member States and the USA (NASA). The XMM-Newton project is supported by the Bundesministerium für Bildung und Forschung/Deutsches Zentrum für Luft- und Raumfahrt (BMBF/DLR), the Max-Planck-Gesellschaft, and the Heidenhain-Stiftung.

References

Breitschwerdt, D.: 2001, *Astrophysics and Space Science* **276**, 163–176.
Breitschwerdt, D. and Schmutzler, Th.: 1994, *Nature* **371**, 774–776.
Breitschwerdt, D. and Schmutzler, Th.: 1999, *Astronomy and Astrophysics* **347**, 650–668.
Berghöfer, Th. and Breitschwerdt, D.: 2002, *Astronomy and Astrophysics* **390**, 299–306.
Burrows, D.N.: 1989, *Astrophysical Journal* **340**, 775–785.
Burrows, D.N. and Mendenhall, J.A.: 1991, *Nature* **351**, 629–631.
Dixon, W.V., Hurwitz, M., Jelinsky, P., Welsh, B.Y., Edelstein, J.E. et al.: 1998, Bulletin of the American Astronomical Society **193**, 12.08.
Freyberg, M.J. and Breitschwerdt, D.: 2003a, *Astronomische Nachrichten* **324**(1-2), 162.
Freyberg, M.J. and Breitschwerdt, D.: 2003b, in: G. Hasinger, Th. Boller and A.N. Parmar (eds.), *XEUS – Studying the Evolution of the Hot Universe*, MPE Report 281, Garching, pp. 255–258.
Freyberg, M.J. and Egger, R.: 1999, in: B. Aschenbach and M.J. Freyberg (eds.), *Highlights in X-ray Astronomy – International Symposium in Honour of Joachim Trümper's* 65th *birthday*, MPE Report 272, Garching, pp. 278–281.
Freyberg, M.J. and Schmitt, J.H.M.M.: 1998, in: D. Breitschwerdt, M.J. Freyberg and J. Trümper (eds.), *The Local Bubble and Beyond. Lyman-Spitzer Colloquium, Proceedings of the IAU Colluquium No. 166*, Springer-Verlag, Lecture Notes in Physics 506, Berlin Heidelberg New York Tokyo, pp. 311–320.
Maíz-Apellániz, J.: 2001, *Astrophysical Journal* **560**, L83–L86.
McCammon, D., Almy, R., Apodaca, E., Bergmann Tiest, E., Cui, W. et al.: 2002, *Astrophysical Journal* **576**, 188–203.
Parmar, A.N.: 2003, in: G. Hasinger, Th. Boller and A.N. Parmar (eds.), *XEUS – Studying the Evolution of the Hot Universe*, MPE Report 281, Garching, pp. 95–111.
Predehl, P., Friedrich, P., Hasinger, G. and Pietsch, W.: 2003, *Astronomische Nachrichten* **324**(1-2), 128–131.
Sanders, W.T., Edgar, R.J., Kraushaar, W.L., McCammon, D. and Morgenthaler, J.P.: 2001, *Astrophysical Journal* **554**, 694–709.
Sfeir, D.M., Lallement, R., Crifo, F. and Welsh, B.Y.: 1999, *Astronomy and Astrophysics* **554**, 694–709.
Shelton, R.L., Kruk, J.W., Murphy, E.M., Andersson, B.G., Blair, W.P. et al.: 2001, *Astrophysical Journal* **560**, 730–741.
Snowden, S.L., Cox, D.P., McCammon, D. and Sanders, W.T.: 1990, *Astrophysical Journal* **354**, 211–219.

Snowden, S.L., Mebold, U., Hirth, W., Herbstmeier, U. and Schmitt, J.H.M.M.: 1991, *Science* **252**, 1529–1532.

Snowden, S.L., McCammon, D. and Verter, F.: 1993, *Astrophysical Journal* **409**, L21–L24.

Snowden, S.L., Freyberg, M.J., Plucinsky, P.P., Schmitt, J.H.M.M., Trümper, J., Voges, W., Edgar, R.J., McCammon, D. and Sanders, W.T.: 1995, *Astrophysical Journal* **454**, 643–653.

Snowden, S.L., Egger, R., Freyberg, M.J., McCammon, D., Plucinsky, P.P., Sanders, W.T., Schmitt, J.H.M.M., Trümper, J. and Voges, W.: 1997, *Astrophysical Journal* **485**, 125–135.

Snowden, S.L., Egger, R., Finkbeiner, D.P., Freyberg, M.J. and Plucinsky, P.P.: 1998, *Astrophysical Journal* **493**, 715–729.

Trümper, J.E., Hasinger, G. and Staubert, R.: 1998, *Astronomische Nachrichten* **319**(1-2), 113–116.

Voges, W., Aschenbach, B., Boller, Th., Bräuninger, H., Briel, U. et al.: 2000, *IAU Circular* **7432**, 2.

GAS AS A TRACER OF THE GALACTIC POTENTIAL

PETER M.W. KALBERLA
Radioastronomisches Institut der Universität Bonn, Auf dem Hügel 71, D-53121 Bonn, Deutschland

Abstract. We consider a disk-like dark matter model for the Milky Way and compare a few predictions with observations. The observed gaseous flaring for HI and molecular gas fits the model predictions in detail. The global HI distribution in the Milky Way needs to be explained by a multiphase medium. The dark matter distribution in the Galactic halo is traced by a low density component of halo gas. High-velocity clouds with distances up to ~ 50 kpc may be explained as condensations which originate from instabilities within the gaseous halo. Our model explains also 'beards' and 'forbidden velocities' as observed in the rotation curves of external galaxies. A disk-like dark matter model is self-consistent and in excellent agreement with observations.

Keywords: Milky Way, ISM, gravitational potential

1. Introduction

For any of the current dark matter models, the density of the gaseous halo is orders of magnitudes below the density of the dark matter distribution. The large scale distribution of the gas in the Milky Way may therefore be considered as a tracer of the gravitational potential.

In a series of papers, Olling & Merrifield have recently discussed the case of an extended spheroidal dark matter distribution in the Milky Way. They have shown that the gaseous flaring is useful to constrain the parameter space for such a mass model.

Here we consider a disk-like dark matter distribution which is closely related to the observable gaseous and stellar disk. A self-consistent solution of the combined Poisson-Boltzmann Equation for the mass distribution and the associated gravitationally potential is possible if both, the massive model components, as well as the tracers of the gravitational potential are assumed to be isothermal. Taking all observational constraints into account, it can be shown that the dark matter disk needs to have a local exponential scale height of 4 kpc (Kalberla, 2002). Such a mass distribution defines a stable disk which extends to a radial distance of 35 kpc. The associated rotation curve is flat, the derived Oort constants, the local surface density, and the Oort limit agree well with the observations. The disk-like mass model leads to some predictions which we discuss briefly, most pronounced the flaring of the disk and the volume density distribution in the halo.

Astrophysics and Space Science **289:** 239–242, 2004.

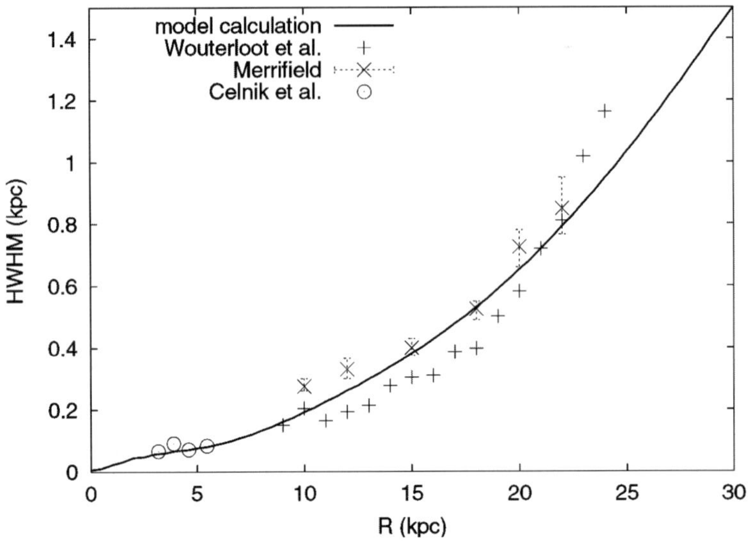

Figure 1. Flaring of the H I disk. The data points represent the HWHM of the H I column density as derived by Wouterloot et al. (1990, +), Merrifield (1992, X), and Celnik et al. 1979, ○). The thick line shows the model.

2. Predictions from the Model

For a gaseous disk in a steady state equilibrium, the turbulent gas pressure needs on average to balance the gravitational attraction of the disk. The scale height of the gas is therefore a sensitive tracer of the gravitational potential. Our model consists of a cold and warm neutral medium (CNM & WNM) with local exponential scale heights of 150 and 400 pc. To explain the global properties of the observed H I distribution, an additional third phase with a velocity dispersion of 60 km s^{-1} and a local scale height of 4 kpc is needed. According to our model about 50% of the gaseous halo exists as a neutral halo medium (NHM), clumpy and embedded in a hot halo medium (NHM) with a temperature of $T \sim 10^{6.2}$ K. The existence of such a neutral gas phase was inferred from the Leiden/Dwingeloo Survey (Kalberla et al., 1998). Recently, Lockman (2002) has used the Green Bank Telescope to resolve this phase into clumps which have diameters of a few tens of pc, temperatures of a few thousand K and densities of $0.1 - 0.9$ cm^{-3}.

In Figure 1 we compare the flaring of the H I gas in the Milky Way according to the model with observational determinations by several authors. For radii $3 < R < 25$ kpc we find an excellent agreement between model and observations. We emphasize that a *multiphase* H I medium is needed to explain the observations in Figure 1.

Our model predicts a flaring of the stellar component. Also, as shown in Figure 2, a flaring of the molecular gas layer can be derived. The model predicts

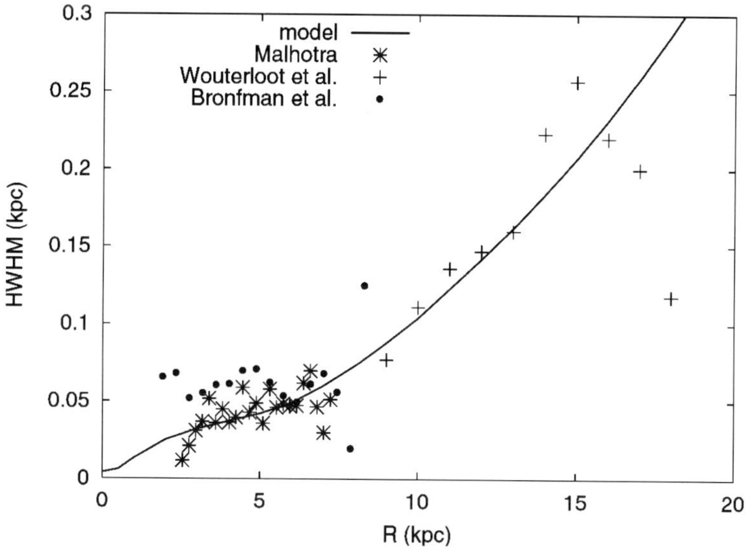

Figure 2. Flaring of the molecular disk. The solid line represents the model, the data points give CO scale heights a derived by Malhotra (1994, $*$), Wouterloot et al. (1990, +) and Bronfman et al. (1988), •).

scale heights which are in excellent agreement with independent observational determinations by several authors.

Flaring exists also within the gaseous halo. The model predicts that the gaseous halo with a local exponential scale height of 4 kpc flares at a radial distance of $R \sim 35$ kpc to a scale height of 35 kpc. The upper halo, locally at distances of $z > 3.3$ kpc, may suffer from thermal-convective or Parker-type instabilities. In consequence, Galactic high velocity clouds (HVCs) may have condensed out of flaring layers in the upper halo. HVC distances according to our model are consistent with direct distance determinations and with estimates derived from optical emission lines by Weiner et al. (2001).

High sensitivity observations of external galaxies at high inclination (NGC 891 & NGC 2403) reveal 'anomalous' gas with high velocity dispersions. This gas appears to lag behind the rotation of the disk (Swaters et al., 1997; Fraternali et al., 2002). We identify this 'anomalous' gas with the NHM phase of our model. To verify, whether our model can explain such a gas phase, we calculated synthetic profiles for the Milky Way, as observed from outside at an inclination of 80^0. Figure 3 shows the de-projected rotation curve. The observable column density distribution for the total H I gas (CNM, WNM & NHM) shows below the –20 dB level a 'beard' similar to the observations of NGC 891 and NGC 2403. Close to the center the beard has a transition to forbidden velocities. Both phenomena are caused by projection effects and are independent from the true rotation velocity of the gas. The thick line in Figure 3 displays the apparent rotational velocity for the

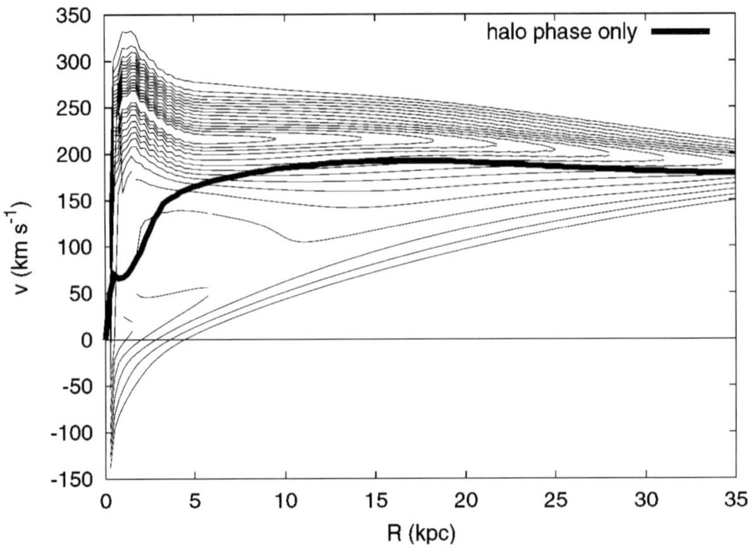

Figure 3. Position velocity diagram along the major axis, observable for a galaxy like the Milky Way with an inclination of 80^0. The thin contours represent normalized column densities in steps of 2 dB, the lowest contour is at –30 dB. The thick lines represent the observable rotation curve for the halo phase only (peak emission).

NHM phase. This halo gas phase with it's high velocity dispersion is reponsible for the 'beard' and the forbidden velocities.

References

Bronfman, L. et al.: 1988, *ApJ* **324**, 248.

Celnik, W., Rohlfs, K. and Braunsfurth, E.: 1979, *A&A* **76**, 24.

Fraternali, F., van Moorsel, G., Sancisi, R. and Oosterloo, T.: 2002, *AJ* **123**, 3124.

Kalberla, P.M.W., Westphalen, G., Mebold, U., Hartmann, D. and Burton, W.B.: 1998, *A&A* **332**, L61.

Kalberla, P.M.W.: 2003, *ApJ* **588**, 805.

Malhotra, S.: 1994, *ApJ* **433**, 687.

Lockman, F.J.: 2002, *ApJ* , in press.

Merrifield, M.R.: 1992, *AJ* **103**, 1552.

Swaters, R.A., Sancisi, R. and van der Hulst, J.M.: 1997, *ApJ* **491**, 140.

Weiner, B.J., Vogel, S.N. and Williams, T.B.: 2001, *ASP Conf. Ser.* **240**: Gas and Galaxy Evolution, 515.

Wouterloot, J.G.A., Brand, J., Burton, W.B. and Kwee, K.K.: 1990, *A&A* **230**, 21.

MOLECULAR CLOUDS: FORMATION AND DISRUPTION

JAVIER BALLESTEROS-PAREDES
Instituto de Astronomía, UNAM; E-mail: j.ballesteros@astrosmo.unam.mx

Abstract. Molecular clouds (MC) are the densest and coldest component of the interstellar gas, and the sites of star formation. They are also turbulent and fractal and their masses and sizes span several orders of magnitude. It is also generally believed that they are close to Virial equilibrium (VE). Since this statement has been questioned by a number of authors, with important implications on molecular clouds' lifetimes, we will review this subject within the context of a turbulent ISM. In this framework, there is significant numerical evidence that MCs are not in VE, that there is a strong exchange of mass, momentum and energy between clouds and their surrounding medium, and that it is difficult (if not impossible) to form quasistatic cores inside MCs, suggesting that they must be transient, short-lived phenomena. Thus, their formation and disruption must be primarily dynamical, and probably not due to just a single mechanism, but rather to the combination of several processes. This picture seems consistent with recent estimates of ages of stars in the solar neighborhood.

Keywords: ISM: clouds, turbulence ISM: kinematics and dynamics, stars: formation

1. Virial Theorem: Are Clouds Actually in Virial Equilibrium?

The Virial theorem can be derived from the momentum equation by dotting it by the position vector and integrating it over the volume of interest (see, e.g., Parker, 1979; Spitzer, 1982; Shu, 1989; Hartmann, 1998). Due to lack of space, we will not write down the equation and their terms. A clear and full description of either the Lagrangian and Eulerian versions can be found in MicKee and Zweibel (1992) (see also Ballesteros-Paredes and Vázquez-Semadeni, 1997; Ballesteros-Paredes et al., 1999a). Here we just mention that it involves the calculation of all the energies of the clouds, as well as the evaluation of pressures at the clouds' surface, and time derivatives of the moment of inertia. In the Eulerian version, additional terms involving mass flux through the clouds' surfaces appear.

It is frequently encountered in the literature that VE is applicable to MCs (e.g., McKee, 1999 and references therein). This statement is generally based on the theoretical assumption that all forces are in balance (Spitzer, 1982), or on the observational fact that clouds exhibit near equipartition between gravity and internal energies. Nevertheless, in a constantly stirred turbulent ISM (see chapter by MacLow, this volume), it is not clear that the forces will balance. Furthermore, equipartition between energies should not be considered evidence for VE. The only probe that VE admits is to measure the second time derivative of the moment of inertia, $\ddot{I}$, and check that it is zero. Such a probe is not possible in an observational way. However, it has been measured in numerical simulations of modeled clouds

Astrophysics and Space Science **289:** 243–254, 2004.

by Ballesteros-Paredes and Vázquez-Semadeni (1997), who have shown that, for an ensemble of clouds in simulations of the ISM, the clouds are *not* in VE. They obey the Virial Theorem, but the second time derivative of the moment of inertia is *not* zero.

The VE assumption ($\ddot{I} = 0$) implies either that clouds are not redistributing their mass inside, or that the variation of their mass distribution is constant with time. Both statements seem highly implausible in a dynamical, nonlinear ISM. McKee (1999) has suggested two possibilities, in order to assume VE for MC: One is that, for a single cloud, $\langle \ddot{I} \rangle = 0$ if the considered averaging time is much larger than the dynamical timescale of the cloud, $t_{\rm avg} \gg t_{\rm dyn}$. The other is that for an ensemble of clouds, some of them may have positive values of $\ddot{I}$, and some others will have negative values, so that VE holds for the ensemble. Regarding the first assumption, there is a hypothesis behind it: the cloud has to be oscillating around a mean shape without a strong redistribution of mass. Nevertheless, in a nonlinear regime, any variable may change in an arbitrary way, and not necessarily the mean values will average to zero. Furthermore, recent numerical and observational evidence (cf. §2) suggests that the lifetimes of the clouds are not significantly larger than their own dynamical times (i.e., they are transient), and averaging over large timescales may be meaningless. Concerning the second assumption, the analysis by Ballesteros-Paredes and Vázquez-Semadeni (1997) shows that variations of the moment of inertia span up to 7 orders of magnitude (in absolute value) between the largest clouds and the smallest. Thus, it is not clear that an average for such large scatter will be representative of the actual dynamics of the clouds. Moreover, even if I averages to zero for the cloud ensemble, this does not alter the fact that clouds are individually *out* of VE.

It is important to mention that, however, for clouds in numerical simulations, there is approximately an energy equipartition between gravitational, kinetic, magnetic and internal energies Ballesteros-Paredes and Vázquez-Semadeni (1995), in agreement with observational determinations (e.g., Myers and Goodman, 1988), suggesting that, in particular 'Virial mass' estimates (which would be better called 'energy equipartition mass') are of the same order of magnitude than the actual mass, in spite of VE probably not being applicable.

It is also widely believed that the turbulent kinetic energy provides support for clouds against their self-gravity (see, e.g., Chandrasekhar, 1951; Bonazzola et al., 1987, Vázquez-Semadeni and Gazol, 1995). However, this is only part of the turbulent motions, and only if they were confined to scales much smaller than the cloud sizes it could be the only effect (Chandrasekhar and Fermi, 1953; Leorat et al., 1990; Padoan, 1995; Ballesteros-Paredes et al., 1999a, Klessen et al., 2000). In reality, interstellar turbulence is a multiscale phenomenon, and there is observational and numerical evidence that turbulence at very different scales is present in MC, with most of the energy residing in the large-scale motions, as in most turbulent flows (Scalo, 1987; Norman and Ferrara, 1996; Falgarone et al., 1998; Ballesteros-Paredes et al., 1999a; Avila and Vázquez-Semadeni, 2001; Ossenkopf and McLow, 2002; Brunt, 2002). Moreover, turbulence in MCs is highly compressible, and compressible turbulent modes have the effect of producing density enhancements at scales smaller than their own, thus promoting, rather than preventing, gravitational collapse. This process is usually referred to as 'turbulent fragmentation' (e.g., Padoan, 1995). Thus, turbulent modes at a given scale are most likely to have a dual role, providing support toward larger scales while simultaneously promoting collapse of smaller ones (Klessen et al., 2000). The precise outcome of collapse or support must be the result of the total energy involved in the compressible modes of turbulence at large scales against the energy involved in the smaller scale modes.

On the other hand, it is found that surface and volumetric terms in the Virial theorem are comparable in order of magnitude (Ballesteros-Paredes and Vázquez-Semadeni, 1997, see also Ballesteros-Paredes et al., 1999a), suggesting that clouds are actually interchanging mass, momentum and energy with the surrounding medium. This result is reasonable because the ISM is highly dynamic, with strong energy injection by stellar outflows, spiral density waves, HII region expansions, etc. In fact, thermal and kinetic pressure confinement has been criticized by Ballesteros-Paredes et al. (1999a), arguing that thermal pressure constancy is relatively meaningless in a medium in which the ram pressure is significantly larger than the thermal one, and that ram pressure 'confinement' is meaningless, because, when the motions are at the scales of the cloud itself, the cloud is actually a (transient) turbulent fluctuation, rather than a persisting entity that somehow requires to be 'confined'. Recent observations confirming this picture are presented by Jenkins and Tripp (2001).

The preceding discussion questioning the assumption of VE for molecular clouds has an analogue at the level of their dense cores. Although in the standard models for star formation, low-mass protostellar cores are assumed to be in quasi-hydrostatic equilibrium (e.g., Mouschovias, 1991, Shu et al., 1987, and references therein), this view has begun to change recently, especially among the proponents of a turbulent scenario for molecular cloud structure and star formation (e.g., Scalo, 1987; Elmegreen, 1994; Padoan, 1995; Ballesteros-Paredes et al., 2000; Klessen et al., 2000; Padoan and Nordlund, 2002; Vázquez-Semadeni et al., 2002a, 2002b). For instance, Ballesteros-Paredes and Vázquez-Semadeni (1998) have presented preliminary results suggesting that the velocity structure of the so-called 'coherent cores' (dense molecular cores with a non-thermal velocity dispersion that becomes constant with size at small – below few $\times$ 0.1 pc – scales; Goodman et al., 1998), may be interpreted as the signature of the collision between gas streams. Similarly, Padoan et al. (2001), have found a velocity dispersion-column density relationship which appears in shocked generated cores in their simulations, as well in observed protostellar cores, suggesting that protostellar cores must be formed by supersonic turbulence.

Furthermore, a particular argument against the possibility that cores within MCs are in hydrostatic regime has been given by Vázquez-Semadeni et al. (2002, see also Tohline et al., 1987; Taylor et al., 1996; Ballesteros-Paredes et al., 1999a, Vázquez-Semadeni et al., 2002b). Molecular clouds are nearly isothermal, and thus a typical Bonnor-Ebert (BE) configuration (Ebert, 1955; Bonnor, 1956), in which a tenuous, hotter medium provides confining pressure, cannot exist, because the core and its surrounding medium are at roughly the same temperature. Thus, cores must be extended, rather than truncated structures. Now, extended equilibrium isothermal structures can be shown to always be gravitationally unstable,[1] and so, if a turbulent, dynamic compression brings the core near a hydrostatic configuration, the core will simply proceed to collapse if it reaches gravitational instability, or else will re-expand back to the average mean density of the cloud. However, cores that re-expand will be delayed by their own self-gravity, spending more time in this process than if they proceed to collapse. This result is consistent with the fact that MC typically contain more starless than star-forming cores (Taylor et al., Lee and Meyers, 1996, 1999; see also Evans, 1999 and references therein).

Of course, the above discussion does not apply if the dense cores are embedded in a hotter, more tenuous medium, which can then confine the core. B68 is probably the best example of a core in this situation (Alves et al., 2001; Hotzel et al., 2002a, 2002b), although there are some pieces of inconsistency: First, the cloud is not round, or oval in a regular way. This makes implausible that it can be in precise hydrostatic equilibrium. Second, their recently reported observations show motions of 0.25−0.5 the sound speed, suggesting that the cloud is near but not precisely in hydrostatic equilibrium. Third, the fitted Bonnor-Ebert profile by Alves et al. (2001) implies a temperature of 16 K, and, even at this temperature, marginal instability. The actual temperature is a factor of 30–50% lower ($\sim$ 10 K, see Hotzel et al. 2002a), and thus the thermal support is even lower. Then, the BE fit implies not hydrostatic equilibrium, but rather, instability. Finally, Ballesteros-Paredes et al. (2002) have found, over a sample of 120 cores on their 3 projections, that 50% of the projections can be fitted succesfully by a BE profile, in spite of not been in hydrostatic equilibrium. This shows that fitting BE-like column density profiles to cores in numerical simulations is not an unambiguous test for hydrostatic equilibrium.

2. Cloud Formation and Destruction

The evidence reviewed in the last section that clouds have important fluxes of the physical quantities (mass, momentum, energy) through their boundaries, and

[1] It is important to mention that the preceding discussion is not in contradiction with the fact that stars are objects in hydrostatic equilibrium within a turbulent medium, since they do not correspond to quasi-isothermal flows. In stars, energy is trapped since the opacity has increased, and the cooling time is about 10^{10} times the free-fall time. In order for a core formed by a compression to reach equilibrium, the isothermalicity of the flow must be released. Under MC conditions, this requirement occurs until scales and densities of protostars are reached (Ballesteros-Paredes et al., 1999a).

that they have important time derivatives of their moment of inertia, suggests that they may be much short-lived than previously thought, in line with suggestions by, among others, Sasao (1973), Hunter (1979), Hunter et al. (1986), Verschuur (1991), Elmegreen (1993) Ballesteros-Paredes et al. (1999a), that clouds are turbulent density fluctuations.

As has been discussed in previous reviews (e.g., Elmegreen, 1993), it is most likely that there are more than a single mechanism for cloud formation. There is more or less agreement that the earlier models of collisional agglomeration of smaller 'cloudlets' by, e.g., Kwan (1979), Scoville and Hersh (1979), Cowie (1980), do not reproduce the observations (Blitz and Shu, 1980). In particular, in those models, the time required for building up a giant MC ($\sim$ 100 Myr) is so long that it is difficult to either support a GMC against collapse, or avoid their disruption by the internal HII regions and SN explosions. In addition to the fact that there is not enough molecular material in the 'chaff' to allow the formation of GMCs via agglomeration (Blitz and Williams, 1999), the picture of 'cloudlets' flying ballistically might be inadequate, since clouds are not a collection of isolated entities of gas, but an interconnected network that has only been made to look like spheres due to the limitations of the observations (Scalo, 1990).

Within the context of a turbulent ISM, clouds may form by the convergence of turbulent flows at large scales. These might be probably produced by different instabilities (Parker, thermal, gravitational, or magnetorational [Sellwood and Balbus, 1999]), the passage of spiral density waves, swept up shells from SN remnants or simply the general action of global turbulence, where no single apparent mechanism is invoked, but the result of streams at different velocities that collide (see Elmegreen, 1990, 1993; Blitz and Williams, 1999 for reviews). An interesting alternative considered by Pringle et al. (2001) is that a large fraction ($\sim$ 50%) of the interstellar medium may be in the form of (non-self gravitating) molecular gas which is too cold to be detected. In their picture, we can only observe the molecular gas that is illuminated by sufficiently nearby heating sources. It is compressed in spiral shocks and only there a substantial fraction of the gas becomes self-gravitating enough to initiate star formation. It is only at this point, they argue, that we are able to see it.

In any event, an important constraint that should be satisfied for any cloud formation model are the observed ages of newborn stars in MCs, since they are the most direct measurement of a time scale in MCs. The other time scale is the dynamical or crossing time, $\tau \sim l/\Delta v$. Although it is applicable only in a statistical sense, there may be strong discrepancies when compared to the ages of stars. For example, for Taurus, ($l \sim 20$ pc, $\Delta v \sim 2$ km s^{-1}), the dynamical timescale is of the order of 10 Myr, while the ages of the young stars are only $\sim$ 2 Myr (see, e.g., Ballesteros-Paredes et al., 1999b and references therein).

It is important to note that there is some controversy on the determination of ages in star-formation regions. For example, Palla and Stahler (2000, 2002), argue that nearby MC have been forming stars in the last 10 Myr or more, with a recent burst of star formation. However, Hartmann (2002) argues that their conclusions are skewed by a statistically small sample of stars with masses larger than $1 M_\odot$, and by biases in their birthline age corrections. He notes that the picture by Palla and Stahler (a) requires the last 1–2 Myr to be a special epoch for most MCs; (b) implies that most MCs are forming stars at extremely low rates, if any; and that (c) the apparently oldest stars are systematically higher in mass, implying that for most of a typical MC's star-forming epoch, the Initial Mass Function was strongly skewed. Item (a) seems implausible, and items (b) and (c) are contradicted by observations (see Hartmann, 2002, and references therein). A more plausible explanation of the observations is that the 'tail' of older stars is really the result of including older foreground stars, as well as problems with the isochrones calibration in the higher mass stars.

Blitz and Williams (1999) argue that giant MC must live for some times 10 Myr based on the ages of stars in Orion quoted by Blaaw (1964). Nevertheless, more recent estimates of the stellar ages (Brown et al., 1994; Briceño et al., 2001) show that in the Orion OB1a association, which has $\sim$ 10 Myr old stars, there is virtually no molecular gas associated. Instead, in Orion OB1b, which contains large amounts of gas, the stars are about 1 Myr old, a value close to the estimated for young stars in Taurus.

Hartmann et al. (2001) tabulate the ages of stars in 13 nearby star-forming regions. For those regions with stars older than $\sim$ 5 Myr, there is no molecular gas associated, suggesting that the time scales for both cloud- and star-formation are shorter even than the dynamical time scale proposed by Elmegreen (2000).

The fact that several molecular regions exhibit synchronized star formation with ages of the newborn stars much smaller than the dynamical time suggests that some kind of external triggering must be involved. Vázquez-Semadeni et al. (1995), Ballesteros-Paredes et al. (1999b) and Harmann et al. (2001) suggest that global turbulence may play a crucial role. In their picture, large-scale flows powered by global events of star formation along the Galaxy may collide, form clouds and stars rapidly, and then dissipate. The idea is not new (Blaaw, 1964; Elmegreen and Lada, 1977; McCray and Kafatos, 1987), but although mechanisms like nearby HII regions in expansion, or SN events, are not discarded, all these processes feed the global turbulence, and thus no single mechanism need be directly responsible for the formation of any particular cloud complex.

How fast can MCs be formed by the general turbulence? Ballesteros-Paredes et al. (1999b) show that clouds can be produced rapidly (in few Myr) by the convergence of large-scale flows, evolving to high densities over scales of tens of parsecs nearly simultaneously. This is because the velocities involved are of the order of the velocity dispersion at the large scales (several km s^{-1}), rather than of the internal velocity dispersion of the clouds. Nevertheless, the exact cloud build-up time may depend on how much mass the streams are carrying, how strong the compression is, the rate of cooling of the compressed (shocked) region, the geometry of the compression, etc. Even with the typical ISM flow velocities $\sim$ 10 km s^{-1}, it can take tens of Myr to accumulate enough mass from the diffuse interstellar medium ($n \sim 1\mathrm{cm}^{-3}$) to form a MC complex. However, a necessary (though not sufficient) condition for the existence of molecular material in the solar neighborhood is that it have sufficient column density to effectively shield H_2 and CO from the dissociating ultraviolet radiation of the diffuse Galactic field. This requires a minimum column density in hydrogen atoms of roughly 10^{21} cm^{-2} (see Franco, 1993 and references therein). Thus, even if the process of building up material from diffuse H I takes a long time, the 'life' time of the MC in the solar neighborhood only begins once this minimum column density is attained Hartmann et al., 2001).

An important point to note is that the column density needed to allow molecular gas to be formed is similar to that required for the MC to become self-gravitating, under solar-neighborhood conditions ($\lesssim 1A_\mathrm{v}$). This may be the main reason why star formation is presently occurring in virtually all MC complexes of significant size within a kiloparsec from Sun (Hartmann et al., 2001), but the situation for the inner or outer galaxy may be quite different (Cox and Franco, 1986).

Concerning the destruction of clouds, studies of other star-forming regions in addition to Taurus (Hartmann, 2002, and references therein) in the solar neighborhood such as Cha I and IC 348 (Lawson et al., 1996; Herbig, 1998) also provide relatively little evidence for large populations of stars older than few Myr, especially when observational biases are eliminated. This suggests that the molecular gas may be dispersed also in a few Myr, a time scale consistent with the cluster survey results of (Leisawitz et al., 1989).

Several mechanisms for cloud-disruption have been proposed: SN explosions, the action of massive stars (UV ionizing radiation and/or winds), the lowering of shielding by an expansion of the cloud, the action of internal and/or external turbulence, etc. As pointed out by Franco et al. (1994), while SN explosions are the products of late stages of evolution, and take some time (6 Myr or more) to be 'turned on', massive stars on their main sequence may power the clouds almost immediately. Thus, even if SN explosions were the main mechanism powering global turbulence in the ISM (see MacLow, this volume), it is likely that the main agent disrupting clouds is the massive OB stars. This view has been recently supported by Matzner (2002), who has argued that the most efficient mechanism for cloud destruction is photoionization in H II regions, even considering the combined effects of winds (from proto-, main-sequence and evolved stars) and SNe. Once HII regions are created, their expansion is responsible for ionizing and photodissociating all the environmental gas.

At this point it is important to mention that a comparison between observational data and numerical simulations with a star formation prescription in which stars are formed in the densest regions (Vázquez-Semadeni et al., 1997) supports the view that clouds are dispersed by OB stars. Indeed, consider the case of Orion OB1 (Figure 1a), mentioned before. OB1a region contains ~10 Myr-old stars but no molecular gas. Instead, the OB1b association contains 1 Myr-old stars, and large amounts of gas. A similar situation is found in the simulations (Figure 1b), where

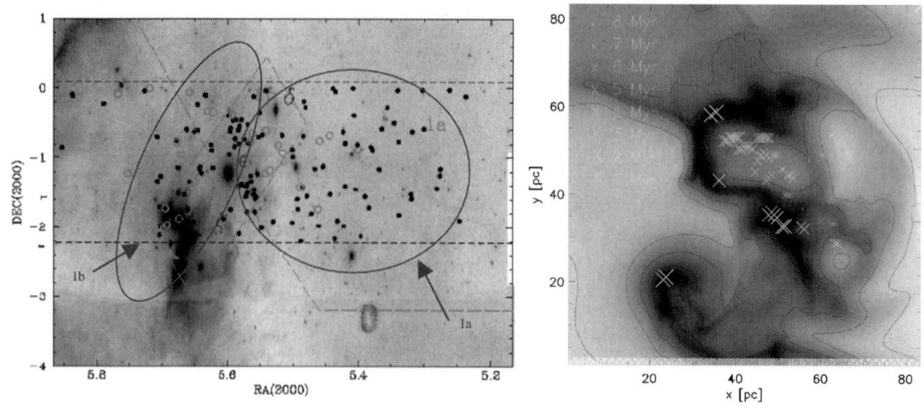

Figure 1. a. Orion OB1 association. b. Cloud in simulations from (Passot et al., 1995). Note that stars of more than 5 Myr old are 10 pc or more far from the dense gas. The simulation was not intended to reproduce the particular behavior of Orion.

the energy input is due to 'O stars' that last 6 Myr. As can be seen, the older, 5 – 8 Myr stars[2] (smaller crosses), are 10 – 20 pc away from the gas.

Other mechanisms may also help in the dispersal of the clouds, and contribute to the low star-formation efficiency, such as ejections from massive stars (Withworth, 1979) and winds and outflows (e.g., Norman and Silk, 1980; Leisawitz et al., 1989; Matzner and McKee, 2000). However, these are not considered as the main mechanisms because they are considerably less energetic than the HII regions and/or SN explosions. Recently, two other mechanisms have been proposed (although not explored in detail): the lowering of shielding (Hartmann et al., 2001), and the same large-scale streams that creates the clouds, may destroy them (Ballesteros-Paredes et al., 1999b). In the first case, if clouds like Taurus (with a mean column density that corresponds to an $A_V \sim 1 - 2$, Arce and Goodman, 1999) suffer an expansion of surface area by a factor of 2 or 3, the column density will be reduced by the same factor, allowing the destruction of molecular gas, with the consequent increase of temperature and, thus, of the Jeans length. In the second case, although a small fraction of the mass in clouds formed by turbulence may be considerably self-gravitating and susceptible for collapse and star formation, most of the mass may be (marginally) unbound and thus easily dispersed (Shadmehri et al., 2001; Vázquez-Semadeni et al., 2002a). These mechanisms, as well as the first mentioned, require further study, in order to quantify their efficiency in disrupting less massive clouds as, e.g., Taurus, where no massive stars or HII regions are available to contribute to the dispersal. In any event, rapid dispersal of gas is required to avoid extended periods of star formation, as stellar ages suggest.

[2] Note that stars older than 6 Myr are fossils, and they are not contributing anymore to the dispersal of the parent cloud.

3. Conclusions

We have discussed the implications of the Virial theorem on MCs' lifetimes. Consisting in two dimensions of spatial information, one dimension of velocity information, and no time evolution information, observations can not probe that clouds are or not in Virial equilibrium. Thus, numerical simulations can show important qualitative features, even if details were missing or incorrect. Numerical works have shown that clouds are not in Virial equilibrium, that there is strong exchanges of mass, momentum and energy with their environment, that time derivatives are non-negligible, and that surface terms can not be neglected, being always comparable to the respective volumetric term. Nevertheless, these simulations show equipartition between the self-gravity and the internal energies of the clouds, just as observations do (e.g., $GM/R \sim \delta v^2$).

If time derivatives and surface terms in the Virial theorem are important, then clouds (and cores) must be transient. We discussed that several mechanisms may be responsible for the formation of clouds, with some emphasis on the global effects of multiscale turbulence. This picture is consistent with recent estimations of ages of stars associated to molecular gas shown in the literature. Regarding destruction of MCs, it is favored the scheme in which HII regions have enough power, and act before SN, to be the dominant factor of cloud dispersal. Nevertheless, some other mechanisms may operate, in order to understand the lack of extended periods of star formation in low-mass starforming regions as Taurus.

Acknowledgements

I want to thank M. de Avillez and D. Breitschwerdt for their invitation. L. Hartmann and E. Vázquez-Semadeni for careful reading of the manuscript and useful comments. C. Briceño for allow me to publish Figure 1a before publication. I acknowledge financial support by CONACYT's grant and I39318-E. This work has made extensive use of NASA's Astrophysics Data System Abstract Service.

References

Alves, J., Lada, C.J. and Lada, E.A.: 2001, *Nature* **409**, 159.
Arce, H.G. and Goodman, A.A.: 1999, *ApJ* **517**, 264.
Avila-Reese, V. and Vázquez-Semadeni, E.: 2001, *ApJ* **553**, 645.
Ballesteros-Paredes, J., Hartmann, L. and Vázquez-Semadeni, E.: 1999, *ApJ* **527**, 285.
Ballesteros-Paredes, J., Klessen, R.S. and Vázquez-Semadeni, E.: 2002, *ApJ* , submitted.
Ballesteros-Paredes, J. and MacLow, M.: 2002, *ApJ* **570**, 734.
Ballesteros-Paredes, J. and Vazquez-Semadeni, E.: 1995, *Revista Mexicana de Astronomia y Astrofisica Conference Series* **3**, 105.
Ballesteros-Paredes, J. and Vázquez-Semadeni, E.: 1997, *Star Formation Near and Far*, 81.

Ballesteros-Paredes, J. and Vazquez-Semadeni, E.: 1998, *Bull. Amer. Astron. Soc.* **30**, 1358.
Ballesteros-Paredes, J., Vázquez-Semadeni, E. and Scalo, J.: 1999, *ApJ* **515**, 286.
Blaauw, A.: 1964, *ARAA* **2**, 213.
Blitz, L. and Williams, J.P.: 1999, *NATO ASIC Proc.* **540**: The Origin of Stars and Planetary Systems, 3
Blitz, L. and Shu, F.H.: 1980, *ApJ* **238**, 148.
Bonazzola, S., Heyvaerts, J., Falgarone, E., Perault, M. and Puget, J.L.: 1987, *AA* **172**, 293.
Bonnor, W.B.: 1956, *MNRAS* **116**, 351.
Briceño, C., Vivas, A.K., Calvet, N., Hartmann, L., Pacheco, R., Herrera, D., Romero, L., Berlind, P., Sánchez, G., S., Jeffrey A. and Andrews, P.: 2001, *Science* **291**, 93.
Brown, A.G.A., de Geus, E.J. and de Zeeuw, P.T.: 1994, *AA* **289**, 101.
Brunt, C.: 2002. *ApJ* , in press.
Carr, J.S.: 1987, *ApJ* **323**, 170.
Caselli, P. and Myers, P.C.: 1995, *ApJ* **446**, 665.
Chandrasekhar, S.: 1951, *Proc. R. Soc. London* **210**, 26.
Chandrasekhar, S. and Fermi, E.: 1953, *ApJ* **118**, 116.
Cowie, L.L.: 1980, *ApJ* **236**, 868.
Dame, T.M., Elmegreen, B.G., Cohen, R.S. and Thaddeus, P.: 1986, *ApJ* **305**, 892.
Cox, D.P. and Franco, J.: 1986, *Revista Mexicana de Astronomia y Astrofisica*, vol. **12**, 12, 295.
Ebert, R.: 1955, *Zeitschrift für Astrophysik* **36**, 222.
Elmegreen, B.G.: 1990, *ASP Conf. Ser.* **12**: *The Evolution of the Interstellar Medium*, 247 p.
Elmegreen, B.G.: 1993, in: *Protostars and Planets III*, 97 p.
Elmegreen, B.G.: 2000, *ApJ* **530**, 277.
Elmegreen, B.G. and Lada, C.J.: 1977, *ApJ* **214**, 725.
Evans, N.J.: 1999, *ARAA* **37**, 311.
Falgarone, E., Panis, J.-F., Heithausen, A., Perault, M., Stutzki, J., Puget, J.-L. and Bensch, F.: 1998, *AA* **331**, 669.
Falgarone, E., Puget, J.-L. and Perault, M.: 1992, *AA* **257**, 715.
Franco, J.: 1993, *Revista Mexicana de Astronomia y Astrofisica*, vol. **26**, 26, 13.
Franco, J., Shore, S.N. and Tenorio-Tagle, G.: 1994, *ApJ* **436**, 795.
Fuller, G.A. and Myers, P.C.: 1992, *ApJ* **384**, 523.
Goodman, A.A., Barranco, J.A., Wilner, D.J. and Heyer, M.H.: 1998, *ApJ* **504**, 223.
Hartmann, L.: 1998, *Accretion Processes in Star Formation*, Cambridge University Press, Cambridge, UK, New York, Cambridge astrophysics series; 32 p.
Hartmann, L., Ballesteros-Paredes, J. and Bergin, E.A.: 2001, *ApJ* **562**, 852.
Hartmann, L.: 2002. *ApJ* , in press.
Herbig, G.H.: 1998, *ApJ* **497**, 736.
Hotzel, S., Harju, J. and Juvela, M.: 2002, *AA* **395**, L5.
Hotzel, S., Harju, J., Juvela, M., Mattila, K. and Haikala, L.K.: 2002, *AA* **391**, 275.
Hunter, J.H.: 1979, *ApJ* **233**, 946.
Hunter, J.H., Sandford, M.T., Whitaker, R.W. and Klein, R.I.: 1986, *ApJ* **305**, 309.
Jenkins, E.B. and Tripp, T.M.: 2001, *ApJS* **137**, 297.
Kegel, W.H.: 1989, *AA* **225**, 517.
Klessen, R.S., Heitsch, F. and MacLow, M.: 2000, *ApJ* **535**, 887.
Kwan, J.: 1979, *ApJ* **229**, 567.
Kwan, J. and Valdes, F.: 1983, *ApJ* **271**, 604.
La Rosa, T.N., Shore, S.N. and Magnani, L.: 1999, *ApJ* **512**, 761.
Larson, R.B.: 1981, *MNRAS* **194**, 809.
Lawson, W.A., Feigelson, E.D. and Huenemoerder, D.P.: 1996, *MNRAS* **280**, 1071.
Lee, C.W. and Myers, P.C.: 1999, *ApJS* **123**, 233.
Leisawitz, D.: 1990, *ApJ* **359**, 319.

Leisawitz, D., Bash, F.N. and Thaddeus, P.: 1989, *ApJS* **70**, 731.
Leorat, J., Passot, T. and Pouquet, A.: 1990, *MNRAS* **243**, 293.
Leung, C.M., Kutner, M.L. and Mead, K.N.: 1982, *ApJ* **262**, 583.
Loren, R.B.: 1989, *ApJ* **338**, 925.
Matzner, C.D.: 2002, *ApJ* **566**, 302.
Matzner, C.D. and McKee, C.F.: 2000, *ApJ* **545**, 364.
McCray, R. and Kafatos, M.: 1987, *ApJ* **317**, 190.
McKee, C.F.: 1999, *NATO ASIC Proc.* **540**: *The Origin of Stars and Planetary Systems*, 29 p.
McKee, C.F. and Zweibel, E.G.: 1992, *ApJ* **399**, 551.
Miesch, M.S. and Bally, J.: 1994, *ApJ* **429**, 645.
Mouschovias, T.C.: 1991, *NATO ASIC Proc.* **342**: *The Physics of Star Formation and Early Stellar Evolution*, 449 p.
Myers, P.C.: 1983, *ApJ* **270**, 105.
Myers, P.C. and Goodman, A.A.: 1988, *ApJL* **326**, L27.
Norman, C.A. and Ferrara, A.: 1996, *ApJ* **467**, 280.
Norman, C. and Silk, J.: 1980, *ApJ* **238**, 158.
Ossenkopf, V. and MacLow, M.-M.: 2002, *AA* **390**, 307.
Ostriker, E.C., Stone, J.M. and Gammie, C.F.: 2001, *ApJ* **546**, 980.
Padoan, P.: 1995, *MNRAS* **277**, 377.
Padoan, P., Juvela, M., Goodman, A.A. and Nordlund, Åke: 2001, *ApJ* **553**, 227.
Padoan, P. and Nordlund, Åke: 2002, *ApJ* **576**, 870.
Palla, F. and Stahler, S.W.: 2000, *ApJ* **540**, 255.
Palla, F. and Stahler, S.W.: 2002. *ApJ* , in press.
Parker, E.N.: 1979, Clarendon Press, Oxford; Oxford University Press, New York, p. 858.
Passot, T., Vazquez-Semadeni, E. and Pouquet, A.: 1995, *ApJ* **455**, 536.
Plume, R., Jaffe, D.T., Evans, N.J., Martin-Pintado, J. and Gomez-Gonzalez, J.: 1997, *ApJ* **476**, 730.
Pringle, J.E., Allen, R.J. and Lubow, S.H.: 2001, *MNRAS* **327**, 663.
Sasao, T.: 1973, *PASJ* **25**, 1.
Scalo, J.M.: 1987, *ASSL* Vol. **134**: *Interstellar Processes*, 349.
Scalo, J.: 1990, *ASSL* Vol. **162**: *Physical Processes in Fragmentation and Star Formation*, 151 p.
Scoville, N.Z. and Hersh, K.: 1979, *ApJ* **229**, 578.
Sellwood, J.A. and Balbus, S.A.: 1999, *ApJ* **511**, 660.
Shadmehri, M., Vazquez-Semadeni, E. and Ballesteros-Paredes, J.: 2001, in: R. Taylor, T. Landecker and T. Willis (eds.), *Seeing Through the Dust: The Detection of HI and the Exploration of the ISM in Galaxies*, ASP, San Francisco.
Shu, F.: 1991, *the Physics of Astrophysics, V2. Gas dynamics*. University Science Books, New York.
Shu, F.H., Adams, F.C. and Lizano, S.: 1987, *ARAA* **25**, 23.
Solomon, P.M., Rivolo, A.R., Barrett, J. and Yahil, A.: 1987, *ApJ* **319**, 730.
Spitzer, L.: 1982, *Physical Processes in the Interstellar Medium*, John Willey & Sons.
Taylor, S.D., Morata, O. and Williams, D.A.: 1996, *AA* **313**, 269.
Tohline, J.E., Bodenheimer, P.H. and Christodoulou, D.M.: 1987, *ApJ* **322**, 787.
Torrelles, J.M., Rodríguez, L.F., Cantó, J., Carral, P., Marcaide, J., Moran, J.M. and Ho, P.T.P.: 1983, *ApJ* **274**, 214.
Vázquez-Semadeni, E., Ballesteros-Paredes, J. and Klessen, R.S.: 2002. *ApJ* , submitted.
Vázquez-Semadeni, E., Ballesteros-Paredes, J. and Rodriguez, L.F.: 1997, *ApJ* **474**, 292.
Vázquez-Semadeni, E. and Gazol, A.: 1995, *AA* **303**, 204.
Vázquez-Semadeni, E., Ostriker, E.C., Passot, T., Gammie, C.F. and Stone, J.M.: 2000, *Protostars and Planets* IV, 3 p.
Vázquez-Semadeni, E., Passot, T. and Pouquet, A.: 1995, *ApJ* **441**, 702.
Vázquez-Semadeni, E., Passot, T. and Pouquet, A.: 1996, *ApJ* **473**, 881.
Vázquez-Semadeni, E., Shadmehri, M. and Ballesteros-Paredes, J.: 2002, *ApJ* , submitted.

Verschuur, G.L.: 1991, Ap&SS **185**, 305.
Williams, J.P., de Geus, E.J. and Blitz, L.: 1994, *ApJ* **428**, 693.
Whitworth, A.: 1979, *MNRAS* **186**, 59.
Wood, D.O.S., Myers, P.C. and Daugherty, D.A.: 1994, *ApJS* **95**, 457.

THE KOSMA LARGE-SCALE CO SURVEY OF CLOUDS IN THE GALACTIC MOLECULAR RING

M. BRÜLL[1], C. KRAMER[1], V. OSSENKOPF[1,2], R. SIMON[3] and J. STUTZKI[1]
[1]*KOSMA, I.Physikalisches Institut, Universität zu Köln, Zülpicher Strasse 77, 50937 Köln, Germany*
[2] *SRON National Institute for Space Research, Postbus 800, 9700 AV Groningen, the Netherlands*
[3] *Boston University, 725 Commonwealth Avenue, Boston, MA 02215, USA*

Abstract. We present the status of the KOSMA large-scale observations of clouds in the Galactic Molecular Ring (GMR) in CO J=3-2, 2-1 and ^{13}CO 2-1 following up on the BU-FCRAO ^{13}CO 1-0 survey. The GMR is one of the prominent large scale structures in the Milky Way and an interesting laboratory to study star formation. Whereas the ^{13}CO data yield a good measure of the overall column density structure, the new observations of CO 2-1 provide a picture of the thin extended gas and the CO 3-2 data show the distribution of the dense and warm molecular gas. The combination of observations of these lines thus allows to distinguish between extended, quiescent gas, dense cold parts, and warm regions, influenced by star formation. In performing large scale observations we are able to analyse the structure of this material.

Keywords: galactic molecular ring, star formation, ISM, cloud structure

1. Introduction

The Milky Way Galaxy is known to be a spiral barred galaxy with a prominent ring between 4 and 7 kpc Galactocentric radius. Ring resonances occur frequently in barred spiral disk galaxies by gas accumulation under the continuous action of gravity torques from the bar pattern and are often sites of enhanced star formation. The Galactic Molecular Ring (GMR) encircles the bar and is thus probably an inner resonance-ring (Buta, 1999). Alternatively, but not necessarily in disagreement, the Milky Way GMR may consist of two pairs of tightly wound spiral arms (Englmaier and Gerhard, 1999).

The GMR was detected already in the distribution of γ-ray emission and HII regions, but measured for the first time in molecular gas by Scoville & Solomon (1975) using the CO 1-0 transition. Based on the available rotation curves they derived a Galactic radius of about 5 kpc for the CO 1-0 peak. The GMR of our Galaxy shows up prominently in the CO J=1-0 Columbia Milky Way survey (Dame et al., 1987) as well as in CI, CII and NII in the COBE survey (Fixsen et al., 1999). It is known for its high star formation rate: Most of the Galactic giant HII regions, diffuse ionized gas, and SN remnants are located in the ring and most of the far IR luminosity is emitted here (Burton, 1976; Robinson et al., 1984).

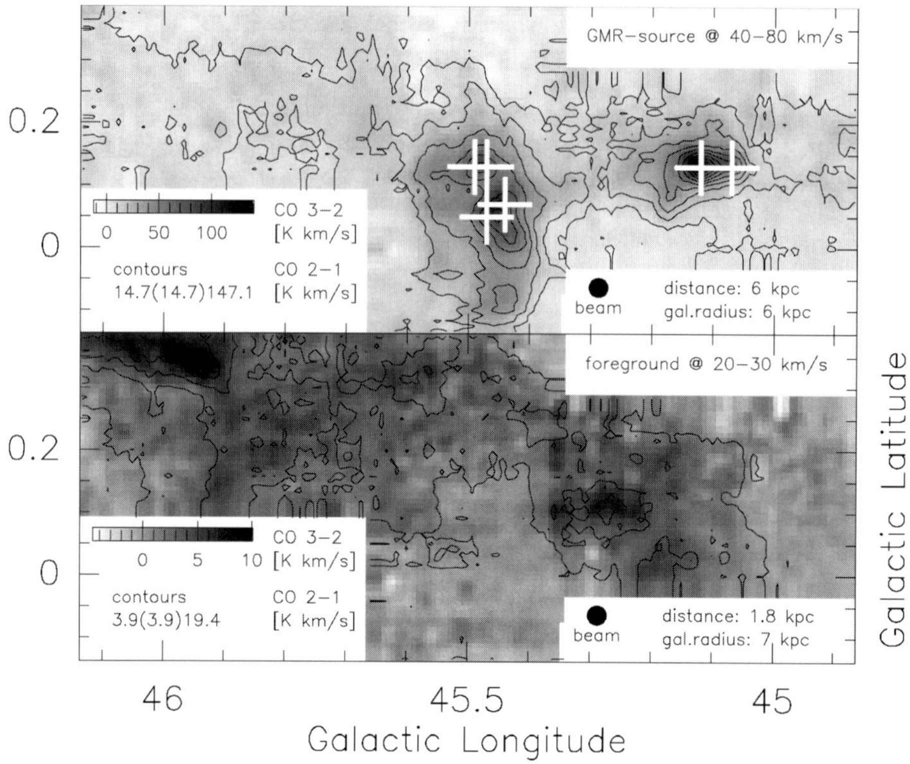

Figure 1. Integrated CO intensities measured in the GMR-source (upper panel, velocity range 40-80 km/s) and the extended foreground source (lower panel, 20-30 km/s). In grey scales one sees the CO 3-2 intensity, in contours the CO 2-1 intensity. The CO 3-2 data are smoothed to 120″, the KOSMA beamsize at 230 GHz. The white crosses mark the positions of known OH-, H_2O- and CH_3OH masers (Braz and Epchtein, 1983; Caswell et al., 1995) indicating star formation and associated compact HII regions (Wood & Churchwell, 1989). These positions coincide well with the peaks in the 8μm-A-band of the Midcourse Space eXperiment (MSX) not shown here (Egan and Price, 1996).

2. The KOSMA survey

To study the excitation conditions and their variation in the GMR, we have started a low-J CO survey of clouds in the GMR with the KOSMA (Kölner Observatorium für Sub-Millimeter Astronomie) 3m telescope on Gornergrat (Brüll et al., 2000). This survey complements the ^{13}CO 1-0 GMR-survey of the Boston University (BU) at the FCARO (Five Colleges Radio Astronomy Observatory, Simon et al., 2001; Jackson et al., 2002). At present, we have observed a 1350 arcmin^2 field on a 30″ grid in CO 3-2, 2-1 and ^{13}CO 2-1. The star forming activity is associated with dense and warm molecular clouds well traced by the CO 3-2 transition ($n_{crit}=10^4$cm^{-3}, E_u/k_B = 33K). The first selected region is shown in Fig.1. In addition to this star forming GMR-source, a quiescent, extended foreground source lies along the lines of sight.

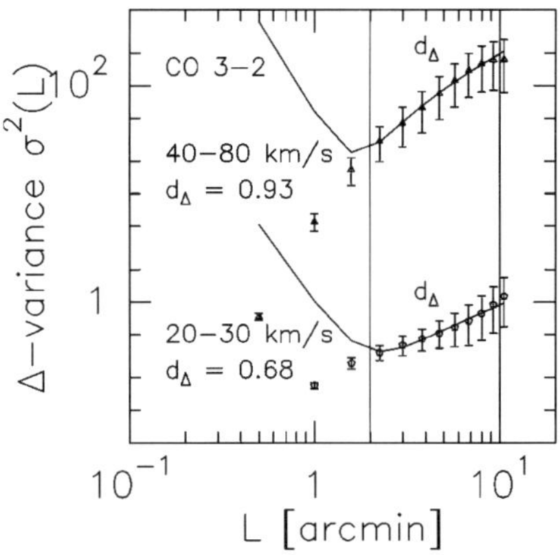

Figure 2. Δ-variance spectra of the CO 3-2 maps. The upper line represents the GMR source, the lower line the foreground source.

3. Results

The morphology of integrated intensities of CO 3-2, 2-1, ^{13}CO 2-1, and 1-0 within each of the two velocity components is quite similar. But peak positions, cloud edges, brightness, and intensity steepness towards the peaks vary between the different tracers. Compared to each other, the two clouds, however, have very different emission characteristics. The foreground source has a narrow velocity range and the spectra show only one Gaussian component. P-v-diagrams of the GMR source appear much more clumpy than the maps of integrated intensity. This behavior is reflected in the spectra, which often show blends of several Gaussian components. This is expected towards a torus, the GMR, with many clouds along the same line of sight, but at different distances and hence velocities. On the other hand the averaged spectrum of the GMR-source looks as smooth as that of the foreground cloud, only having a much wider line width and higher line intensity.

Using $N(H_2)[cm^{-2}] = 2.6{\cdot}10^{20}{\cdot}W_{CO}[Kkm/s]$ we obtain masses of $M_{CO} \approx 5.8 \cdot 10^5 M_\odot$ and $M_{vir} \approx 1.2 \cdot 10^6\ M_\odot$ for the GMR-source and $M_{CO} \approx 1.1 \cdot 10^4 M_\odot$ and $M_{vir} \approx 5.5 \cdot 10^4 M_\odot$ for the foreground-source. We therefore conclude that both clouds are gravitationally bound. For the star forming cloud, assuming LTE and the same dilution factors, the CO3-2/CO2-1 ratio indicates the presence of warm gas close to the HII regions and temperatures of 10-20 K for the colder gas.

4. Structure of the Clouds

Visual inspection shows that the CO 3-2 intensity map (grey scales in Figure 1) of the foreground source is more fragmented than the one of the GMR-source. The GMR-source is more dominated by large components. To quantify this description, we used the Δ-variance analysis (Stutzki et al., 1998; Bensch et al., 2001) measuring the relative amount of structure in a map depending on the structure size. This method distinguishes the scaling behaviour of the real cloud structure from observational noise and the telescope beam influence (right and left from the vertical lines in Figure 2, respectively). The Δ-variance spectra of the CO 3-2 maps in both sources are shown in Figure 2. The slopes d_Δ differ significantly reflecting the concentration of the emission to a few warm cores in the GMR source. The slopes of the Δ-variance are not as discrepant for the CO 2-1 line. Tracers of the warm dense gas thus appear to show differences in cloud structure stemming from the influence of massive star formation.

5. Conclusion

The current status of the KOSMA survey of clouds in the Galactic molecular ring allows a first analysis of the density and temperature structure in the observed clouds. Comparing the structural scaling behaviour in maps taken in different tracers, we study the effect of star formation on cloud structure in the Galactic molecular ring. The Δ-variance spectra of the CO 3-2 maps clearly show the difference in the distribution of warm gas between the GMR-source and the non star forming foreground source.

Acknowledgements

We thank the European Astronomical Society for the financial assistance of M.Brüll at the Joint European and National Astronomical Meeting 2002 in Porto, Portugal.

References

Bensch, F., Stutzki, J. and Ossenkopf, V.: 2001, *A&A* **366**, 636–650.
Braz, M.A. and Epchtein, N.: 1983, *A&AS* **54**, 167–185.
Brüll, M., Kramer, C. and Stutzki, J.: 2000, *YERAC conference* **E**, 9.
Burton, W.B.: 1976, *ARA&A* **14**, 275–306.
Buta, R.: 1999, *Ap&SS* **269**, 79–99.
Caswell, J.L. et al.: 1995, *MNRAS* **272**, 96–138.
Dame, T.M. et al.: 1987, *ApJ* **322**, 706–720.
Egan, M.P. and Price, S.D.: 1996, *AJ* **112**, 2862+.
Englmaier, P. and Gerhard, O.: 1999, *MNRAS* **304**, 512–534.
Fixsen, D.J., Bennett, C.L. and Mather, J.C.: 1999, *ApJ* **526**, 207–214.
Jackson, J.M. et al.: 2002, *ApJ* **566**, L81–L84.
Robinson, B.J. et al.: 1984, *ApJ* **283**, L31–L35.
Scoville, N.Z. and Solomon, P.M.: 1975, *ApJ* **199**, L105–L109.
Simon, R. et al.: 2001, *ApJ* **551**, 747–763.
Stutzki, J. et al.: 1998, *A&A* **336**, 697–720.
Wood, D.O.S. and Churchwell, E.: 1989, *ApJS* **69**, 831–895.

THE STRUCTURE OF MOLECULAR CLOUDS FROM PC TO AU SCALE

A new approach to the problem

J. ALVES
European Southern Observatory; E-mail: jalves@eso.org

Abstract. We present results from powerful new method to study the structure of dense molecular clouds on an unprecedent column density range of $0.5 < A_V < 50$ magnitudes ($10^{21} < N < 10^{23}$ cm^{-2}). The application of the method to Barnard 68 leads to the discovery of the best example in nature of a Bonnor-Ebert sphere.

1. The Search for the Initial Conditions to Star Formation

Stars and planets form within dark molecular clouds. However, despite 30 years of study little is understood about the internal structure of these clouds and consequently the initial conditions that give rise to star and planet formation. This is largely due to the fact that molecular clouds are primarily composed of molecular hydrogen, which is virtually inaccessible to direct observation. There is a need for a less complicate and more robust tracer of H_2 to access not only the physical structure of these objects but also to accurately calibrate molecular abundances and dust emissivity inside these clouds. The deployment of sensitive, large format infrared array cameras on large telescopes has, however, fullfilled this need by enabling the direct measurement of the dust extinction toward thousands of individual background stars observed through a molecular cloud. Such measurements are free from the complications that plague molecular-line or dust emission data and enable detailed maps of cloud structure to be constructed.

2. The Method and Results So Far

We have developed a new powerful technique for measuring and mapping the distribution of dust through a molecular cloud using data obtained in large-scale, multi-wavelength, infrared imaging surveys (Lada et al., 1994; Alves et al., 1998; Lombardi and Alves, 2001). This method combines measurements of near-infrared color excess to directly measure extinctions and map the dust column density distribution through a cloud. *It is the most straightforward and unambiguous way of determining the density structure in dark molecular clouds*. Moreover, the measurements can be made at significantly higher angular resolutions and substantially greater optical depths than previously thought possible. We have conclusively

Astrophysics and Space Science **289:** 259–263, 2004.

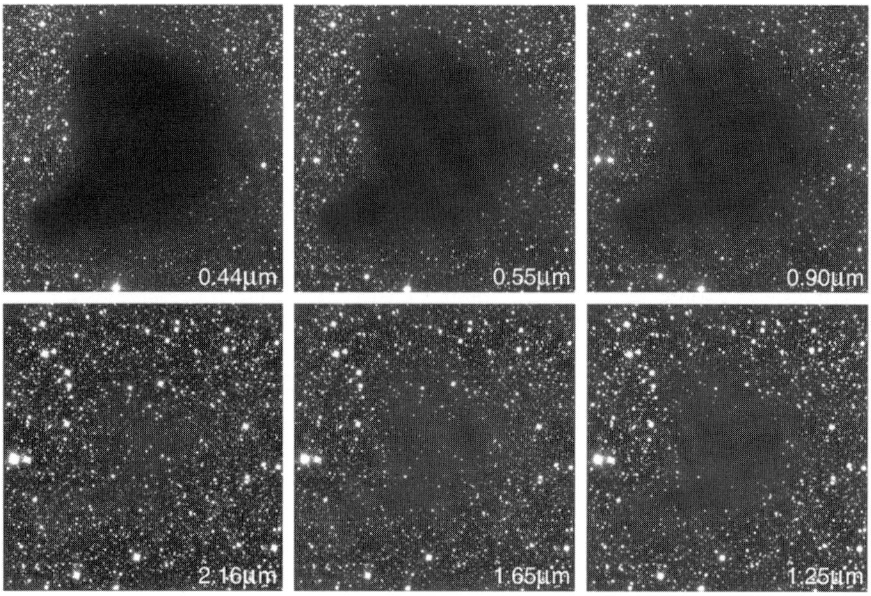

Figure 1. Deep BVIJHK-band (0.44μm, 0.55μm, 0.9μm, 1.25μm,1.65μm, 2.16μm) images of dark molecular cloud Barnard 68 taken with ESO's Very Large Telescope (VLT) and New Technology Telescope (NTT). The wavelength dependence of interstellar dust extinction in Barnard 68 is clearly depicted in these images. The analysis of the near-infrared colors of the stars seen through the dark cloud allow the construction of a very high resolution map (5 arcsec or about 500 AU) of the distribution of mass as traced by dust extinction, and a finely sampled and higher S/N density profile ever obtained for a cold dark cloud.

demonstrated the efficacy of this technique with our study of the dark cloud complex IC 5146 (Lada et al., 1994, Lada et al., 1999), L 977 (Alves et al., 1998), and in particular Barnard 68 (Alves, Lada and Lada, 2001, see Figure 1), where we detected more then a 1000 infrared sources background to this cloud and produced detailed maps of the extinction across the cloud to optical depths and spatial resolution an order of magnitude higher than previously possible (Av $\sim$ 40 magnitudes, spatial resolution $\sim$ 5 arcsec).

We have used our extinction observations to measure the masses, density structure, extinction laws, and distances to these objects. We found the radial density profiles of filamentary clouds (IC 5146 and L 977) to be well behaved and smoothly falling with a power-law index of $\alpha = -2$, significantly shallower than predicted by early theoretical calculations of Ostriker (1964) ($\alpha = -4$). Moreover because we are using pencil beam measurements of dust column density along the line of sight to background stars we were able to demonstrate that the small-scale structure of the clouds is surprisingly smooth with random density fluctuations ($\delta A_V/A_V$) present at very small levels ($< 3\%$!). This result is in very good agreement with optical studies of the small scale structure of the diffuse Interstellar Medium (ISM) (Thoraval, Boissé and Duvert, 1999).

When convolved to the appropriate spatial resolution our maps showed structure in the dust distribution which was strikingly well correlated with millimeter-wave CO and CS emission maps of the cloud, although showing crucial differences at high optical depths where these other tracers of column density become unreliable. These comparisons enabled us to, for the first time, directly derive CO, CS, and N_2H^+ abundances, and variations of, over a visual extinction range of 1–30 magnitudes, a range nearly an order of magnitude greater than achieved previously with optical star counting techniques. In a recent experiment we were able to make a direct measurement of molecular depletion in a cold cloud core (Kramer et al., 1999, Bergin et al., 2002). Finally, a comparison between our extinction data and millimeter continuum emission data, allowed a most accurate measurement of the ratio of dust absorption coefficients at millimeter and near-infrared wavelengths (Kramer et al., 1998, Kramer et al., 2003).

3. Barnard 68 is a Bonnor-Ebert Sphere

The extinction profile of Barnard 68 is the projection of the cloud volume density profile function, and therefore provides an exquisite view of the internal structure of this dense dark cloud. As early as 1948 Bart Bok pointed out that roughly spherical homogenous looking clouds, such as Barnard 68, resemble single dynamical units much like the polytropic models of Lane and Emden used to describe stellar structure (Lane, 1870; Emden, 1907). Can Barnard 68 be described as a self-gravitating, polytropic sphere of molecular gas? To investigate the physical structure of the cloud we begin with the assumptions of an isothermal equation of state and spherical symmetry.

The fluid equation that describes a self-gravitating, isothermal sphere in hydrostatic equilibrium is the following well-known variant of the Lane-Emden equation (Lane, 1870; Emden, 1907):

$$\frac{1}{\xi^2}\frac{d}{d\xi}\left(\xi^2\frac{d\psi}{d\xi}\right) = e^{-\psi} \tag{1}$$

where ξ is the non-dimensional radius,

$$\xi = \frac{r}{a}\,(4\pi\, G\, \rho_c)^{\frac{1}{2}} \tag{2}$$

while $\psi(\xi)$ equals,

$$\psi(\xi) = -ln(\rho/\rho_c) \tag{3}$$

for which r is the distance from the center of the sphere, a is the isothermal sound speed inside the gas cloud ($a = (KT/m)^{\frac{1}{2}}$), ρ is the density, and ρ_c is the central density. For an isothermal sphere bounded by a fixed external pressure there is a family of solutions characterized by a single parameter (Ebert, 1955; Bonnor, 1956):

$$\xi_{max} = \frac{R}{a} \, (4\pi \, G \, \rho_c)^{\frac{1}{2}} \tag{4}$$

Here ξ_{max} is the value of ξ at the outer boundary, R. Each of these solutions corresponds to a unique cloud mass density profile. Bonnor demonstrated that for $\xi_{max} > 6.5$ such a gaseous configuration would be unstable to gravitational collapse (Bonnor 1956). The high quality of our extinction data permits a detailed comparison with the Bonnor-Ebert predictions and we find that there is a particular solution, $\xi_{max} = 6.9 \pm 0.2$, that fits the data extraordinarily well. The close correspondence of the observed extinction profile with that predicted for a Bonnor-Ebert sphere strongly suggests that Barnard 68 is indeed an isothermal, pressure confined, and self-gravitating cloud. It is also likely to be in a state near hydrostatic equilibrium with thermal pressure primarily supporting the cloud against gravitational contraction. For Barnard 68, ξ_{max} is very near and slightly in excess of the critical radial parameter and the cloud may be only marginally stable and on the verge of collapse. If this is the case we should expect molecular radio-spectroscopy of this cloud to reveal a quiet, non-turbulent cloud with narrow molecular emission lines. Indeed, preliminary results from our radio-spectroscopy observing campaign of Barnard 68 (with the IRAM 30 m Radio Telescope at Pico Veleta, Granada) reveal that the linewidth of the $C^{18}O$ line in this cloud is $\Delta v \sim 0.18$ kms^{-1}, one of the narrowest lines ever observed in molecular clouds, in perfect agreement with the Bonnor-Ebert sphere nature of Barnard 68.

3.1. Reverse Engineering: Distance and Gas to Dust Ratio to Few Percent

The exact physical state of the Barnard 68 cloud is further constrained by the fact that ξ_{max}, which is solely derived from the shape of the observed column density distribution, uniquely specifies the combination of central density, sound speed and physical size that characterizes the cloud (i.e., Equation 4). Independent knowledge of any two of these parameters directly determines the value of the third. For example, in Barnard 68 we independently measure the dust extinction (which is directly related to the mass column density, via the gas-to-dust ratio in the cloud) and the angular size of the cloud (which is directly related to its physical size via the cloud's distance). Thus, if we know the temperature of the cloud, our measurement of its extinction and angular size (combined with the constraint that $\xi_{max} = 6.9 \pm 0.2$) independently gives the distance to the cloud, provided the gas-to-dust ratio is assumed. A characterization of dense molecular cores as Bonnor-Ebert spheres, perhaps modified Bonnor-Ebert spheres to account for turbulence and magnetic field, promises to be a fruitful approach in the understanding dense molecular cores and hence the initial conditions to star formation.

References

Alves, Lada and Lada: 2001, *Nature* **409**, 159.
Alves, Lada and Lada: 1999, *ApJ* **515**, 265.
Alves et al.: 1998, *ApJ* **506**, 292.
Bergin et al.: 2002, *ApJ* **570**, 101.
Bonnor: 1956, *MNRAS* **116**, 351.
Ebert: 1955, *Zeitschrift für Astrophysik* **37**, 217.
Emden: 1907, *Gaskugeln*, Teubner, Leipzig.
Kramer et al.: 1999, *A&A* **342**, 257.
Kramer et al.: 2003, *A&A*, in press.
Lada et al.: 1994, *ApJ* **429**, 694.
Lada, Alves and Lada: 1999, *ApJ* **512**, 250.
Lane: 1870, *American Journal of Science and Arts*, Series 2, **4**, 57.
Lombardi and Alves: 2001, *A&A* **377**, 1023.
Ostriker: 1964, *ApJ* **140**, 1056.
Thoraval, Boissé and Duvert: 1999, *A&A* **351**, 1051.

MOLECULAR CLOUD FORMATION IN DISK GALAXIES: THE CASE OF M33

ERIK ROSOLOWSKY and LEO BLITZ
University of California Berkeley

Abstract. We present results from the BIMA survey of $^{12}CO(1 \rightarrow 0)$ emission from M33. The survey resolves this emission into individual objects making this study the first unbiased, complete census of giant molecular clouds (GMCs) in a spiral galaxy. The GMCs are clearly associated with dense atomic gas, suggesting formation from the HI. Moreover, high resolution observations of these clouds show that they appear the same as Milky Way GMCs when viewed through the same observational filters. The GMCs show significantly less angular momentum than predicted by simple formation theories, suggesting the importance of magnetohydrodynamic effects in their formation.

1. The BIMA Survey of M33

We used the BIMA millimeter interferometer at Hat Creek Radio observatory to study $^{12}CO(1 \rightarrow 0)$ emission in the nearby Sc spiral M33. The study consisted of two phases: an all-disk survey with a synthesized beam of 14″ (50 pc) (Engargiola et al., 2002) and high resolution follow-up at 6″ (20 pc) on the brightest sources in the galaxy (Rosolowsky et al., 2002). Both sets of observations had spectral resolutions of 2 km s^{-1}, and the high resolution observations had sufficient spatial resolution to measure radii of many of the observed clouds. These studies represent the first unbiased census of molecular clouds in a spiral galaxy.

We identified 148 giant molecular clouds (GMCs) throughout the galaxy with masses ranging from $2.5 \times 10^4\ M_\odot$ to $8 \times 10^5\ M_\odot$. The clouds are distributed with a power-law mass spectrum with $dN/dM \propto M^{-1.6\pm0.2}$, comparable to the Milky Way mass spectrum. The GMCs are found almost exclusively in regions of dense atomic gas, as is seen in Figure 1. This, coupled with the fact that the HI is significantly more massive than the H_2 globally and locally by at least an order of magnitude, suggests that the atomic gas is the progenitor material for the molecular gas.

High resolution observations of these molecular clouds show that they appear to be drawn from the same molecular cloud population as is found in the Milky Way. Using similar characterizations of cloud properties as used by Solomon et al. (1987), we find the molecular clouds to have comparable masses, radii and line-widths to GMCs in the Milky Way, and these clouds follow similar relations between these three properties as first stated by Larson (1981). In using the same

Astrophysics and Space Science **289:** 265–268, 2004.

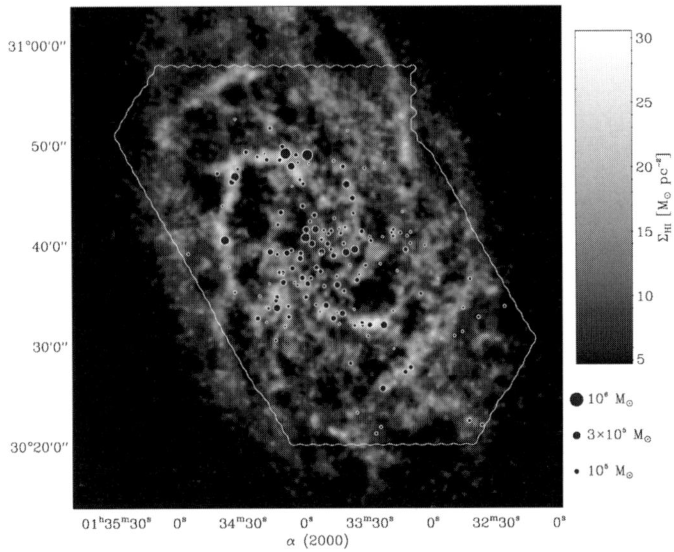

Figure 1. Greyscale image of HI 21 cm emission from Deul and van der Hulst (1987) with catalog sources overlayed. All molecular clouds lie in regions of HI overdensity. The area of the molecular cloud has been scaled to represent the relative masses of the clouds. The coincidence of molecular clouds with HI overdensity is evidence that clouds form out of the atomic gas. The extent of the survey is shown as the white contour.

observational tools to quantify cloud properties, we find that the clouds appear to be drawn from the same statistical population as those found in the Milky Way.

2. Angular Momentum Constraints on Cloud Formation

Our high resolution observations allow measurement of the velocity gradient across the molecular clouds. Many studies of Milky Way molecular clouds note systematic velocity gradients across the molecular clouds (e.g. Phillips, 1999; and Blitz, 1993, and references therein). The clouds in M33 also show significant velocity gradients with magnitudes comparable to Milky Way clouds. Using these velocity gradients, coupled with the observed intensity distribution of the molecular gas, we can calculate the specific angular momentum of the GMCs of M33. The values found are similar to those found in the Milky Way (Phillips, 1999).

We compared the observed values of the angular momentum to those predicted by common molecular cloud formation theories. In all cases, the theories predict angular momentum values significantly higher than the observed values Figure 2. The angular momentum is imparted by accumulating HI from different galactic radii which carries angular momentum from shear in the velocity field of the rotating HI disk. The theories studied include collapse via the Toomre Instability, which is implicated in many studies of galaxy-wide star formation (Kennicutt et

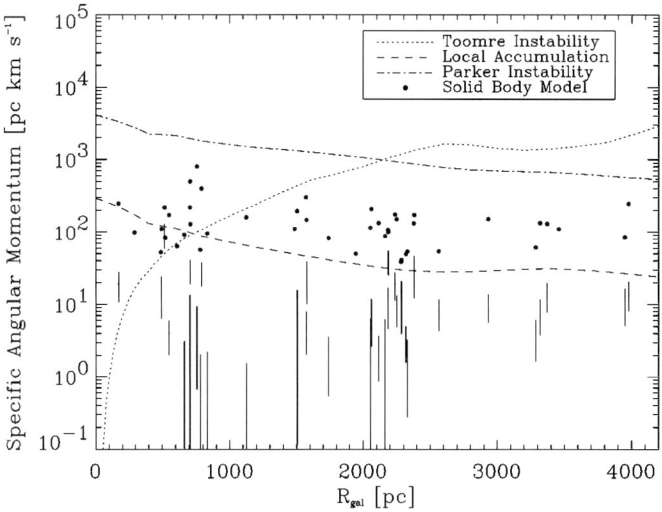

Figure 2. Specific angular momenta imparted by different collapse mechanisms as a function of galactic radius. The measurements are represented by $\pm 1\sigma$ error bars. The three curves are the imparted angular momenta as a function of R_{gal} for three simple accumulation mechanisms. The Jeans collapse curve assumes a cloud of $10^5 M_\odot$. The filled points are the predicted values of the specific angular momentum for the solid body rotator discussed in the text. All measurements of the specific angular momentum lie well below the predictions for all models.

al., 1998), the Parker Instability (Giz and Shu, 1993), and two models based on accumulating the local HI (Local Accumulation and Solid Body Rotators). The latter two models measure the angular momentum in accumulated HI based on different accumulation geometries. For the Local Accumulation model, we assumed the progenitor HI arose from cylinders reaching to infinity in z with radius large enough to include HI mass equal to that of the GMC. The Solid Body Model includes projection effect assuming the clouds are aligned with the galaxy's rotation and only accumulates HI from a cylinder with height equal to its diameter. In all cases, the angular momentum measured is significantly smaller than than predicted by any of the models. *If the clouds form by one of these formation mechanisms, there must be significant braking of cloud rotation in their formation.*

3. Conclusions

The dominant formation mechanism of molecular clouds remains unknown, despite many conjectures as to the important physical mechanism(s) (Elmegreen, 1996). Angular momentum studies show that the most naive treatments fail to explain the small angular momentum values for the M33 GMCs. Molecular clouds may simply form through other mechanisms, specifically those that do not gather material over a large range of galactic radii. Other options include magnetohydrodynamic braking, such as is important later in the star formation process. The measured value of the magnetic field in M33 is dynamically significant (Beck, 2000), therefore MHD braking is a viable option. These studies do not specifically address turbulent cloud formation theories, since these models do not lend themselves to simple comparisons with observations.

Extragalactic molecular clouds provide a unique trial population for testing molecular cloud formation theories. The GMCs in M33 show similarity to those found in the remainder of the Local Group. The clouds are shown to correlate well with dense atomic gas and their angular momentum is shown to be small compared to simple predictions from formation theories and observation of the galactic shear.

References

Beck, R.: 2000, preprint (astroph-0009455).
Blitz, L.: 1993, *Protostars and Planets* **III**, 125.
Deul, E. and van der Hulst, J.: 1987, *A&ApS* **67**, 509.
Elmegreen, B.G.: 1996, *IAU Symp.* **169**: *Unsolved Problems of the Milky Way* **169**, 551.
Engargiola, G., Plambeck, R., Rosolowsky, E. and Blitz, L.: 2002 *ApJ*, in preparation.
Giz, A.T. and Shu, F.H.: 1993, *ApJ* **404**, 185.
Kennicutt, R.C. et al.: 1998, *ApJ* **498**, 181.
Larson, R.B.: 1981, *MNRAS* **194**, 809.
Phillips, J.P.: 1999, *A&ApS* **134**, 241.
Rosolowsky, E., Plambeck, R., Engargiola, G. and Blitz, L.: 2002, *ApJ*, in preparation.
Solomon, P.M., Rivolo, A.R., Barrett, J. and Yahil, A.: 1987, *ApJ* **319**, 730.

SUPERBUBBLE ACTIVITY IN STAR-FORMING GALAXIES

M.S. OEY
Lowell Observatory, 1400 W. Mars Hill Rd., Flagstaff, AZ 86001, USA

Abstract. Mechanical feedback from massive stars, primarily from supernovae, can dominate ISM structuring and phase balance, thereby profoundly affecting galactic evolutionary processes. Our understanding of mechanical feedback is based on the adiabatic, wind-driven bubble model, applied on size scales ranging over three decades. Tests of the model, and our consequent understanding of feedback, are reviewed. While the model is broadly successful, critical unknowns still prevent a comprehensive understanding of the consequences of feedback.

Keywords: galaxies: evolution, intergalactic medium, ISM: bubbles, ISM: general, supernova remnants

1. Introduction

Mechanical feedback from massive stars is a dominant driver of evolutionary processes in galaxies, and takes place on scales ranging from individual wind-driven bubbles to galactic superwinds. Our understanding of the feedback process is based on the standard evolutionary model for stellar wind- and supernova-driven bubbles (Pikel'ner 1968; Weaver et al., 1977): hot ($\log T/\mathrm{K} \sim 6-7$), low-density ($n \sim 0.01\ \mathrm{cm}^{-3}$) gas is generated within a double-shock structure, and the pressure of this hot gas, chemically enriched by the stellar products, drives the growth of the thin, radiatively-cooled shell. In the adiabatic model, energy loss from the hot gas is negligible, yielding simple analytic expressions for the shell radius R and expansion velocity v as a function of time t:

$$\begin{aligned} R &\propto (L/n)^{1/5}\, t^{3/5} \quad , \\ v &\propto (L/n)^{1/5}\, t^{-2/5} \quad . \end{aligned} \tag{1}$$

For a stellar wind-driven bubble, the mechanical luminosity $L = 1/2\ \dot{M} v_\infty^2$, where $\dot{M}$ and v_∞ are the wind mass-loss rate and terminal velocity, respectively. For OB associations, supernovae (SNe) quickly dominate over winds, in which case $L = N_* E_{51}/t_e$ (e.g., MacLow and McCray, 1988), where N_* and E_{51} are the total number of SNe and SN energy, respectively, and t_e is the time over which the SNe occur.

The fate of the interior hot gas is crucial to the phase balance and enrichment of the interstellar and intergalactic media. Is stellar feedback indeed the source of the diffuse, hot gas in the ISM? Do galactic superwinds from starbursts eject metal-enriched gas from galaxies? How does mechanical feedback affect the structure of

Astrophysics and Space Science **289:** 269–277, 2004.

the ISM and porosity, for example, to ionizing radiation? Do superbubbles trigger renewed star formation?

Determining the relevance of the standard, adiabatic model for shell evolution is clearly critical in answering these fundamental questions. Several tests can be applied: 1) Comparing the observed vs predicted dynamics of individual bubbles and superbubbles; 2) Comparing the observed vs predicted statistical properties of superbubble populations, for example size, velocity, and energy distributions; 3) Identifying spatial correlations of superbubbles with the progenitor OB associations and their relics; and 4) Testing the observed vs predicted dynamics and properties of galactic superwinds. The last is presently more difficult and the subject of entire reviews in its own right (e.g., Heckman 2002). Therefore I will discuss here only the first three tests.

2. Individual Shell Systems

2.1. Single Star Bubbles

Few studies exist of single star bubbles from isolated OB stars. Oey and Massey (1994) examined two nebular examples in M33, and spectroscopically classified the parent O stars. The inferred stellar masses and ages implied wind parameters that were consistent with the observed sizes and shell ages predicted by the adiabatic model. However, the parameters were loosely constrained. H I shells with radii of several tens of pc have been identified as wind-blown bubbles around a number of Galactic O and Of stars (Cappa and Benaglia, 1998; Benaglia and Cappa, 1999). These are largely consistent with the standard model, and probe a specific subset of fairly evolved stars with old shells that have essentially stopped expanding.

Studies of Wolf-Rayet ring nebulae suggest shells that are too small, equivalent to an overestimate in L/n by an order of magnitude (e.g., Treffers and Chu, 1982; García-Segura and MacLow 1995; Drissen et al., 1995). However, the progenitor star produces several wind phases, including both fast and slow winds, with extreme changes in L. Their cumulative effect on shell morphology is complex and poorly understood. It is therefore unsurprising to find significant discrepancies between the predictions and observations of shell parameters. Hence, more studies of the simpler, single OB star bubbles are needed.

2.2. Superbubbles

Superbubbles around OB associations are more prominent than single-star bubbles, and thus have been studied more extensively. Soft X-ray emission has been detected within many objects, which is qualitatively consistent with the adiabatic evolution model. Two classes of X-ray emission have been identified: objects with X-ray luminosity L_x in excess of the model's prediction (Chu and MacLow, 1990; Wang and Helfand, 1991), and objects that remain undetected in X-rays (Chu et al., 1995). The X-ray–bright objects are thought to be overluminous because of SNR impacts on the shell walls. Upper limits on the X-ray–dim objects remain consistent with L_{X} predicted by the adiabatic model. It will thus be of great interest to determine L_{X} for these objects with *XMM-Newton* or *Chandra*. Also, an interface region between the hot gas and cooler shells should generate intermediate temperatures and ions. Chu et al. (1994) searched a number sightlines through LMC superbubbles and confirmed the existence of C IV and Si IV absorption in all cases.

A stringent test of the adiabatic model is to compare the predicted and observed shell kinematics in cases where the input mechanical power and other parameters are well-constrained. This was carried out for eight, young, wind-dominated LMC superbubbles by Oey and Massey (1995), Oey (1996), and Oey and Smedley (1998). The predicted growth rate for the shells was higher than implied by their observed R and v, equivalent to an overestimate in L/n by an order of magnitude. However, even after adjusting L/n in the models, over half the objects still showed observed expansion velocities that were typically a factor of two higher than predicted for the given R. Similar discrepancies were reported for Galactic objects by Saken et al. (1992) and Brown et al. (1995).

SNR impacts on the shell wall are the favored explanation for the high-velocity shells, since these also exhibit the anomalously high X-ray emission and elevated [S II]/Hα ratios. However, a sudden drop in the ambient density can induce a 'mini-blowout' with shell kinematics that can easily reproduce the anomalous velocities (Oey and Smedley, 1998; MacLow et al., 1998; Silich and Franco, 1999). Indeed, were it not for the X-ray and nebular diagnostics, it would be impossible to distinguish the shell acceleration mechanism from the kinematics alone.

Thus we see that the ambient properties are critical in determining the shell evolution. An underestimate in n could, for example, contribute to the growth rate discrepancy described above, that is seen in all the objects. To clarify the ambient gas distribution, Oey et al. (2001) mapped the H I distribution within a $\sim 40'$ radius of three nebular LMC superbubbles at $30''$ resolution. The results show neutral environments that vary to an extreme, despite morphologically similar optical nebulae. It is therefore essentially impossible to infer properties of the ambient material without direct, multi-wavelength observations.

Another vital parameter for shell evolution is the ambient pressure, P_0, which determines whether and when the superbubble growth becomes pressure-confined. While P_0 is usually unimportant in young, high-pressure superbubbles like the nebular objects mentioned above, it is of vital importance in the mid- to late-stage evolution. It may also be relevant in high-pressure, ionized environments like dense star-forming regions (e.g., García-Segura and Franco, 1996). Ultimately, P_0 determines the final size of the shells, and conditions relative to blowout. The uniformity and distribution of P_0 in the multiphase ISM is therefore especially relevant to a global understanding of superbubbles in galaxies (see below).

Finally, if the hot gas within superbubbles does not blow out and merge into the hot, ionized medium (HIM), it is likely that the objects will cool and depart from energy conservation. Indeed, whether and how the hot interior cools has long been a major question for superbubble evolution and the fate of the hot gas. Thermal conduction at the interface between the cool shell wall and hot gas should cause a high rate of mass-loading into the interior. The evaporated shell material dominates the mass in the hot region, which could be further supplemented by evaporation and ablation from small clouds overrun by the expanding shocks (e.g., Cowie and McKee, 1977; McKee et al., 1984; Arthur and Henney, 1996). If the interior density is sufficiently increased, radiative cooling will dominate, and the shells will no longer grow adiabatically. In addition, Silich et al. (2001) point out the importance of enhanced metallicity in the superbubble interiors, caused by the stellar and SN yields. Preliminary investigation for individual objects by Silich and Oey (2001) shows enhancement in L_{x} by almost an order of magnitude for low-metallicity ($Z = 0.05\ Z_{\odot}$) objects. This increase in the cooling rate could facilitate a transition from adiabatic to momentum-conserving evolution.

2.3. Supergiant Shells

The very largest H I shells, having sizes of order 1 kpc, emphasize some of the problems with the mechanical feedback model, and also highlight possible alternative shell-creating mechanisms.

The existence of infalling high-velocity clouds (HVCs) suggests that the impact of these objects could be an important contributor to supergiant shell populations. This suggestion is consistent with galactic fountain models for disk galaxies (e.g., Shapiro and Field, 1976), which are ultimately also powered by mechanical feedback in the disk. A number of hydrodynamical simulations of infalling HVCs confirm that these impacts result in shell-like structures (e.g., Tenorio-Tagle et al., 1986; Rand and Stone, 1996; Santillán et al., 1999).

In addition, tidal effects, which dominate energetics and structure formation at the largest length scales, could also create H I features that resemble shells. Note that many SN-driven shells will not exhibit expansion velocities if they have become pressure-confined by the ambient medium, thus a lack of observed expansion velocities cannot distinguish between the feedback model and other models. It has been suggested that some of the largest holes in, e.g., M33 are simply morphologically-suggestive inter-arm regions (Deul and den Hartog, 1990). The same may be true of the giant hole identified by de Blok and Walter (2000) in NGC 6822. Simple self-gravity effects have also produced shell- and hole-like structures in numerical simulations (Wada et al., 2000), although morphologically these structures appear more filamentary than the observations.

While such alternative mechanisms for creating shell-like structures undoubtedly contribute to the supergiant shell population, the conventional mechanical feedback model nevertheless also appears to apply in many situations. Meaburn's (1980)

LMC-4 is a well-known example that is unambiguously linked to Shapley's Constellation III, a large, extended complex of young stars. Kim et al. (1999) are able to identify an evolutionary sequence for supergiant shells in the LMC, based on the relative loci of Hα and H I emission. In addition, Lee and Irwin (1997) considered formation mechanisms for supergiant shells in the edge-on SBc galaxy NGC 3044. They found no evidence of HVCs, and since the galaxy is isolated, tidal interactions are also unable to explain the supergiant shells. They therefore conclude that the active star formation seen in NGC 3044 most likely explains its supergiant shell structures.

Thus, probably both mechanical feedback and other mechanisms form supergiant shell structures. Presumably different processes dominate under different circumstances, and these remain to be understood.

3. Statistical Properties of Superbubble Populations

The statistical properties of the superbubble populations offer another test of the standard evolutionary model for the shells. Oey and Clarke (1997) derived expressions for the differential size distribution $N(R)\,dR$ of superbubbles in a uniform ISM, using the analytic expressions for adiabatic evolution (equation 1). We considered a power-law mechanical luminosity function for the parent OB associations,

$$\phi(L)\,dL = AL^{-\beta}\,dL \quad , \tag{2}$$

with $\beta \simeq 2$, which is robustly associated with the H II region luminosity function (e.g., Kennicutt et al., 1989; Oey and Clarke, 1998a). The superbubble growth is taken to be pressure-confined when the interior pressure $P_i = P_0$. Star formation is assumed to be coeval within each OB association, with SNe therefore exploding over a period $t_e = 40$ Myr, the lifetime of the lowest-mass SN progenitors. For constant star-formation rate ψ and power-law $\phi(L)$, we found that:

$$N(R) \propto R^{1-2\beta} \quad , \tag{3}$$

effectively yielding $N(R) \propto R^{-3}$ for $\beta = 2$. Oey and Clarke (1997) also derive $N(R)$ for other combinations of ψ and $\phi(L)$.

This result agrees well with the H I shell catalog for the Small Magellanic Cloud (SMC) compiled by Staveley-Smith et al. (1997). This is by far the most complete sample of H I shells obtained for any galaxy, as evidenced by the fact that the relative number counts of H II regions and H I shells are in excellent agreement with their relative life expectancies. For shells having $R \geq 100$ pc, the fitted power-law slope $\alpha = 1-2\beta$ is 2.7 ± 0.6, in excellent agreement with the general prediction of $\alpha = 3$.

We note that different models for ISM structure yield different predictions for $N(R)$. For example, Stanimirović et al. (1999) suggest a possible fractal structure for the neutral ISM. From the same H I dataset of the SMC, they find a fractal dimension implying a size distribution for H I holes of $\alpha = 3.5$. It is difficult to empirically differentiate this from our model, having $\alpha = 3$; but it is worth noting that the predictions are intrinsically different.

However, the superbubble size distribution presently is not a sensitive test in determining whether or not the objects evolve adiabatically. If all the internal energy is radiated away, the objects are predicted to follow the momentum-conserving law given by Steigman et al. (1975):

$$R \propto (L/nv_{\infty})^{1/4}\, t^{1/2} \quad . \tag{4}$$

The stall radius $R_{\rm f}$ in this case is only 1.3 times larger than for the adiabatic model, and the size distribution follows the same law $N(R) \propto R^{1-2\beta}$ (Oey and Clarke, 1997). The observations of hot gas are therefore vital confirmation that the adiabatic model applies to a significant fraction of superbubbles.

We can also derive the distribution of expansion velocities $N(v)\ dv$, which describes only the growing objects (Oey and Clarke, 1998b):

$$N_{\rm grow}(v) \propto v^{-7/2}\ , \qquad \beta > 1.5 \quad . \tag{5}$$

This again compares well with the SMC H I shell catalog: the fitted power-law slope is 2.9 ± 1.4. Thus, despite the crude assumptions in deriving the shell size and velocity distributions, the data suggest that the neutral ISM in the SMC is fully consistent with superbubble activity dominating the structure. Although most other available H I shell catalogs are highly incomplete, preliminary results for a few other galaxies also show agreement with our model for the size distribution (Kim et al., 1999; Mashchenko et al., 1999; Oey and Clarke, 1997).

4. Correspondence with Star-Forming Regions

One of the most obvious global tests of mechanical feedback is to identify the parent stellar populations, or their remains, with the superbubbles. M31 (Brinks and Bajaja, 1986) and M33 (Deul and den Hartog, 1990) both show correlations of OB assocations with H I holes. However, Ho II shows contradictory results, based on the H I hole catalog compiled by Puche et al. (1992). Tongue and Westpfahl (1995) found that the SN rate implied by radio continuum emission is consistent with the hole energetics in that galaxy. However, Rhode et al. (1999) carried out a direct, BVR search for remnant stellar populations within the H I holes, and found little evidence for the expected stars. But Stewart et al. (2000) used far-UV data from the *Ultraviolet Imaging Telescope* and Hα images to conclude that a significant correlation between the H I holes and recent star formation does indeed support a feedback origin for the holes.

It is perhaps unsurprising that studies of Ho II yield these confusing results in view of that galaxy's distance of 3 Mpc. The LMC, which is 60 times closer, presents much better spatial resolution and should therefore yield less ambiguous results. Kim et al. (1999) examined the correspondence between their H I shell catalog, catalogued H II regions (Davies et al., 1976), and Hα imaging. Not only do they find a correspondence, but they are also able to identify an evolutionary sequence with respect to the relative sizes, expansion velocities, and Hα emission. Further investigation of the Magellanic Clouds should confirm and reveal more quantitative details of the mechanical feedback process (Oey, Gerken, and Walterbos, in preparation).

5. ISM Porosity and Galactic Superwinds

The consequences of feedback for the global ISM can be evaluated quantitatively in terms of the interstellar porosity parameter Q, which is the ratio: (total area or volume occupied by superbubbles) / (total area or volume of the galaxy). Thus it is essentially the filling factor of hot gas, assuming hot gas is contained within all of the superbubbles. Values of Q near unity indicate the HIM dominates the multiphase ISM, and values $\gg 1$ imply an outflow, with the galaxy generating more hot gas than it can contain.

It is straightforward to use the analytic expression for $N(R)$ (equation 3) to derive Q in terms of a galaxy's star-formation rate, Ψ (Oey et al., 2001; see also Clarke and Oey, 2002):

$$Q \simeq 16 \frac{\Psi(\mathrm{M}_\odot\ \mathrm{yr}^{-1})}{h R_g^2(\mathrm{kpc}^3)} \qquad \propto \frac{1}{P_0}\,, \tag{6}$$

for $\beta = 2$, a Salpeter (1955) IMF for stellar masses $0.1 \leq m \leq 100\ \mathrm{M}_\odot$, and $P_0/k = 9500$. R_g and h are the radius of the gaseous star-forming disk and gas scale height, respectively. We caution that Q depends on ambient interstellar parameters, for example, P_0^{-1} as indicated.

Oey et al. (2001) estimated Q for all the galaxies in the Local Group. The Milky Way yields $Q \sim 1$ for some methods and $Q < 1$ for others, consistent with the ambiguous results found in the past (e.g., McKee and Ostriker, 1977; Slavin and Cox, 1993). The LMC yields $Q \sim 1$, implying that hot gas dominates the ISM volume. The remainder of the Local Group galaxies all show $Q \ll 1$, with the sole exception of IC 10, a starburst galaxy for which $Q \sim 20$, thereby unambiguously predicting an outflow. Oey et al. (2001) crudely estimate the mass-loss rate in this outflow $\dot{M}_{\mathrm{out}}$, assuming that the material is largely evaporated from shell walls by thermal conduction. We find that $\dot{M}_{\mathrm{out}} \sim \Psi$; indeed, absorption-line studies of local starburst galaxies by Heckman et al. (2000) also show that empirically, the outflow and star-formation rates have the same order of magnitude for that sample.

Since $Q \sim 1$ represents a rough threshold for the escape of superwinds from the galactic disk, this also implies the simultaneous escape of newly-synthesized metals, which are contained in the hot gas. Likewise, the shredding of the ISM into filaments facilitates the escape of ionizing radiation, thus $Q \sim 1$ also represents an escape threshold for ionizing photons (Clarke and Oey, 2002). We finally note the extensive body of numerical work on superbubbles and blowout conditions. MacLow (1999) and Strickland and Stevens (2000) provide overviews of this field. The details of the numerical predictions are presently difficult to confirm empirically, but observations with *XMM-Newton* and *Chandra* are beginning to constrain the dominant processes.

6. Summary

Observations of mechanical feedback ranging from individual stellar wind bubbles to galactic superwinds are all largely consistent with the conventional adiabatic model for shell evolution. Presently none of the discrepancies are of a magnitude that suggest any need for major revision of the conventional understanding. The existence and properties of multiphase gas and filamentary structure are broadly consistent with the adiabatic model. It is also a remarkable strength that the model succeeds across size scales ranging over at least three orders of magnitude.

However, characterizing the dominant parameters and their effects on the shell evolution is still highly problematic. For example, critical ambient ISM conditions like density, pressure, and ionization distributions remain elusive. The mechanisms and conditions for cooling of the interior energy need to be identified, and energy budgets reliably determined. Perhaps the most fundamental question is the fate of the hot gas generated within the superbubbles: Does it escape to constitute the HIM? Does it escape from starburst galaxies, and from their gravitational potentials? These issues have crucial consequences for galactic evolutionary processes, and our understanding depends on further clarifying the mechanical feedback process.

Acknowledgements

I gratefully acknowledge support from the WS-ISM session organizers.

References

Arthur, S.J. and Henney, W.J.: 1996, *ApJ* **457**, 752.
Benaglia, P. and Cappa, C.E.: 1999, *A&A* **346**, 979.
Brown, A.G.A., Hartmann, D. and Burton, W.B.: 1995, *A&A* **300**, 903.
Cappa, C.E. and Benaglia, P.: 1998, *AJ* **116**, 1906.
Chu, Y.-H., Chang, H.-W., Su, Y.-L. and MacLow, M.-M.: 1995, *ApJ* **450**, 157.
Chu, Y.-H. and MacLow, M-M.: 1990, *ApJ* **365**, 510.

Chu, Y.-H., Wakker, B., MacLow, M.-M. and García-Segura, G.: 1994, *AJ* **108**, 1696.
Clarke, C.J. and Oey, M.S.: 2002, *MNRAS*, in press.
Cowie, L.L. and McKee, C.F.: 1977, *ApJ* **211**, 135.
Drissen, L., Moffat, A.F.J., Walborn, N.R. and Shara, M.R.: 1995, *AJ* **110**, 2235.
García-Segura, G. and Franco, J.: 1996, *ApJ* **469**, 171.
García-Segura, G. and MacLow, M.-M.: 1995, *ApJ* **455**, 145.
Heckman, T.M.: 2002, in: J.S. Mulchaey and J. Stocke (eds.), *Extragalactic Gas at Low Redshift*, ASP, San Francisco, p. 292.
Heckman, T.M., Lehnert, M.D., Strickland, D.K. and Armus, L.: 2000, *ApJS* **129**, 493.
Kennicutt, R.C., Edgar, B.K. and Hodge, P.W.: 1989, *ApJ* **337**, 761.
Kim, S., Dopita, M.A., Staveley-Smith, L. and Bessell, M.S.: 1999, *AJ* **118**, 2797.
MacLow, M.-M., in: A.R. Taylor, T.L. Landecker and G. Joncas (eds.), *New Perspectives on the Interstellar Medium*, Astron. Soc. Pacific, San Francisco, p. 303.
MacLow, M.-M., Chang, T.H., Chu, Y.-H., Points, S.D., Smith, R.C. and Wakker, B.P.: 1998, *ApJ* **493**, 260.
MacLow, M.-M. and McCray, R.: 1988, *ApJ* **324**, 776.
Mashchenko, S.Y., Thilker, D.A. and Braun, R.: 1999, *A&A* **343**, 352.
McKee, C.F. and Ostriker, J.P.: 1977, *ApJ* **218**, 148.
McKee, C.F., Van Buren, D. and Lazareff, B.: 1984, *ApJ* **278**, L115.
Oey, M.S.: 1996, *ApJ* **467**, 666.
Oey, M.S. and Clarke, C.J.: 1997, *MNRAS* **289**, 570.
Oey, M.S. and Clarke, C.J.: 1998a, *AJ* **115**, 1543.
Oey, M.S. and Clarke, C.J.: 1998b, in: J. Franco and A. Carramiñana (eds.), *Interstellar Turbulence*, Cambridge Univ. Press, Cambridge, p. 112.
Oey, M.S., Clarke, C.J. and Massey, P.: 2001, in: K.S. de Boer, R.-J. Dettmar and U. Klein (eds.), *Dwarf Galaxies and Their Environment*, Shaker Verlag, p. 181.
Oey, M.S., Groves, B., Staveley-Smith, L. and Smith, R.C.: 2001, in preparation.
Oey, M.S. and Massey, P., 1994, *ApJ* **425**, 635.
Oey, M.S. and Massey, P., 1995, *ApJ* **452**, 210.
Oey, M. S. and Smedley, S. A.: 1998, *AJ* **116**, 1263.
Pikel'ner, S. B.: 1968, *Astrophys. Lett.* **2**, 97.
Saken, J.M., Shull, J.M., Garmany, C.D., Nichols-Bohlin, J. and Fesen, R.A.: 1992, *ApJ* **397**, 537.
Salpeter, E.E.: 1955, *ApJ* **121**, 161.
Silich, S.A. and Franco, J.: 1999, *ApJ* **522**, 863.
Silich, S.A. and Oey, M.S.: 2001, in: E.K. Grebel, D. Geisler and D. Minniti (eds.), *Extragalactic Star Clusters*, ASP, San Francisco, in press.
Silich, S.A., Tenorio-Tagle, G., Terlevich, R., Terlevich, E. and Netzer, H.: 2001, *MNRAS* **324**, 191.
Slavin, J.D. and Cox, D.P.: 1993, *ApJ* **417**, 187.
Stanimirović, S., Staveley-Smith, L., Dickey, J.M., Sault, R.J. and Snowden, S.L.: 1999, *MNRAS* **302**, 417.
Staveley-Smith, L., Sault, R.J., Hatzidimitriou, D., Kesteven, M.J. and McConnell, D.: 1997, *MNRAS* **289**, 225.
Steigman, G., Strittmatter, P.A. and Williams, R.E.: 1975, *ApJ* **198**, 575.
Strickland, D.K. and Stevens, I.R.: 2000, *MNRAS* **314**, 511.
Treffers, R.R. and Chu, Y.-H.: 1982, *ApJ* **254**, 569.
Wang, Q. and Helfand, D.J.: 1991, *ApJ* **373**, 497.
Weaver, R., McCray, R., Castor, J., Shapiro, P. and Moore, R.: 1977, *ApJ* **218**, 377.

HI SHELLS IN THE LEIDEN-DWINGELOO HI SURVEY

SOŇA EHLEROVÁ, JAN PALOUŠ and RICHARD WÜNSCH
Astronomical Institute, Academy of Sciences of the Czech Republic, Prague; E-mail: sona@ig.cas.cz

Abstract. An algorithm of an automatic search for HI shells is presented and results of identifications in the Leiden-Dwingeloo HI survey are shown. An identification algorithm searches for local minima in channel maps, then creates 3-dimensional objects, HI holes. There is no *a priori* constraint on the shape of an HI hole, only the dimension criterion: dimensions of the structure must lie in a preselected interval. When this algorithm is applied to the Leiden-Dwingeloo HI survey, it finds about 1000 structures, which are seen in 4 or more consecutive velocity channels ($\Delta v \geq 4\ kms^{-1}$). A total energy contained in all structures in the Milky Way is of the order of $10^{55} erg$, which is less than the energy corresponding to the observed SNR. 80% of the energy contained in holes resides at low galactic z-coordinates, $|z| < 2\ kpc$.

Keywords: HI shells, HI holes, an automatic identification, Leiden-Dwingeloo HI survey

1. HI Shells in Galaxies

HI shells (or holes) exist in many galaxies including our own (for a review on this topic see Brinks and Walter, 1998, or the contribution of S. Oey in this volume). The identification of shells is, in most cases, based on the subjective (i.e. by-eye) methods, though sometimes efforts of several people are combined or the observer is supported by e.g. the `kshell` routine in the KARMA visualisation software. However, human eyes can be cheated, not only parts of shells can be missing, but eyes predominantly identify structures at certain positions (most notably edges of a gaseous disc) and at certain T_B levels. `kshell` is not an identification tool and it is sensitive to only spherical structures.

There were efforts to identify HI shells automaticaly. Most notably there is a series of papers (Thilker et al., 1998; Mashchenko et al., 1999) describing an automatic code for the shell recognition. The code is based on a model of the shell (either a model in the thin shell approximation or a hydrodynamical one), it searches imprints of the model (or rather of a family of different models) in the data. The algorithm was used for the galaxy NGC 2403. The difficulty of this approach is the dependance on the model. In our opinion the turbulent and irregular nature of the ISM may prevent the usability of universal models.

Our goal was to write a model-independent algorithm identifying automaticaly HI shells in datacubes. Therefore we abstain from searching for complex shapes and velocity profiles, but rather we search for simple HI holes. The basic strategy

Astrophysics and Space Science **289:** 279–282, 2004.

used in our identifying code resembles de Heij's et al. (2002) approach (an HVCs search), only reversed.

2. HOLMES, an Automatic Identification of HI Holes

Our identifying code HOLMES detects local minima in maps[1]. An identified structure is a continuous region around a minimum, which is equal to or larger than a required minimum dimension Δ_{min} and smaller than a maximum dimension Δ_{max}. A defining contour of the structure is the maximum one, for which the enclosed structure does not exceed the dimension Δ_{max}. There are no *a priori* constraints on the shape or velocity profile of the structure, only dimensions must fall into a selected interval of values. Identifications in subsequent channel maps are then glued together. We discard all structures, which appear in 3 or less velocity channels ($\Delta v \leq 3\ kms^{-1}$), as they are probably not real HI holes or hole-like structures.

There are problems with certain types of shells, which the code cannot solve. If the wall of the shell is fragmented to pieces and the interior is directly or with a low contrast connected to the surrounding medium, the code does not locate any structure. Therefore we think that structures with fragmented walls and no clear holes are missing in HOLMES identifications. Secondly, complex shapes of HI shells – like two lobes, which are not connected by a low-density tunnel, or where this tunnel is hidden by the intervening emission – are not found by the code or are identified as two or more separate structures. A comparison of our results for the Leiden-Dwingeloo HI survey (LDS) of the Milky Way and the list of HI shells by Heiles (1979) confirm this superstition. A human eye easily extrapolates and finds structures with partly missing walls. On the other hand, a computer is systematic and is not biased to structures at certain temperatures or at certain positions and therefore finds more shells than were previously known in the Milky Way.

3. Leiden-Dwingeloo HI survey

The code HOLMES was applied to the Leiden-Dwingeloo HI survey (Hartmann and Burton, 1997). Nearly 4000 structures were identified, about 1000 of them were found in more than 3 consecutive velocity channels. Positions of these 1000 structures are shown at Figure 1. As expected, holes are concentrated towards the galactic plane. The extrapolated number of shells in the whole Milky Way is 1600.

Using the rotation curve of Wouterloot et al. (1990) we calculate kinematic distances of HI holes. A majority of shells is found close to the galactic equator, the halfthickness is about 800 *pc*. The size distribution of holes is shown at Figure 2a. There is a clear trend for shells closer to the galactic equator (i.e. a denser

[1] In 2D channel maps, though the algorithm works in 3D as well. However, 3D identifications take significantly more CPU time than 2D identifications and therefore were not used.

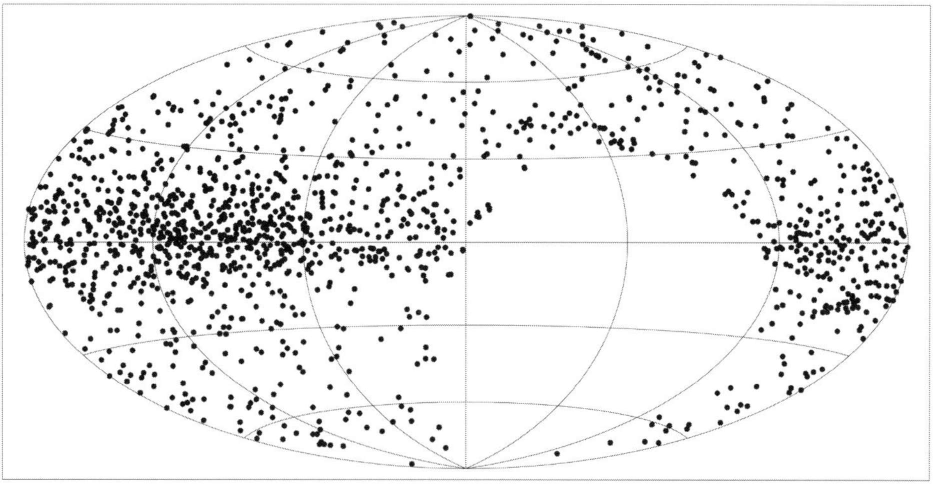

Figure 1. Positions of HI shell-like structures in the Milky Way.

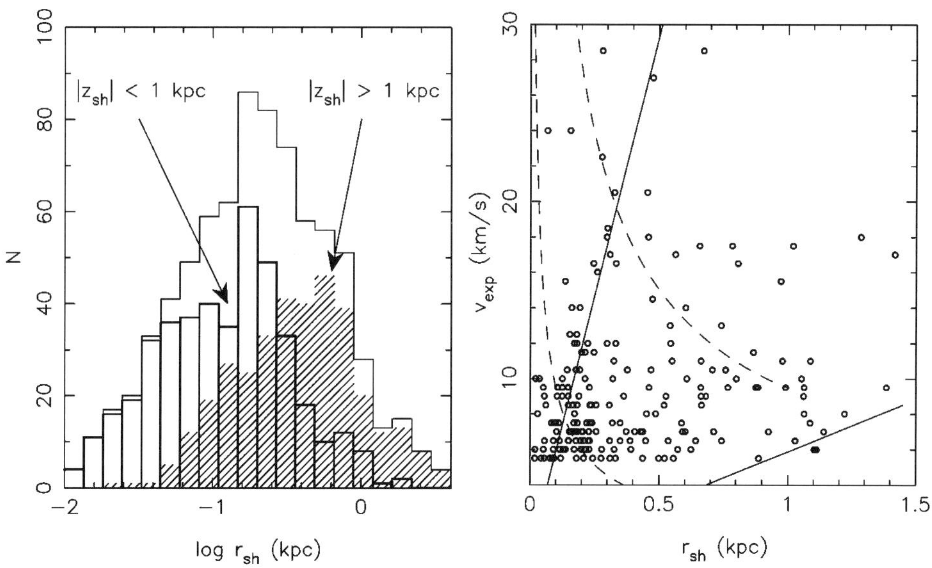

Figure 2. The size distribution (left) and r_{sh} vs v_{exp} graph (right) of identified structures. Solid lines in the right panel show an analytical solution for 10 and 100 Myr, dashed lines show solutions for two different $\left(\frac{L}{n}\right)$, 10^{50} and $10^{52} erg/Myr/cm^{-3}$.

region) to be smaller than shells at a higher z. A greater part of identified structures have expansion velocities lower than 10 kms^{-1}, see Figure 2b.

A typical energy of the identified structure (Chevalier's formula, 1974) is $5 \times 10^{50} erg$, about 80 structures have energies $E_{chev} \geq 5 \times 10^{51} erg$. We extrapolate the total energy contained in all identified shells to the whole Milky Way and reach the value $E_{tot} \simeq 10^{55} erg$. Assuming the typical life-time of an HI hole ($20 - 30$ Myr)

we calculate the needed energy input, which we consider to come mostly from supernovae, to be $\dot{E} = 1.4 \times 10^{40} erg s^{-1}$, i.e. $SNR = 1\ SN$ per 2000 yr. This is much smaller than the observed SNR or than the SNR corresponding to the total SFR in our Galaxy. The discrepancy probably comes from three effects: 1) an underestimation of dimensions and the energy; 2) an underrepresentation of evolved and fragmented shells in our sample; 3) a part of the energy released by stellar winds and supernovae escapes from shells to the galactic halo and therefore cannot be derived by our method: energies released by SF and contained in shells are different. Currently we work on the way how to get rid of the first two effects and how to estimate their influence on results.

Acknowledgements

This work was supported by the grant of the Grant Agency of the Academy of Sciences of the Czech Republic, No. B3003106. The participation of SE at the JENAM conference was supported by the grant of the European Astronomical Society.

References

Brinks, E. and Walter, F. 1998: in: T. Richtler and J.M. Braun (eds.), *The Magellanic Clouds and Other Dwarf Galaxies*, Shaker Verlag, p. 1–10.
Chevalier, R.: 1974: *ApJ* **188**, 501.
de Heij, V., Braun, R. and Burton, W.B.: 2002: *A&A* **391**, 159.
Hartmann, D. and Burton, W.B.: 1997: *Atlas of Galactic Neutral Hydrogen*, Cambridge University Press.
Heiles, C.: 1979, *ApJ* **229**, 533.
Mashchenko, S.Y., Thilker, D.A. and Braun, R.: 1999, *A&A* **343**, 352.
Thilker, D.A., Braun, R. and Walterbos, R.A.M.: 1998, *A&A* **332**, 429.
Wouterloot, J.G.A., Brand, J., Burton, W.B. and Kwee, K.K.: 1990, *A&A* **230**, 21.

SNR SURFACE DENSITY DISTRIBUTION IN NEARBY GALAXIES

MANAMI SASAKI, DIETER BREITSCHWERDT and RODRIGO SUPPER
Max-Planck-Institut für extraterrestrische Physik, Giessenbachstraße, Postfach 1312, 85741 Garching, Germany

Abstract. Since supernova remnants (SNRs) are believed to be the primary sources of Galactic cosmic rays (CRs), their distribution in galaxies is an important basis for modelling and understanding the distribution of the CRs and their γ-ray spectrum. We analysed the radial surface density of X-ray and radio selected SNRs in the Large Magellanic Cloud (LMC) and M 33. Both in X-rays and in radio, the surface densities of the SNRs are in excellent agreement in both galaxies, showing an exponential decay in radius. The results were compared to the SNR distribution in the spiral galaxies M 31 and NGC 6946 as well. The radial scale length of the distribution is $\frac{1}{4} - \frac{1}{3}$ of the radius of the galaxies, fully consistent with values derived for the Milky Way, the LMC, and M 33. Therefore, not only the radio SNRs, but also the X-ray detected SNR sample can be interpreted to be representative for the CR sources within a galaxy.

Keywords: X-rays: galaxies, ISM: supernova remnants, cosmic rays

1. Introduction

Diffusive shock acceleration in supernova remnants (SNRs) is thought to play a dominant role in producing Galactic cosmic rays (CRs) below $\sim 10^{15}$ eV. Therefore, the distribution of SNRs (and pulsars) in the Galaxy is a crucial basis for the understanding of CR distribution. Based on Galactic SNRs, Stecker and Jones (1977) showed that the surface density can be described as $f(x) = C\, x^{\alpha} \exp(-\beta x)$ (x: radial distance) with a maximum at a distance of several kpc from the Galactic centre. Case and Bhattacharya (1998) used the function $f(x) = C\, sin(\pi x + \theta) \exp(-\beta x)$, noticing that the distribution seems to be larger than zero at $x = 0$. However, since the sample of Galactic SNRs is flux limited, older and distant remnants are missed systematically. The analysis of *supernovae* (SNe) in different galaxies by Dragicevich et al. (1999) showed that the average distribution can be written as $f(x) = C \exp(-\beta x)$. While the observation of SNRs reveals us the population in a time interval of more than 1000 years, the sample of SNe only gives information about the last few decades. Therefore, we analysed the radial distribution of the X-ray selected and radio selected SNRs of the Large Magellanic Cloud (LMC) which is a dwarf galaxy with spiral arms and bar-like structure. In order to compare the results with the Milky Way, we also selected the grand-design spiral galaxy M 31 and the face-on spiral galaxy M 33. Moreover, we analysed NGC 6946, which is a spiral starburst galaxy.

Astrophysics and Space Science **289:** 283–286, 2004.

TABLE I

Fit results for $f(x) = C \exp(-x/R)$. For abbreviations see references

Data		C	$R = 1/\beta$	Red. χ^2
LMC (Radio)	[FHW98a], [FJW98b]	3.74±1.59	0.29±0.06	1.16
LMC (X-rays)	[HP99], [SHP00]	3.25±1.17	0.27±0.05	1.03
M 33 (Radio)	[GDK99]	2.84±0.71	0.28±0.03	1.06
M 33 (X-rays)	[HP01]	1.52±0.84	0.23±0.06	1.23
M 31 (Radio)	[DD84], [BW93]	0.10±0.05	0.37±0.08	1.65
M 31 (X-rays)	[SHL01]	0.12±0.07	0.33±0.09	1.82
NGC 6946 (Radio)	[LD01]	1.32±0.50	0.24±0.04	1.42

2. Radial SNR Surface Density Distribution

After the SNR positions were corrected for projection and the radial distances to the centre of the galaxies were normalised to $D_{25}/2$, the surface density of the SNRs was calculated for equidistant radial binning. In Figure 1, the SNR surface densities are plotted against the normalised radial distance together with model distributions. The best fit parameters for the exponential distributions are listed in Table I.

The radial dependences of the SNR surface densities of the X-ray selected and radio selected SNRs in the *LMC* are comparable with a scale length of $\sim 0.3 \times D_{25}/2 \approx 1.4$ kpc. In *M 33*, the surface density distribution of H I is known to peak at $\sim 10' = 0.3 \times D_{25}/2 \approx 1.9$ kpc. Also in the SNR distribution, there is a maximum at the corresponding distance within the statistical errors. As for *M 31*, the radio and X-ray selected SNRs have similar surface densities with a maximum at $\sim 0.3 \times D_{25}/2 \approx 5.8$ kpc distance from the centre of the galaxy. This distribution is in perfect agreement with the SNR distribution in the Milky Way. The SNR surface density in *NGC 6946* increases steeply towards the centre, probably due to the starburst activities in the central part. In total, the radial surface density distribution of the radio selected SNRs in NGC 6946 has a scale length of $0.24 \times D_{25}/2 \approx 2.2$ kpc.

3. Discussion

Flat radial distribution of the diffuse Galactic γ-ray emission has been observed by COS-B (Strong et al., 1988) and EGRET (Strong and Mattox, 1996; Digel et al., 1996). This weak radial gradient of the γ-ray emissivity in our Galaxy can be reproduced in theoretical models with a CR source distribution similar to the SNR distribution: e.g. Bloemen et al. (1993) presented a model in which diffusion is

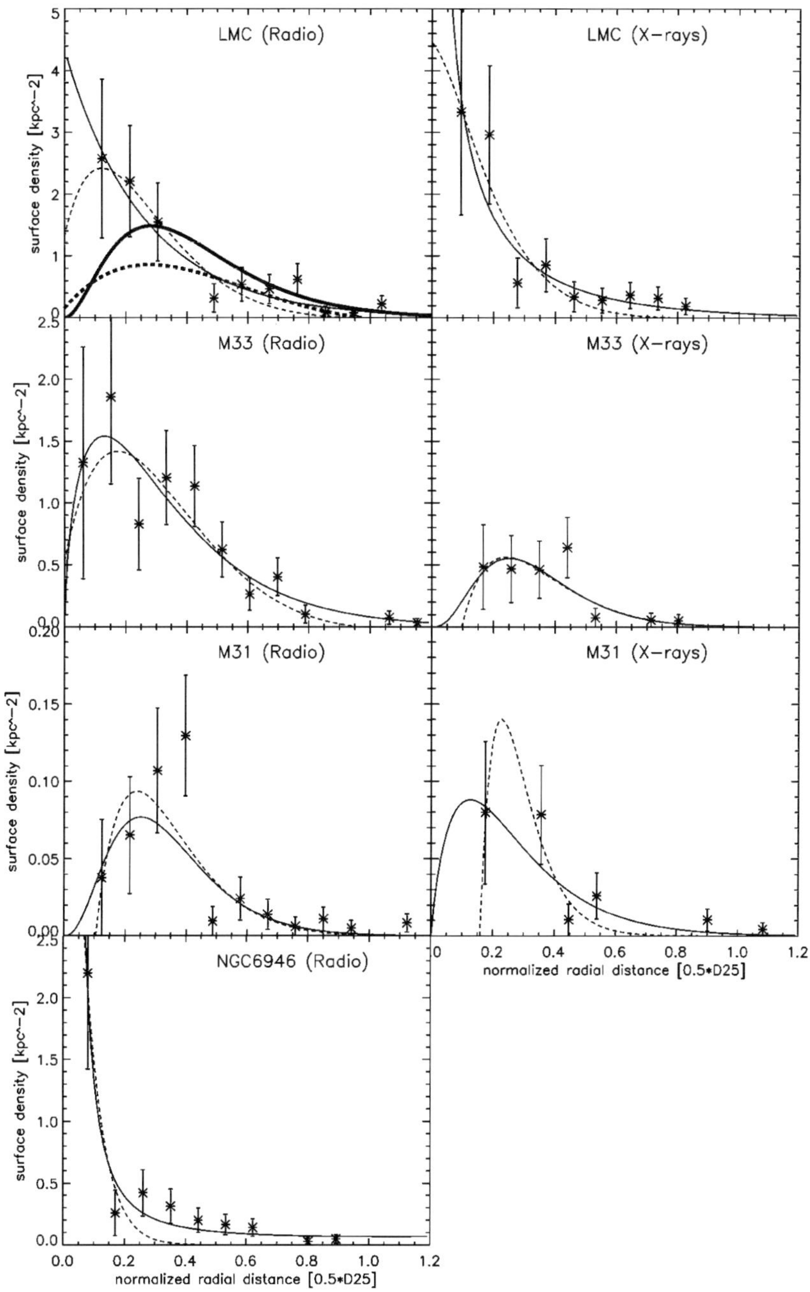

Figure 1. Surface densities and fit curves for radio and X-ray selected SNRs. Solid lines are used for the model function $f(x) = C\,x^{\alpha}\exp(-\beta x)$ and dashed lines for $f(x) = C\,sin(\pi x + \theta)\exp(-\beta x)$. The models used for the Galactic SNRs by Case and Bhattacharya (1998) are plotted in the first diagram with thick lines.

the dominant process of the transport, but with a huge diffusion halo which is inconsistent with CR secondary measurements. Therefore, a model considering dynamical coupling of the CR particles and the thermal plasma was developed by Breitschwerdt et al. (2002). Our analysis has shown that the radio and X-ray selected SNRs in M 31 and M 33 have distributions similar to the Galactic SNRs, with a scale length of $\sim (0.25 - 0.35) \times D_{25}/2$ and a maximum at $(0.2 - 0.4) \times D_{25}/2$. The perfect correlation between radio and X-ray selected SNRs indicates that both types are good candidates for CR sources. Although the SNR surface densities of the LMC (active star formation) and NGC 6946 (starburst nucleus) differ significantly from that of the Milky Way, M 31, and M 33, the scale lengths of the exponential fit are comparable for all of these galaxies.

References

Bloemen, J.B.G.M., Dogiel, V.A., Dorman, L. and Ptuskin, V.S.: 1993, *A&A* **267**, 372.
Braun, R. and Walterbos, R.A.M.: 1993, *A&AS* **98**, 327. [BW93]
Breitschwerdt, D., Dogiel, V.A. and Völk, H.J.: 2002, *A&A* **385**, 216.
Case, G.L. and Bhattacharya, D.: 1998, *ApJ* **504**, 761.
Dickel, J.R. and Dodorico, S.: 1984, *MNRAS* **206**, 351. [DD84]
Digel, S.W., Grenier, I.A., Heithausen, A., Hunter, S.D. and Thaddeus, P.: 1996, *ApJ* **463**, 609.
Dragicevich, P.M., Blair, D.G. and Burman, R.R.: 1999, *MNRAS* **302**, 693.
Filipovic, M.D., Haynes, R.F., White, G.L. and Jones, P.A.: 1998a, *A&AS* **130**, 421. [FHW98a]
Filipovic, M.D., Jones, P.A., White, G.L. and Haynes, R.F.: 1998b, *A&AS* **130**, 441. [FJW98b]
Gordon, S.M., Duric, N., Kirshner, R.P., Goss, W.M. and Viallefond, F.: 1999, *ApJS* **120**, 247. [GDK99]
Haberl, F. and Pietsch, W.: 1999, *A&AS* **139**, 277. [HP99]
Haberl, F. and Pietsch, W.: 2001, *A&A* **373**, 438. [HP01]
Lacey, C.K. and Duric, N.: 2001, *ApJ* **560**, 719. [LD01]
Sasaki, M., Haberl, F. and Pietsch, W.: 2000, *A&AS* **143**, 391. [SHP00]
Stecker, F.W. and Jones, F.C.: 1977, *ApJ* **217**, 843.
Strong, A.W., Bloemen, J.B.G.M., Dame, T.M., Grenier, I.A., Hermsen, W., Lebrun, F., Nyman, L.-A., Pollock, A.M.T. and Thaddeus, P.: 1988, *A&A* **207**, 1.
Strong, A.W. and Mattox, J.R.: 1996, *A&A* **308**, L21.
Supper, R., Hasinger, G., Lewin, W.H.G., Magnier, E.A., van Paradijs, J., Pietsch, W., Read, A.M. and Trümper, J.: 2001, *A&A* **373**, 63. [SHL01]

KINEMATICS OF GIANT H II REGIONS IN NEARBY GALAXIES

H. PLANA[1], J. MAÍZ-APELLÁNIZ[2] and E. TELLES[1]
[1] *Observatorio Nacional – MCT*
[2] *Space Telescope Science Institute*

Abstract. We present optical high-spectral-resolution long-slit observations of massive star formation regions in the nearby galaxies M101, NGC 2403, and M33 obtained with the WHT. We have constructed velocity, line dispersion and intensity maps using between 6 and 12 parallel slits in each case. We will present the data for seven regions: NGC 5455, NGC 5461, NGC 5471 in M101; three regions in NGC 2403; and NGC 595 in M33. We will also present a detailed study of the different profiles found across the regions, paying special attention to the presence of supersonic profiles and multiple components.

Keywords: Giant H II regions, Kinematics

1. Introduction

In the early 60's Ambartsumian et al. (1963) first investigated superassociations (SA) of young stars. He observed a number superassociations brighter than M_V= -14 and with a projected diameter on the sky of ~500pc. In his study, the luminosity threshold was arbitrary. Petrosyan et al. (1983) did a more quantitative study of Ambartsumian sample using morphology criteria and noted that these criteria changes when $M_{pg} = -11$. This study defined the concept of superassociations and Giant H II Regions (GHRs). Quantitative observables can be define as followed: a large nebula ionized by massive cluster of approximately coeval stars with a diameter 100 pc < D < 1 kpc, a gas mass of $M_{gas} \sim 10^6$ $M_\odot$, a mass stellar of the same order of magnitude and a Star Formation Rate, SFR < 10 $M_\odot$ / yr (late type galaxy) but, the SFR / volume unit is orders of magnitude higher in comparison with that is found in a late type galaxy.

Smith and Weedman (1972) found that the gas motions of GHRs were highly supersonic (velocity dispersion $\sigma_{gas} > c_{H\,II} \sim 13$ km·s^{-1}). In comparison, the σ of Galactic H II regions is subsonic or sonic. Collisions between gas clouds at supersonic speed will dissipate kinematic energy in a time scale shorter than the life time of an H II region.

Two important correlations have been found linking the Luminosity ($L_{H\beta}$), the velocity dispersion σ_{gas} and the radius of the region (R). Terlevich & Melnick 1981 found that $L_{H\beta}\ \alpha\ \sigma^4$ and Melnick et al. 1987 found that R $\alpha\ \sigma^{2.5}$. These correlations have been found to be true in a wide range of objects, from H_2 clouds to ellipticals

TABLE I
Journal of observations

Mother Galaxy	Region	Date	Nb Slits	Slits Aper. & Separation	Expo.Time
NGC 2403	Regions II - IV	Apr. 18th, 1992	10 slits	1" - 1"	900s
M101	Region NGC 5455	Apr. 18th, 1992	6 slits	1" - 1"	900s
	Region NGC 5461	Apr. 18th, 1992	6 slits	1" - 1"	900s
	Region NGC 5471	Apr. 18th, 1992	10 slits	1" - 1"	900s
M33	Region NGC 595	Aug. 19th, 1992	9 slits	1"- 2"	1000/1200s

nuclei, going through GHRs and H II galaxies. This led Melnick et al. (1988) to think that virial equilibrium plays an important role in gas motion.

Several mechanisms have been proposed to explain how to maintain a high velocity dispersion. Dyson (1979) and others proposed the action of thousands of strong stellar winds from coeval massive stars to explain these observations. These stellar winds would form giant 'bubbles' with supersonic thin shells around them. The 'champagne' flow model (Tenorio-Tagle, 1979) can also explain the supersonic expansion. However these mechanisms fail to explain the position of different sources in the (σ - r) plane.

2. Observations and Data

2.1. Observations

In the context of the GEFE program (Grupo de Estudios de Formación Estelar), high resolution observations have been carried out with the ISIS Spectrograph attached to the 4.20 m William Hershel Telescope. These observations provide high spectral resolution (0.4 Å/px) in the blue (4660 Å– 5060 Å) region and in the red (6390 Å– 6840 Å) region, allowing us to measure several emission lines (Hβ, [OIII]5007 Å, Hα, [NII]6584 Å, [SII]6717 Å and [SII]6731 Å. The spatial resolution of the instrument was $0.3''$/px.

In order to produce 2D maps (velocity, monochromatic intensity, and dispersion), observations of several parallel slits have been made. Table I shows the journal of observations:

2.2. Objects description

The five GHRs have been observed in three nearby late type galaxies: NGC 2403 (SAB(s)cd type, V_{syst} = 141 km·s^{-1}), M101 (SA(s)cd type, V_{syst}=-179 km·s^{-1}) and M33 (SAB(rs)cd type, V_{syst} = 241 km·s^{-1}). The five GHRs observed have diameters between 400 pc and 1000 pc. For all these regions we obtained monochromatic in all emission lines mentioned before and used only Hα for the velocity and velocity dispersion maps.

Recent observations of the NGC 2403 Giant H II Regions by Drissen et al. (1999) show that NGC 2403 II has an ionizing cluster within 43 pc radius but a ionized nebula extension of more than 300pc. For Region II and IV $\sim$ 10 WR stars candidates have been found. The velocity field we obtained first shows the rotation of the galaxy (with an amplitude of $\sim$ 150 km·s^{-1}). Position-Velocity plots show large velocity variations inside both Regions II and IV, between 5 to 20 km·s^{-1}. A more careful analysis shows the presence of double (or more) component profiles in both regions. Double profiles are located in the center of Region IV where the intensity is maximum while double profiles also appear in a less intense area for Region II. The analysis of the velocity dispersion map also revealed that high intensity regions have low velocity dispersion ($\sim$ 12 km·s^{-1}) while low intensity regions (around them) show high dispersion (20 km·s^{-1}).

The three regions NGC 5455, NGC 5461 and NGC 5471 are part of M101.

NGC 5471 is formed by six clusters (Skillman et al., 1984). Our velocity map shows that the six clusters have a higher velocity than areas of lower monochromatic intensity. In fact a circular region of high velocity is present including the six clusters. The anti correlation between velocity dispersion and monochromatic emission is not as strong as it it was for NGC 2403 II-IV. Here, double component profiles are present for cluster B and regions of lower monochromatic intensity.

NGC 5461 is formed by five clusters (Muñoz-Tuñón et al., 1995). Clusters A and B show signs of multiple components profiles in their centers which will explain the high velocity dispersion they show, $\sim$ 30 km·s^{-1}. The others clusters (C,D,E) show low dispersion ($\sim$ 16 km·s^{-1}).

With NGC 604, NGC 595 is one of the largest H II complex of M33. Like NGC 2403, $\sim$ 10 WR stars have been detected in NGC 595. In Figure 1 we present, as an example the kind of maps we have derived. The Hα monochromatic map is superimposed to the dispersion velocity map. Here too, we can see clearly the anti correlation between regions with high velocity dispersion and low monochromatic emission. We also show the regions where multiple component profiles are present.

NGC 2403 II and NGC 595 show kinematical signatures of expanding shells into the ISM. The detailled study of the velocity and dispersion maps will explain the evolution state of these GHRs.

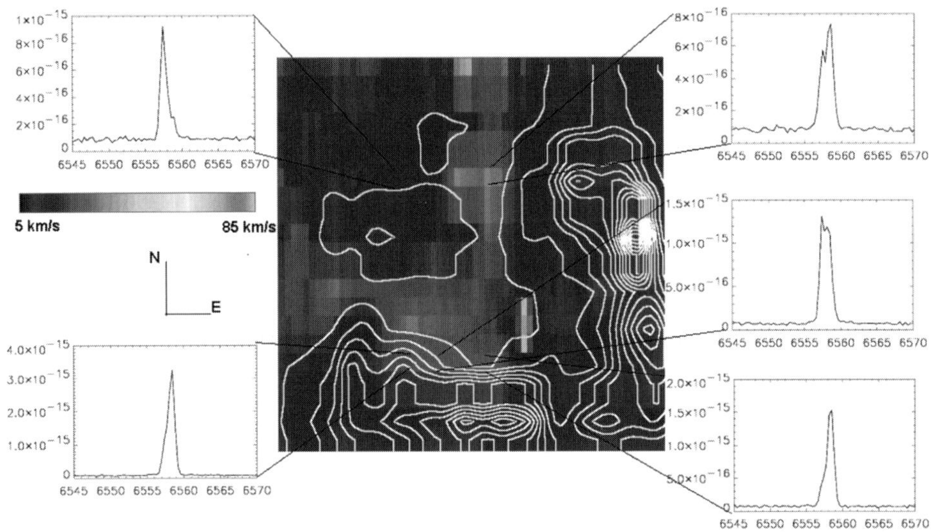

Figure 1. We show here the main clusters of NGC 595. Isophotes represent Hα intensity line superimposed on the velocity dispersion maps.

Acknowledgements

H.P. wants to thank the PRONEX program and the LOC of JENAM2002 for their financial support.

References

Ambartsumian et al.: 1963, *Soobshch. Byur. Obs.* **33**, 3.
Drissen et al.: 1999, *AJ* **117**, 1249.
Dyson: 1979, *A&A* **73**, 132.
Melnick et al.: 1987, *MNRAS* **226**, 849.
Melnick et al.: 1988, *MNRAS* **235**, 297.
Muñoz-Tuñón et al.: 1995, *AJ* **110**, 1630.
Petrosyan et al.: 1983, *Astrofisika* **19**, 631.
Tenorio-Tagle: 1979, *A&A* **71**, 59.
Terlevich and Melnick: 1981, *MNRAS* **195**, 839.
Skillman et al.: 1984, *ApJ* **280**, 580.
Smith and Weedman: 1972, *ApJ* **172**, 307.

TURBULENCE, MAGNETIC FIELDS AND CRs

THE ROLE OF MAGNETIC FIELDS IN SPIRAL GALAXIES

RAINER BECK
MPI für Radioastronomie, Auf dem Hügel 69, 53121 Bonn, Germany

Abstract. Interstellar magnetic fields are strong: up to 25μG in spiral arms and 40μG in nuclear regions. In the spiral galaxy NGC 6946 the average magnetic energy density exceeds that of the thermal gas. Magnetic fields control the evolution of dense clouds and possibly the global star formation efficiency in galaxies. Gas flows and shocks in spiral arms and bars are modified by magnetic fields. Magnetic forces in star-forming circumnuclear regions are able to drive mass inflow towards the active nucleus. Magnetic fields are essential for the propagation of cosmic rays and the formation of galactic winds and halos.

Keywords: Galaxies: spiral, Galaxies: magnetic fields, ISM: magnetic fields

1. Interstellar Magnetic Fields Are Hard To Observe

... *but it's easier than most people think.* In radio continuum the typical degrees of polarization are much higher than in the other spectral ranges, and we benefit from the development of large instruments and sensitive receivers. This is why most of our knowledge on interstellar magnetic fields in our Galaxy (Reich, 1994; Heiles, 1996; Beck, 2001) and in external galaxies is based on polarized radio emission and its Faraday rotation (see also Beck, 2000; Beck, 2002). To detect extended features and achieve high resolution, data from interferometric (synthesis) and single dish telescopes have to be combined.

2. Radio Waves Are Tracers of Magnetic Fields

... *simply the best ones.* Interstellar magnetic fields are illuminated by cosmic-ray electrons emitting synchrotron radiation, the dominant contribution to the diffuse radio continuum emission at centimeter and decimeter wavelengths. Synchrotron emission is intrinsically highly linearly polarized, 70–75% in a completely regular magnetic field. The observable degree of polarization in galaxies is reduced by a contribution of unpolarized thermal emission which may dominate in star-forming regions, by Faraday depolarization (Sokoloff et al., 1998) and by geometrical depolarization within the beam. A map of the *total* nonthermal intensity (Figure 1) reveals the *strength of the total interstellar magnetic fields* in the plane of the sky, while the *polarized intensity and polarization angle* (Figures 2 and 3) reveal the

Astrophysics and Space Science **289:** 293–302, 2004.

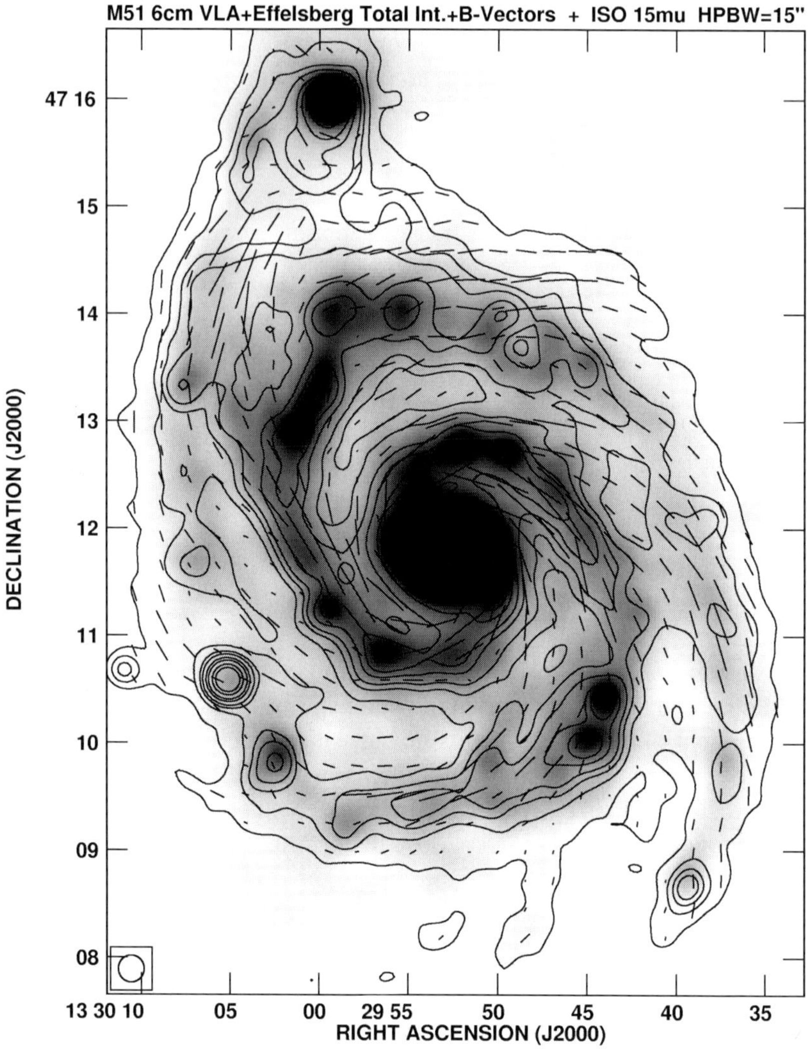

Figure 1. Total radio continuum intensity (contours) and **B**-vectors of polarized intensity of M 51 at 15″ resolution, combined from VLA and Effelsberg observations at λ6 cm. The grey-scale ISOCAM image shows the λ15 μm emission (kindly provided by M. Sauvage), smoothed to 15″ resolution. (Beck, unpublished)

strength and structure of the resolved regular fields in the plane of the sky. The orientation of polarization vectors is changed in a magneto-ionic medium by *Faraday rotation* which is generally small below about λ6 cm so that the **B**-vectors (i.e. the observed **E**-vectors rotated by 90°) directly trace the *orientation* of the regular fields in the sky plane.

3. Magnetic Fields *Are* Strong

... *much stronger than most people think.* The average strength of the total $\langle B_{t,\perp} \rangle$ and the resolved regular field $\langle B_{reg,\perp} \rangle$ in the plane of the sky can be derived from the total and polarized radio synchrotron intensity, respectively, if energy-density *equipartition* between cosmic rays and magnetic fields is assumed (Beck et al., 1996).

In our Galaxy the accuracy of the equipartition assumption can be tested, because we have independent information from in-situ measurements about the local cosmic-ray energy density and from γ-ray data about their distribution. Combination of the radio synchrotron emission, the local cosmic-ray electron density and diffuse continuum γ-rays yields a local strength of the total field of $6 \pm 1\,\mu$G (Strong et al., 2000), the same value as derived from energy equipartition (Berkhuijsen in Beck, 2001). Even the scale length of the radial variation of the equipartition field strength is similar to that in Strong et al. (2000).

The mean equipartition strength of the total field for a sample of 74 spiral galaxies (derived from their total radio flux densities) is $\langle B_t \rangle = 9\,\mu$G, ranging between $\langle B_t \rangle \simeq 6\,\mu$G in radio-faint galaxies like M 31 and M 33, and $\simeq 15\,\mu$G in grand-design galaxies like M 51, M 83 and NGC 6946. In prominent spiral arms the total field strength is 20–25μG, up to 40μG in nuclear regions (Sect. 7.3), and even 1 mG in molecular clouds compressed by supernova shocks (Brogan et al., 2000), in compact cloud cores (Myers and Goodman, 1988; Uchida et al., 2001) and in filaments near the Galactic center (Reich, 1994; Yusef-Zadeh et al., 1996).

4. Magnetic Fields *Are* Dynamically Important

... *more important than most people think.* A standard approach to estimate the dynamical importance of various competing forces is to compare the corresponding energy densities (Table I). Note, however, that the dynamical effects of magnetic fields are anisotropic due to their vector nature. NGC 6946 (Figure 2) is a useful standard spiral galaxy because it has no companion, no strong density waves, no active nucleus and has been observed in many spectral ranges. The average total equipartition field strength in the inner disk (within 4 kpc radius) is $B_t \simeq 18\,\mu$G (using the map of synchrotron emission by Walsh et al., 2002) and the average density of the warm ionized gas ($T \simeq 10^4$ K) is $n_e \simeq 0.5$ cm^{-3} (using their map of thermal radio emission and a filling factor of 5%). The map of the total neutral gas (molecular + atomic, where the molecular gas dominates until about 5 kpc radius) yields a mean number density (within a disk of 4 kpc radius and 100 pc height) of $n \simeq 20$ H atoms cm^{-3}. The turbulent velocity of the neutral gas is assumed to be $v_{turb} = 7.5$ km/s, as for the cold neutral gas in our Galaxy (Kalberla and Kerp, 1998), and the global rotation velocity $v_{rot} = 170$ km/s. For the cosmic rays we know the slope of the energy spectrum ($N[E] \propto N_0 E^{-2\alpha-1} dE$) from the observed

TABLE I

Average energy densities ϵ (in 10^{-12} erg cm^{-3}) in the ISM of the inner disk ($\leq$4 kpc radius) of NGC 6946

Magnetic field (equip.)	$\epsilon_{B_t^2}$	$B^2/8\pi$	13
Cosmic rays (equip.)	ϵ_{CR}	$C \int E^{-2\alpha} dE$	13
Warm ionized medium	ϵ_{WIM}	$\frac{3}{2} n_e k T$	1
Turbulent cloud motion	ϵ_{turb}	$\frac{1}{2} \rho v_{turb}^2$	10
Global gas rotation	ϵ_{rot}	$\frac{1}{2} \rho v_{rot}^2$	5000

synchrotron spectral index α, but no information on the absolute particle number N_0 is available so that we have to rely on the validity of energy equipartition with the magnetic fields.

The ratio β between the thermal and magnetic energy densities of the warm ionized medium (WIM) is only $\simeq 0.1$. The hot ionized gas and the warm neutral gas are believed to be in pressure equilibrium with the WIM so that their β should be similar. (The thermal energy of the cold gas is negligible.) A low thermal contribution to the total gas pressure was also found in the local Galactic ISM (Jenkins and Tripp, 2001; Jenkins, this volume). A significant fraction of the diffuse ISM must be unstable, giving rise to gas flows. From X-ray and radio continuum observations in the halo of the spiral galaxy M 83, Ehle et al. (1998) derived a similarly low value of β. The magnetic field dominates thermal processes in the disk and in the halo of galaxies.

5. Magnetic Fields Drive Star Formation

Turbulent motion of neutral gas and total magnetic fields in the inner disk of NGC 6946 are roughly in energy equipartition (Table I) (but see Sect. 6). This fact supports the idea that magnetic fields are anchored in the (partly ionized) envelopes of gas clouds (Beck, 1991). This also explains the tight radio-infrared correlation (Niklas and Beck, 1997; Walsh et al., 2002). Between the clouds, field loops can be generated by the Parker instability which can drive the galactic dynamo (Hanasz and Lesch, 1998; Moss et al., 1999; Hanasz et al., this volume).

Interstellar magnetic fields affect the motion of small gas clouds (Elmegreen, 1981) and their collision rate. The importance of internal fields for the evolution of clouds is generally accepted, though not understood (Mestel and Paris, 1984; Myers and Goodman, 1988; Crutcher, 1999; Heitsch et al., 2001). However, little is known about the influence of magnetic fields in the ISM on general properties of star formation.

Stability of cloud cores depends also on the fractional ionization by cosmic rays, and the number density of cosmic rays in the ISM is controlled by the interstellar magnetic fields. A larger fractional ionization allows a better coupling of the fields to the gas (weaker ambipolar diffusion) and thus stronger internal fields. As a result, the field strength inside and outside of the clouds should be correlated.

The star formation rate SFR is known to relate non-linearly with gas density ρ, the *Schmidt law* $SFR \propto \rho^{\mathrm{n}}$, with $n \simeq 1.5$ (Wong and Blitz, 2002). Hence, the star formation efficiency ($SFE \propto SFR/\rho$) increases with gas density. As the total field strength B_t in galaxies is known to scale globally with ρ as $B_t \propto \rho^{0.5}$ (Niklas and Beck, 1997), a relation $SFE \propto B_t^{\mathrm{x}}$ could well be the cause of the non-linearity. The observational data are still poor (Vallée, 1994). The supernova-driven ISM model including magnetic fields by Gazol-Patiño and Passot (1999) indeed predicts a significant increase of the average star-formation rate with field strength. Furthermore, strong fields may shift the stellar mass spectrum towards the more massive stars (Mestel, 1994).

6. Do Magnetic Fields Dominate in Outer Galaxies ?

Most observable quantities in galactic disks have exponential distributions. For NGC 6946 the scale length of the synchrotron emission is $l_{\mathrm{syn}} = 3.9 \pm 0.1$ kpc (radial range 3–8 kpc). The synchrotron disk is a combination of the distributions of cosmic-ray electrons and total magnetic fields. In case of equipartition between these two components, the scale length of the total magnetic field is $l_{\mathrm{B}} = (3 + \alpha)l_{\mathrm{syn}} \simeq 4l_{\mathrm{syn}} \simeq 16$ kpc, and the scale lengths of the magnetic energy density and of the cosmic rays are $l_{\mathrm{B}^2} = l_{\mathrm{CR}} \simeq$ **8 kpc**. All other energy densities in Table I fall off with the scale length of the neutral gas $l_\rho = 3.2 \pm 0.1$ kpc if T, v_{turb} and v_{rot} are radially constant, or with a smaller scale length if one of these parameters decreases with radius.

The scale length of the distribution of supernova remnants (the sources of cosmic rays) in NGC 6946 is only $\simeq$2 kpc (Sasaki et al., this volume) so that a large diffusion length is needed to reach equipartition also at large radii (see the cosmic-ray propagation model by Breitschwerdt et al., 2002). A large value of l_{CR} is also required to explain the distribution of Galactic γ-rays (Strong et al., 2000).

As a consequence of the different scale lengths $\beta = \epsilon_{\mathrm{turb}}/\epsilon_{\mathrm{B}^2}$ decreases with radius: The dominance of the magnetic field increases towards large radii. It is however hard to understand why the magnetic field energy density is larger that of the turbulent motions. This can be in conflict with turbulent generation of interstellar magnetic fields which predicts energy equipartition between turbulent motions and magnetic fields, and hence $l_{\mathrm{B}^2} = l_\rho$ (if $v_{\mathrm{turb}} \simeq$ const) and $l_{\mathrm{syn}} < l_\rho$. Radial diffusion of the magnetic field (Priklonsky et al., 2000), field connections through the wind-driven halo (Breitschwerdt et al., 2002) or a supra-equipartition turbulent dynamo (Belyanin et al., 1993) are possible explanations.

In the outermost parts of galaxies the magnetic field energy density may even reach the level of global rotational gas motion. If so, the rotation curves at large radii could be affected by magnetic fields, as discussed by Battaner and Florido (2000). Field strengths in the outer parts of galaxies are difficult to measure. Synchrotron emission is weak, but Faraday rotation of polarized background sources can be measured out to large radii. Han et al. (1998) found evidence for regular fields in M 31 at 25 kpc radius of similar strength as in the inner disk. More detailed studies in a number of galaxies are required.

7. Magnetic Fields Act Everywhere

7.1. Spiral Arms

Maps of the total radio emission and ISOCAM maps of the mid-infrared dust emission (Figure 1) reveal a surprisingly close connection (Frick et al., 2001; Walsh et al., 2002), suggesting a coupling of the *total* magnetic field to the warm dust mixed with cool gas ($B_t \propto \rho^{0.5}$, Niklas and Beck, 1997). Strongest total fields generally coincide with highest emission from dust and gas in the spiral arms (Figure 1), whereas the polarized emission is weak in the arms due to field tangling by turbulent motions or by supernova shock fronts. Radio polarization observations show that in most galaxies the *regular* field follows the spiral structure seen in the stars and the gas, though generally offset, sometimes forming *magnetic spiral arms* between the gaseous arms (Figure 2). Here the energy density of the regular field alone ($\simeq 10\,\mu$G) is comparable to that of the neutral interarm gas ($\rho \simeq 5$cm^{-3}).

In galaxies with strong density waves (like M 51) the regular field is strong at the positions of the dust lanes on the inner edge of the spiral arms. However, the arm-interarm contrast of the regular field is low. The strong field may modify the gas flow and the compression ratio.

The observation of large-scale patterns in Faraday rotation measures proves that the regular field in galaxies has a coherent direction and is not generated by compression or stretching of irregular fields in gas flows. The *dynamo* mechanism is able to generate and preserve coherent magnetic fields of spiral shape (Beck et al., 1996; Moss et al., 1999). The observed alignment of magnetic pitch angles with those of the gaseous arms can be achieved by inclusion of the shear flow around spiral arms (von Linden et al., 1998). Future dynamo models should address the interaction between dynamo wave and density wave.

7.2. Bars

Average radio intensity, radio luminosity and star-formation activity in barred galaxies correlate with *relative bar length* (Beck et al., 2002). In the best-studied case, NGC 1097, the general similarity of the B-vectors (Figure 3) and gas streamlines around the bar as obtained in simulations is striking (Beck et al., 1999). The field

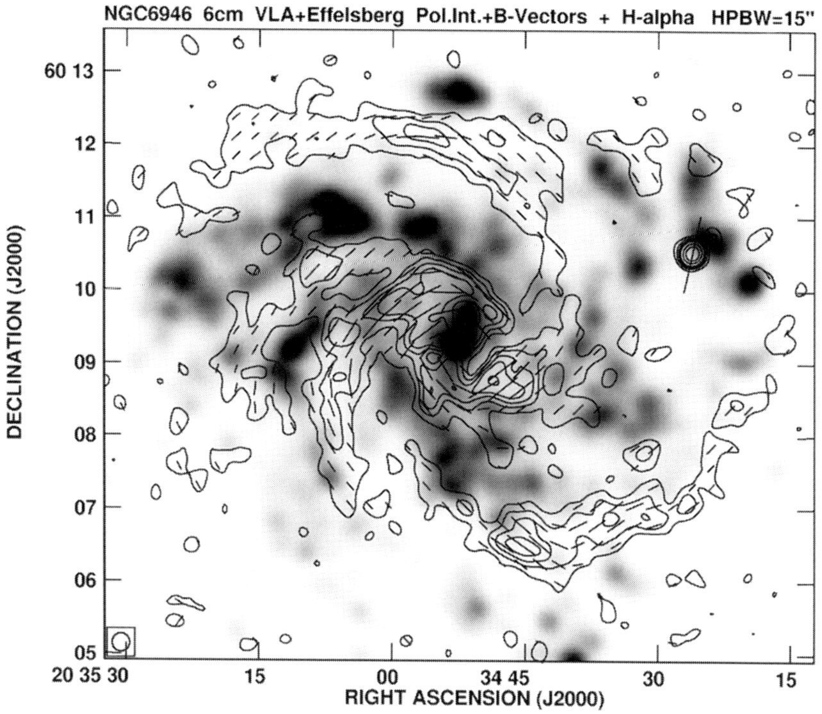

Figure 2. Polarized radio intensity (contours) and **B**-vectors of polarized intensity of NGC 6946 at $15''$ resolution, combined from VLA and Effelsberg observations at $\lambda 6$ cm. The grey-scale image shows the Hα emission (kindly provided by A. Ferguson), smoothed to $15''$ resolution. (From Beck and Hoernes, 1996)

lines are a tracer of the sheared flow pattern in the sky plane which is otherwise unobservable. Remarkably, the optical image of NGC 1097 shows dust filaments outside the bar which are aligned with the field.

The bar's major dust lane is believed to indicate the shock front. However, the region of strongest shear in NGC 1097 (where the field changes its direction abruptly) is located $\simeq$ 1 kpc in front of the dust lane (Figure 3), in conflict with the hydrodynamical models. Furthermore, the regular field is only weakly compressed in the bar. Either the field decouples from the gas (at least from its diffuse component) and thus avoids the gas shock, or there is no shock at all. Detailed observations of the velocity field are required. Moss et al. (2001) used velocity shear and dynamo action to model the field configuration, but a MHD model including the back-reaction of the field onto the gas flow is still needed.

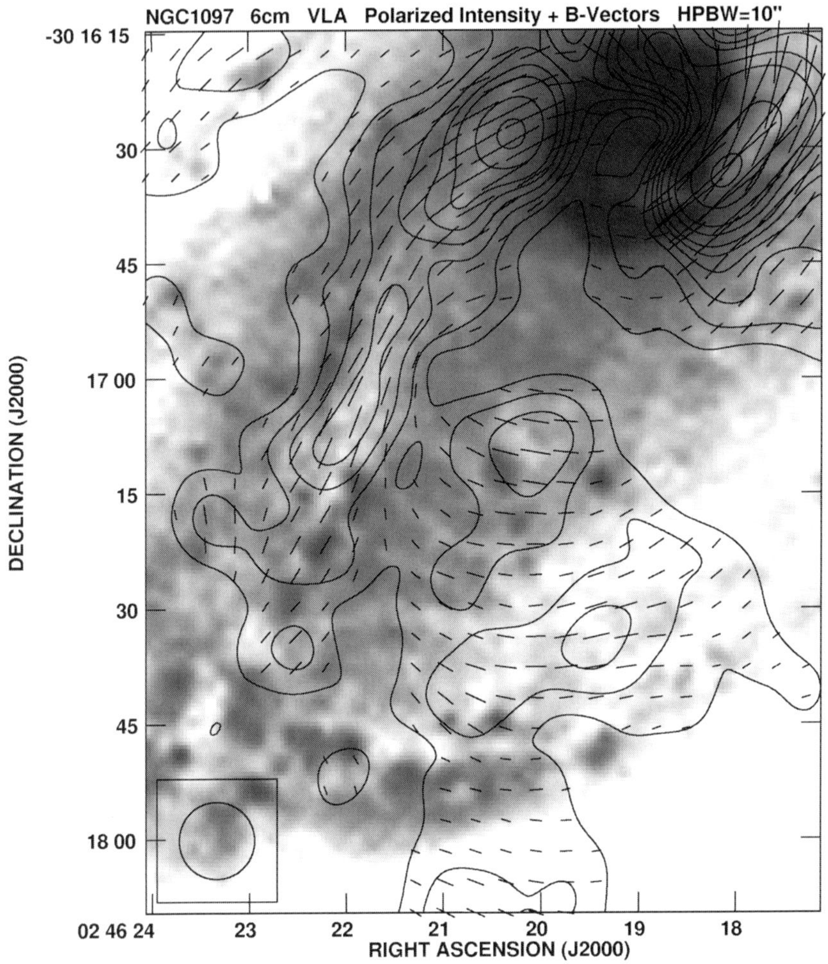

Figure 3. Polarized radio intensity of NGC 1097 (contours) and **B**-vectors of polarized intensity at $10''$ resolution, observed with the VLA at $\lambda 6$ cm. The grey-scale image shows an optical image from the Cerro Tololo Observatory (kindly provided by H. Arp). (From Moss et al., 2001)

7.3. Nuclear regions

Many barred galaxies have circumnuclear rings, sites of ongoing intense star formation, sometimes with an active nucleus in the center. Radio polarization maps of these inner regions (Beck et al., 2002) revealed strong regular fields (up to 40μG) with spiral patterns and large pitch angles, accompanied by spiral dust filaments visible on optical images. Magnetic stress in the circumnuclear ring can drive mass inflow to feed the active nucleus, as suggested for NGC 1097 (Beck et al., 1999).

7.4. Halos

In edge-on galaxies the observed field orientations are mainly parallel to the disk (Dumke et al., 1995). A prominent exception is NGC 4631 with the brightest and largest halo observed so far (Beck, 2000), composed of vertical magnetic spurs connected to star-forming regions in the disk (Golla and Hummel, 1994). The field is probably dragged out by a strong galactic wind. The magnetic energy density in the halo of, e.g. M 83, exceeds that of the hot gas (Ehle et al., 1998). Halo magnetic fields are important for the propagation of cosmic rays, the formation of a galactic wind (Breitschwerdt et al., 2002) and the stability of gas filaments (Tüllmann et al., 2000; Dettmar, this volume).

8. Outlook

Much has been learnt about the global properties of interstellar magnetic fields, but their detailed structure on scales below about 100 pc is still unclear. Polarization observations in our Galaxy trace structures on pc and sub-pc scales (e.g. Uyanıker et al., 1999; Gaensler et al., 2001). Future radio polarization observations with increased sensitivity and resolution (Extended VLA, Square Kilometer Array) will show the full wealth of magnetic structures in galaxies.

References

Battaner, E. and Florido, E.: 2000, *Fund. Cosmic Phys.* **21**, 1–154.

Beck, R.: 1991, *Astron. Astrophys.* **251**, 15–26.

Beck, R.: 2000, *Phil. Trans. R. Soc. Lond. A* **358**, 777–796.

Beck, R.: 2001, *Space Science Reviews* **99**, 243–260.

Beck, R.: 2002, in: E. Athanassoula, A. Bosma and R. Mujica (eds.), *Disks of Galaxies*, ASP Conf. Ser. **275**, pp. 331–342.

Beck, R. and Hoernes, P.: 1996, *Nature* **379**, 47–49.

Beck, R., Brandenburg, A., Moss, D., Shukurov, A. and Sokoloff, D.: 1996, *Ann. Rev. Astron. Astrophys.* **34**, 155–206.

Beck, R., Ehle, M., Shoutenkov, V., Shukurov, A. and Sokoloff, D.: 1999, *Nature* **397**, 324–327.

Beck, R., Shoutenkov, V., Ehle, M. et al.: 2002, *Astron. Astrophys.* **391**, 83–102.

Belyanin, M., Sokoloff, D. and Shukurov, A.: 1993, *Geophys. Astrophys. Fluid Dyn.* **68**, 227–261.

Breitschwerdt, D., Dogiel, V.A. and Völk, H.J.: 2002, *Astron. Astrophys.* **385**, 216–238.

Brogan, C.L., Frail, D.A., Goss, W.M. and Troland, T.H.: 2000, *Astrophys. J.* **537**, 875–890.

Crutcher, R.M.: 1999, *Astrophys. J.* **520**, 706–713.

Dumke, M., Krause, M., Wielebinski, R. and Klein, U.: 1995, *Astron. Astrophys.* **302**, 691–703.

Ehle, M., Pietsch, W., Beck, R. and Klein, U.: 1998, *Astron. Astrophys.* **329**, 39–54.

Elmegreen, B.G.: 1981, *Astrophys. J.* **243**, 512–525.

Frick, P., Beck, R., Berkhuijsen, E.M. and Patrickeyev, I.: 2001, *Mon. Not. R. Astron. Soc.* **327**, 1145–1157.

Gaensler, B.M., Dickey, J.M., McClure-Griffiths, N.M. et al.: 2001, *Astrophys. J.* **549**, 959–978.

Gazol-Patiño, A. and Passot, T.: 1999, *Astrophys. J.* **518**, 748–759.

Golla, G. and Hummel, E.: 1994, *Astron. Astrophys.* **284**, 777–792.
Han, J.L., Beck, R. and Berkhuijsen, E.M.: 1998, *Astron. Astrophys.* **335**, 1117–1123.
Hanasz, M. and Lesch, H.: 1998, *Astron. Astrophys.* **332**, 77–87.
Heiles, C.: 1996, in: W.G. Roberge and D.C.B. Whittet (eds.), *Polarimetry of the Interstellar Medium*, ASP Conf. Ser. **97**, pp. 457–474.
Heitsch, F., Mac Low, M. and Klessen, R.S.: 2001, *Astrophys. J.* **547**, 280–291.
Jenkins, E.B. and Tripp, T.M.: 2001, *Astrophys. J. Suppl.* **137**, 297–340.
Kalberla, P.M.W. and Kerp, J.: 1998, *Astron. Astrophys.* **339**, 745–758.
von Linden, S., Otmianowska-Mazur, K., Lesch, H. and Skupniewicz, G.: 1998, *Astron. Astrophys.* **333**, 79–91.
Mestel, L.: 1994, in: D. Lynden-Bell et al. (eds.), *Cosmical Magnetism*, Kluwer, Dordrecht, pp. 181–211.
Mestel, L. and Paris, R.B.: 1984, *Astron. Astrophys.* **136**, 98–120.
Moss, D., Shukurov, A. and Sokoloff, D.: 1999, *Astron. Astrophys.* **343**, 120–131.
Moss, D., Shukurov, A., Sokoloff, D., Beck, R. and Fletcher, A.: 2001, *Astron. Astrophys.* **380**, 55–71.
Myers, P.C. and Goodman, A.A.: 1988, *Astrophys. J.* **326**, L27–L30.
Niklas, S.: 1995, Ph.D. Thesis, University of Bonn.
Niklas, S. and Beck, R.: 1997, *Astron. Astrophys.* **320**, 54–64.
Priklonsky, V.I., Shukurov, A., Sokoloff, D. and Soward, A.: 2000, *Geophys. Astrophys. Fluid Dyn.* **93**, 97–114.
Reich, W.: 1994, in: R. Genzel and A.I. Harris (eds.), *The Nuclei of Normal Galaxies*, Kluwer, Dordrecht, pp. 55–62.
Sokoloff, D.D., Bykov, A.A., Shukurov, A., Berkhuijsen, E.M., Beck, R. and Poezd, A.D.: 1998, *Mon. Not. R. Astron. Soc.* **299**, 189–206 and *Mon. Not. R. Astron. Soc.* **303**, 207–208 (Erratum).
Strong, A.W., Moskalenko, I.V. and Reimer, O.: 2000, *Astrophys. J.* **537**, 763–784.
Tüllmann, R., Dettmar, R.J., Soida, M., Urbanik, M. and Rossa, J.: 2000, *Astron. Astrophys.* **364**, L36–L41.
Uchida, K.I., Fiebig, D. and Güsten, R.: 2001, *Astron. Astrophys.* **371**, 274–286.
Uyanıker, B., Fürst, E., Reich, W., Reich, P. and Wielebinski, R.: 1999, *Astron. Astrophys. Suppl.* **138**, 31–45.
Vallée, J.P.: 1994, *Astrophys. J.* **433**, 778–779.
Walsh, W., Beck, R., Thuma, G., Weiss, A., Wielebinski, R. and Dumke, M.: 2002, *Astron. Astrophys.* **388**, 7–28.
Woltjer, L.: 1967, in: H. van Woerden (ed.), *Radio Astronomy and the Galactic System*, Reidel, Dordrecht, pp. 479–485.
Wong, T. and Blitz L.: 2002, *Astrophys. J.* **569**, 157–183.
Yusef-Zadeh, F., Roberts, D.A., Goss, W.M., Frail, D.A., and Green, A.J.: 1996, *Astrophys. J.* **466**, L25–L29.

BUILDING THE FAST GALACTIC DYNAMO

M. HANASZ, R. KOSIŃSKI and H. LESCH
Centre for Astronomy, Nicholas Copernicus University, PL-87-148 Piwnice/Toruń, Poland
E-mail: mhanasz@astri.uni.torun.pl

Abstract. In this paper we demonstrate the importance of cosmic rays for the dynamics of the interstellar medium. We present the first 3D-MHD numerical simulations of the Parker instability triggered by cosmic rays accelerated in randomly distributed supernova remnants. We show that in the presence of galactic rotation a net radial magnetic field is produced as a result of the cosmic ray injection and Coriolis force. This process provides a possibility of very efficient magnetic field amplification within the general frame of so called *fast galactic dynamo* proposed by Parker (1992).

Keywords: Interstellar medium, magnetic field, cosmic rays

1. Amplification of Galactic Magnetic Fields

The dynamo theory applied to galactic magnetic fields successfully reproduces majority of observational features of large scale galactic magnetic fields (Ruzmaikin et al., 1988; Beck et al., 1996). The dynamo process is a combination of differential rotation, helical turbulence and a dissipative process. The exponential growth time of galactic magnetic fields, characteristic for classical mean field dynamos relying on turbulent motions powered by supernova explosions, is typically estimated to be of the order of $t_{\rm dynamo} \sim (5 - 10) \times 10^8$ yr.

Despite successes of the dynamo theory, serious criticism against it has been raised by different authors (see the review by Widrow (2002) and references therein). On the other hand galactic magnetic fields are subject to the action the Parker (magnetic buoyancy) instability, which evacuates magnetic flux from galactic disks in timescales of $t_{\rm PI} \sim 5 \times 10^7$, about 10 times faster than classical dynamos can regenerate them (Hanasz and Lesch, 2001).

An attractive idea of *fast galactic dynamo* offering a solution the classical dynamo problems has been proposed by Parker (1992). The idea relies on two new fundamental ingredients: (1) cosmic rays (CR) continuously supplied to the disk by SN remnants and (2) fast magnetic reconnection which operates in current sheets and makes it possible to dissipate and relax the random magnetic field component even in the limit of vanishing resistivity.

Astrophysics and Space Science **289:** 303–306, 2004.

2. The Cosmic Ray Component of the ISM

The diffusive cosmic ray transport in galactic disks can be described by the diffusion-advection equation. Following Schlickeiser and Lerche (1985) we apply the following form of the transport equation

$$\frac{\partial e_{\rm cr}}{\partial t} + \nabla \left(e_{\rm cr}\boldsymbol{v}\right) = \nabla \left(\hat{K}\nabla e_{cr}\right) - p_{\rm cr}\left(\nabla \cdot \boldsymbol{v}\right) + Q, \tag{1}$$

where $e_{\rm cr}$ and $p_{\rm cr} = (\gamma_{\rm cr} - 1)e_{\rm cr}$ are the cosmic ray energy density and cosmic ray pressure, $\gamma_{\rm cr} = 4/3$ is the ratio of the specific heats of the relativistic cosmic ray gas, $\boldsymbol{v}$ is velocity of the thermal gas and Q is the source term for the CR energy density resulting from the cosmic ray injection in SNR. In the present approach we assume that 10 % of the 10^{51} erg of kinetic energy output from a supernova is converted into the cosmic ray energy.

The cosmic ray component described by the diffusion equation is supplemented in the Zeus-MP MHD code in addition to the standard set of ideal MHD equations. We add also an extra term related to the gradient of cosmic ray pressure $-\nabla p_{cr}/\rho$ in the equation of motion for the gaseous component (see Berezinski et al., 1990). The CR diffusion along the magnetic field lines is characterized by the diffusion coefficients $K_{\parallel} = 3 \times 10^{26}{\rm cm}^2{\rm s}^{-1}$ and $K_{\perp} = 0$. The applied magnitude of the parallel diffusion coefficient is less then the estimated realistic value for the ISM, a few $10^{28}{\rm cm}^2{\rm s}^{-1}$ (see eg. Jokipii, 1999) due to limitations of a simple explicit numerical algorithm applied for the cosmic ray transport.

We consider a 3D region of size 600pc $\times$ 1200pc $\times$ 1200pc in corotating Cartesian coordinates (x, y, z) corresponding to radial, azimuthal and vertical directions, respectively. We first construct a magnetohydrostatic equilibrium permeated by purely azimuthal magnetic field $B_y(z)$ in a realistic vertical gravity including a background cosmic ray pressure equal to the magnetic pressure. We assume that the disk equilibrium is of the Parker type, ie. $p_{\rm mag} = \alpha p_{\rm gas}$ and $p_{\rm cr} = \beta p_{\rm gas}$ with $\alpha = \beta = 1$, where $p_{\rm gas}$ corresponds to 1 H atom per cm^3 and the sound speed of 7km s^{-1}. We take into account Coriolis force assuming the angular velocity of the disk matter $\Omega = 2 \times 10^{-15}{\rm s}^{-1}$. Such an equilibrium is perturbed by cosmic rays injected in SNRs (of radius 50 pc) appearing in random locations. The SNR distribution is uniform in x and y-coordinates and is represented by a Gaussian distribution in the z-coordinate. The half width of the Gaussian profile is 100 pc and frequency of explosions is 0.72 SN per squared kpc of the disk per Myr (1 SN explosion per Myr in the full computational domain).

3. Results

Cosmic rays injected in random SNRs initially inflate bubbles in magnetic field and subsequently diffuse along magnetic field lines, in a way described by Hanasz

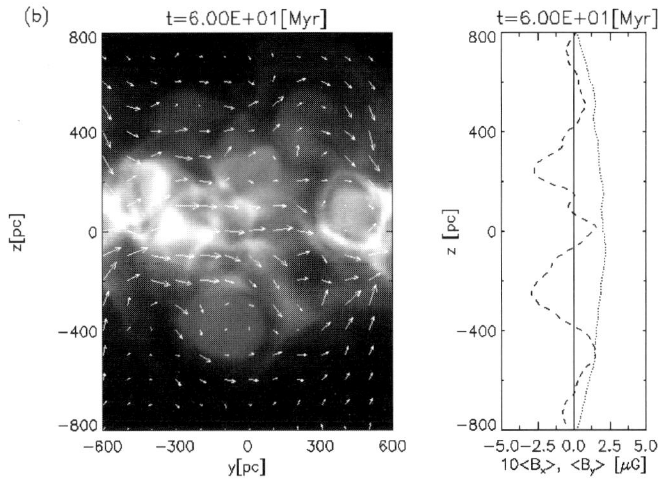

Figure 1. Evolution of cosmic rays injected to ISM by randomly distributed supernova remnants. Left panel displays cosmic ray energy density after 60 Myr of evolution of the system. Right panel displays the mean azimuthal magnetic field B_y (dotted line) and the mean radial magnetic field B_x ($\times 10$, dashed line) produced by Coriolis force as a result of cosmic ray injection.

and Lesch (2000, 2002). Contrary to our former papers, in the present paper we impose multiple and random SN explosions.

A snapshot from simulations at $t = 60$Myr is shown in Figure 1 (left panel), where magnetic vectors are displayed on the background of cosmic ray energy density. The initial equilibrium disk is inflated by the supply of cosmic rays. The azimuthal magnetic field is transported out of the disk in vertical direction by buoyant motions. On the other hand horizontal motions of gas, induced by the cosmic ray pressure gradient and by a slope of magnetic lines in gravitational field, are influenced by the Coriolis force. We compute the mean magnetic field through averaging over planes $z =$ const . The averaged horizontal components of magnetic field are displayed in Figure 1 (right panel), where dotted line represents mean azimuthal field B_y, and dashed line represents mean radial field B_x.

We note that the mean radial magnetic field raises from 0 to about 10 % of the azimuthal magnetic field within 60 Myr of evolution of our system. The profile of mean redial magnetic field B_x represents a wavelike pattern which slowly propagates in vertical direction outward of the disk plane. This behaviour demonstrates very well the separation of positive and negative radial magnetic fluxes desired by the dynamo process.

4. Conclusions

The main conclusion from the MHD simulations with an active cosmic ray component is that cosmic rays are very important for the dynamics of interstellar medium. Second, the powerful contribution of cosmic rays provides a very efficient way of generation of the radial magnetic field component in galactic disks in timescales shorter than 10^8yr . Such an efficient generation of radial magnetic field together with the effect of differential rotation and magnetic reconnection is expected to provide amplification of large scale galactic magnetic fields on time scales comparable or even shorter than 10^8 yr.

Acknowledgements

The presented work was supported by Polish Committee for Scientific Research (KBN) through the grant PB 404/P03/2001/20 and is a continuation of research work supported by Alexander von Humboldt Foundation. We thank organizers of WS-ISM workshop and EAS for financial support to our participation.

References

Beck, R., Brandenburg,A., Moss, D., Shukurov, A. and Sokoloff, D.: 1996, *ARAA***34**, 155.
Berezinskii, V.S., Bulanov, S.V., Dogiel, V.A., Ginzburg, V.L. and Ptuskin, V.S.: 1990, *Astrophysics of cosmic rays*, Amsterdam, North-Holland.
Giacalone, J. and Jokipii, R.J.: 1999, *ApJ***520**, 204–214.
Hanasz, M. and Lesch, H.: 2000, *ApJ***543**, 235–244.
Hanasz, M. and Lesch, H.: 2001, *Space Sci. Rev.***99**, 231–242.
Hanasz, M. and Lesch, H.: 2002, Ap&SS **281**, 289.
Jokipii, J.R.: 1999, in: J. Franco and A. Carraminana (eds.), *Interstellar Turbulence*, Cambridge University Press, pp. 70–78.
Parker, E.N.: 1992, *ApJ***401**, 137–145.
Ruzmaikin, A.A., Shukurov, A.M. and Sokoloff, D.D.: 1988, *Magnetic Fields in Galaxies.*, Kluwer, Dordrecht.
Schlickeiser, R. and Lerche, I.: 1985, *A&A***151**, 151–156.
Widrow, L.M.: 2002, *Rev. Mod. Phys.* **74**, 775.

MAGNETIC RECONNECTION AND TURBULENT MIXING: FROM ISM TO CLUSTERS OF GALAXIES

A. LAZARIAN
Dept. of Astronomy, UW-Madison; E-mail: lazarian@astro.wisc.edu

J. CHO
Dept. of Astronomy, UW-Madison; E-mail: cho@astro.wisc.edu

Abstract. Magnetic reconnection, or the ability of the magnetic field lines that are frozen in plasma to change their topology, is a fundamental problem of magnetohydrodynamics (MHD). We briefly examine the problem starting with the well-known Sweet-Parker scheme, discuss effects of tearing modes, anomalous resistivity and the concept of hyperresistivity. We show that the field stochasticity by itself provides a way to enable fast reconnection even if, at the scale of individual turbulent wiggles, the reconnection happens at the slow Sweet-Parker rate. We show that fast reconnection allows efficient mixing of magnetic field in the direction perpendicular to the local direction of magnetic field. While the idea of stochastic reconnection still requires numerical confirmation, our numerical simulations testify that mixing motions perpendicular to the local magnetic field are up to high degree hydrodynamical. This suggests that the turbulent heat transport should be similar to that in non-magnetized turbulent fluid, namely, should have a diffusion coefficient $\sim V_L L$, where V_L is the amplitude of the turbulent velocity and L is the scale of the turbulent motions. We present numerical simulations which support this conclusion. The application of this idea to thermal conductivity in clusters of galaxies shows that this mechanism may dominate the diffusion of heat and may be efficient enough to prevent cooling flow formation.

Keywords: interstellar medium, intergalactic medium, magnetic fields

1. Introduction

Interstellar plasma has high conductivity and therefore magnetic field and fluid move together. As the result magnetic field influences fundamental properties of interstellar plasma and must be accounted in star formation processes, support and evolution of molecular clouds, transfer of mass and energy in the interstellar space etc. Magnetic fields control cosmic rays propagation, induce Parker (see Parker, 1979) and magneto-rotational (see Balbus and Hawley, 1998) instabilities. Transport processes in the interstellar medium and beyond it, e.g. clusters of galaxies, depend on the ability of magnetic field lines to change their topology or to reconnect. It is difficult to overestimate the role of magnetic reconnection as it is also a necessary component of magnetic dynamo (see Parker, 1979, and also Vishniac, Lazarian and Cho, 2003 for a recent review), and it is likely to affect magnetic flux removal from hotbeds of star formation etc. More generally, it is impossible to

Astrophysics and Space Science **289:** 307–317, 2004.

claim that we understand MHD unless we can predict whether crossing magnetic flux tubes will reconnect or bounce from one another.

Mixing motions perpendicular to magnetic field lines are an essential part of the Goldreich-Sridhar model (1995, henceforth GS95) of MHD turbulence. Should those motions proceed in the hydrodynamic-type fashion or magnetic field would accumulate unresolvable knots that would affect the motion of the fluid? The answer to this question has far reaching consequences for heat transport in the interstellar medium and beyond. Indeed, magnetic fields are known to suppress heat conductivity (see Chandran and Cowley, 1998) in the direction perpendicular to **B**. Transport of heat along wondering magnetic field lines (Narayan and Medvedev, 2001; Zakamska and Narayan, 2002) partially alleviates the problem. If reconnection proceeds efficiently, the mixing motions should allow an efficient transport of heat in magnetized turbulent fluids. This would have big implications for problems from cooling flows in the clusters of galaxies (see Fabian, 1994) to mixing layers in our and other galaxies (see Slavin, Shull and Begelman, 1993).

In §2 and §3, we discuss reconnection and, in §4 and §5, we consider turbulent mixing of magnetized plasma.

2. Sweet-Parker, Petschek Models and their Modifications

The literature on magnetic reconnection is rich and vast (see, for example, Priest and Forbes (2000) and references therein). We start by discussing a robust scheme proposed by Sweet and Parker (Parker, 1957; Sweet, 1958). In this scheme oppositely directed magnetic fields are brought into contact over a region of length L_x (see Figure 1). In general there will be a shared component, of the same order as the reversed component. However, this has only a minor effect on our discussion. The gradient in the magnetic field is confined to the current sheet, a region of vertical size Δ, within which the magnetic field evolves resistively. The velocity of reconnection, V_r, is the speed with which magnetic field lines enter the current sheet, and is roughly $\eta \approx V_r \Delta$. Arbitrarily high values of V_r can be achieved (transiently) by decreasing Δ. However, for sustained reconnection there is an additional constraint imposed by mass conservation. The plasma initially entrained on the magnetic field lines must escape from the reconnection zone. In the Sweet-Parker scheme this means a bulk outflow, parallel to the field lines, within the current sheet. Since the mass enters along a zone of width L_x, and is ejected within a zone of width Δ, this implies

$$\rho V_{rec} L_x = \rho' V_A \Delta \quad , \tag{1}$$

where we have assumed that the outflow occurs at the Alfvén velocity V_A. This is actually an upper limit set by energy conservation. If we ignore the effects of compressibility $\rho = \rho'$ and the resulting reconnection velocity allowed by Ohmic diffusivity and the mass constraint is

$$V_{rec,sweet-parker} \approx V_A \mathcal{R}_L^{-1/2}, \tag{2}$$

where $\mathcal{R}_L$ is the Lundquist number using the current sheet *length*. Depending on the specific astrophysical context, this gives a reconnection speed which lies somewhere between 10^{-3} (stars) and 10^{-10} (the galaxy) times V_A.

Attempts to accelerate Sweet-Parker reconnection are numerous. We start by considering schemes to broaden the current sheet. Anomalous resistivity is known to broaden current sheets in laboratory plasmas. But this is microphysical broadening which is not important when the thickness of the current sheet that in the Sweet-Parker scheme scales as $L_x^{1/2}$ is much larger than the Larmor radius of the thermally moving proton. The latter scale is of the order of 200 km in the interstellar medium and it would be naive to expect to be able to squeeze the interstellar gas out through such a narrow slot (see eq (1).

Tearing modes are a robust instability connected to the appearance of narrow current sheets (Furth, Killeen and Rosenbluth, 1963). Their importance for reconnection by many authors (see Strauss, 1988; Zweibel, 1989). The tearing modes broaden the reconnection layer and enhance the reconnection speed. A treatment of the problem in Lazarian and Vishniac (1999; hereafter LV99) of 3D reconnection involving tearing modes provided an estimate

$$V_{rec,tearing} = V_A \left(\frac{\eta}{V_A L_x} \right)^{3/10}, \tag{3}$$

which is substantially faster than the Sweet-Parker rate, but still very slow in any astrophysical context. Note that unlike anomalous effects, tearing modes do not require any special conditions and therefore should constitute a generic scheme of reconnection.

Hyperresistivity was introduced in the literature (see Hameiri and Bhattacharjee, 1987; Strauss, 1988) to describe the ability of current sheet instabilities to drive plasma turbulence that would affect the reconnection rates. Tearing instability is one of the major current sheet instabilities and therefore the concept of hyperresistivity is related to the previously discuss issue. If the reconnecting magnetic fields are nearly anti-parallel, the turbulent mixing of magnetic field lines becomes possible and the reconnection can go much faster. However, this assumption of very nearly parallel field lines substantially limits the application of the concept.

The failure to find fast reconnection speeds due to broadening the current sheet has stimulated interest to fast reconnection through radically different global geometries. Petschek (1964) conjectured that reconnecting magnetic fields would tend to form structures whose typical size in all directions is determined by the resistivity ('X-point' reconnection). This results in a reconnection speed of order $V_A / \ln \mathcal{R}_L$. However, attempts to produce such structures in numerical simulations of reconnection have been disappointing. Typically the X-point region collapses toward the Sweet-Parker geometry as the Lundquist number becomes large (Biskamp, 1984, 1986, 1996; Wang, Ma and Bhattacharjee, 1996; Ma and Bhattacharjee, 1996).[1]

One may invoke collisionless plasma effects to stabilize the X-point reconnection (for collisionless plasma). For instance, a number of authors (Shay et al., 1998; Shay and Drake, 1998; Shay et al., 1999) have reported that in a two fluid treatment of magnetic reconnection, a standing whistler mode can stabilize an X-point with a scale comparable to the ion plasma skin depth, $c/\omega_{pi} \sim (V_A/c_s)r_L$. The resulting reconnection speed is a large fraction of V_A, and apparently independent of L_x, which would suggest that something like Petschek reconnection emerges in the collisionless regime. However, these studies have not yet demonstrated the possibility of fast reconnection for generic field geometries, since they assume that there are no bulk forces acting to produce a large scale current sheet. Similarly, those studies do not account for fluid turbulence. Magnetic fields embedded in a turbulent fluid will give fluctuating boundary conditions for the current sheets. On the other hand, boundary conditions need to be fine tuned for a Petschek reconnection scheme (Priest and Forbes, 2000).

3. Stochastic Reconnection

The scheme of the stochastic reconnection is presented in Figure 1. We consider the case in which there exists a large scale, well-ordered magnetic field, of the kind that is normally used as a starting point for discussions of reconnection. This field may, or may not, be ordered on the largest conceivable scales. However, we will consider scales smaller than the typical radius of curvature of the magnetic field lines, or alternatively, scales below the peak in the power spectrum of the magnetic field, so that the direction of the unperturbed magnetic field is a reasonably well defined concept. In addition, we expect that the field has some small scale 'wandering' of the field lines. On any given scale the typical angle by which field lines differ from their neighbors is $\phi \ll 1$, and this angle persists for a distance along the field lines $\lambda_{\|}$ with a correlation distance $\lambda_{\perp}$ across field lines (see Figure 1). Unlike the traditional Sweet-Parker scheme it allows many magnetic field lines to enter simultaneously into the reconnection region. The reconnection therefore happens simultaneously at many surfaces and over substantially smaller scales. Therefore the reconnection velocity increases substantially. The limitation for the scheme is the outflow of the matter, which in this case is controlled by magnetic field line diffusivity. More discussion of the scheme is provided below.

[1] Recent plasma reconnection experiments (Yamada et al., 2000) do not support Petschek scheme either.

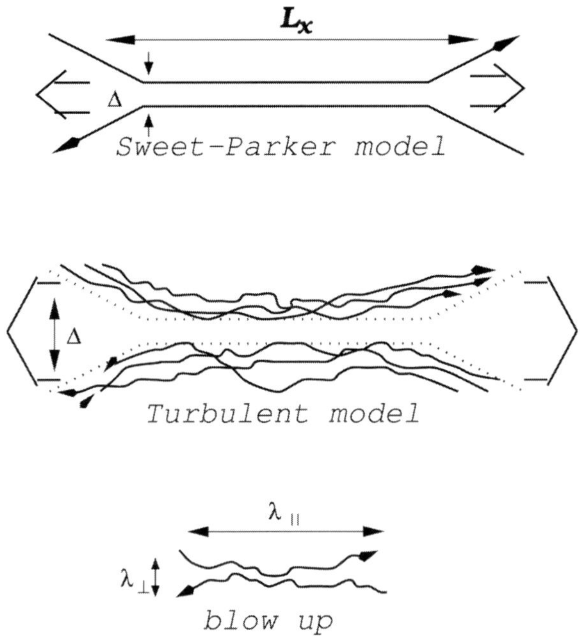

Figure 1. Upper plot: Sweet-Parker scheme of reconnection. Middle plot: scheme of stochastic reconnection that accounts for field line stochasticity. Lower plot: a blow up of the contact region. Thick arrows depict outflows of plasma. From Lazarian and Vishniac (2000)

Note that the model of stochastic reconnection is 3 dimensional[2]. Our result cannot be obtained by considering two dimensional turbulent reconnection (cf. Matthaeus and Lamkin, 1985). In fact, it does not arise from the turbulent transport of magnetic flux, as it incorrectly understood in Kim and Diamond (2001), but is a geometric effect arising from the appearance of stochastic field line wandering in three dimensions.

The modification of the mass conservation constraint in the presence of a stochastic magnetic field component is self-evident. Instead of being squeezed from a layer whose width is determined by Ohmic diffusion, the plasma may diffuse through a much broader layer, $L_y \sim \langle y^2 \rangle^{1/2}$ (see Figure 1), determined by the diffusion of magnetic field lines. This suggests an upper limit on the reconnection speed of

$$V_{rec,\ mass\ constr} \sim V_A(\langle y^2 \rangle^{1/2}/L_x) \tag{4}$$

[2] The Sweet-Parker scheme can easily be extended into three dimensions, in the sense that one can take a cross-section of the reconnection region such that the shared component of the two magnetic fields is perpendicular to the cross-section. In terms of the mathematics nothing changes, but the outflow velocity becomes a fraction of the total V_A and the shared component of the magnetic field will have to be ejected together with the plasma. This result has motivated researchers to do most of their calculations in 2D, which has obvious advantages for both analytical and numerical investigations.

To make further progress one should adopt a model of magnetic turbulence. LV99 considered the whole range of possible models. Using GS95 scalings and assuming that the localized reconnection goes at the Sweet-Parker rate it is possible to get

$$V_{rec,\ SP-Alfven,\ local} \sim V_A(\eta/V_A L)^{1/4} \tag{5}$$

where the turbulent velocity at the scale L is assumed to be Alfvenic. The isotropy of the fast modes has been proved numerically (Cho and Lazarian, 2002,; also review by Cho and Lazarian, 2003). Our work suggests that both in magnetic pressure dominated and gas pressure dominated environments fast modes scaling is similar to that of acoustic turbulence. Assuming that the spectrum of such waves is truncated because of resistivity, LV99 obtained

$$V_{rec,\ SP-fast,\ local} \sim V_A(\eta/V_A L)^{1/6}, \tag{6}$$

which is faster than the Alfvenic turbulence can provide.

If we assume that resistivity presents the bottleneck for the reconnection rate, the global reconnection rate can be estimated as $V_{rec\ local} L_x/\lambda_{\|}$. If the plasma is fully ionized λ gets very small and the limit for the global reconnection rate provided by the resistivity scales as $V_A Rm^{1/4} \gg V_A$ for Alfven modes and $V_A Rm^{1/2} \gg V_A$ for fast modes. Similarly, considering other possible bottlenecks, LV99 concluded that the mass conservation (eq. (4)) presents the most stringent constraint on the reconnection rate for the reconnection of the fully ionized gas. For Alfvenic turbulence this gives (see LV99)

$$V_{rec,global} \approx V_A min\left[(L_x/l)^{1/2}, (l/L_x)^{1/2}\right](V_L/V_A)^2 \tag{7}$$

The case of the partially ionized gas is considered in Lazarian, Vishniac and Cho (2003). There it is found that in the assumption of the local Sweet-Parker rates, the reconnection is slower because the turbulent cascade gets modified by ambipolar drag. However, the resulting reconnection rates are sufficiently high to be important for removing flux from star-forming clouds.

4. Mixing and Turbulent Diffusion in Magnetized Media

Let us start with a note of warning. The scheme of stochastic reconnection that we have discussed is very different from the so-called 'turbulent diffusivity of magnetic field'. The latter is a wrong concept that was used at earlier stages of the mean dynamo theory (see recent review by Vishniac, Lazarian and Cho, 2003) to justify some of its assumptions. According to 'turbulent diffusivity' idea, the magnetic fields of opposite polarity can be mixed up by turbulence and then dissipate at small scales. This picture is wrong as magnetic fields are dynamically important and they get to dominate the small scale motions of the magnetized fluid (see Parker, 1992).

In the stochastic reconnection scheme magnetic field wondering happens within each layer of the opposite polarity, but no small scale mixing of the magnetic field of opposite polarities is assumed.

Mixing of magnetic field lines happens in the direction perpendicular to local direction of the magnetic field. GS95 picture of turbulence can be understood in terms of the eddies in the planes perpendicular to magnetic field lines (see discussion in Cho, Lazarian and Vishniac, 2002). LV99 showed that the eddies will not be forming magnetic knots if the reconnection is as fast as the stochastic reconnection scheme suggests. This means that the motions of the magnetized fluid will be very similar to the hydrodynamic motions in the planes perpendicular to the local direction of magnetic field.

Cho, Lazarian and Vishniac (2002) scrutinized the statistics of the motions of the magnetized turbulent fluid in the direction perpendicular to the local direction of magnetic field. Using numerical simulations they found that for sufficiently strong magnetic field the turbulent motions perpendicular to magnetic field are identical to the hydrodynamic motions. *Although this fact cannot be used as a proof of the stochastic reconnection but it can be used as the evidence that, in the absence of any anomalous effects, the change of magnetic topology does not constrain fluid motions.*

While the 'turbulent diffusivity of magnetic field' is a faulty concept, turbulence may provide efficient diffusivity for heat and matter. Indeed, the turbulent motions perpendicular to magnetic field should transport heat and other passive scalar field in the same way as the hydrodynamic turbulence does. Below we will be mostly talking about heat transfer, as this is the problem of largest astrophysical significance. The corresponding diffusion coefficient is expected to be of the order of LV_L.

Do we expect the turbulent heat transport to be strongly anisotropic? As the large scale field wanders, the direction of the local magnetic field that determines the mixing motions of hydrodynamic-like eddies changes as well. Therefore no strong anisotropy in heat conduction is expected.

To test our theoretical considerations related to the diffusion of heat we performed numerical simulations using our hybrid ENO MHD code described, for instance, in Cho and Lazarian (2002). We use a passive scalar $\psi(\mathbf{x})$ to trace thermal particles. We inject a passive scalar with a Gaussian profile:

$$\psi(\mathbf{x}, t = t_0) \propto \exp^{-(\mathbf{x}-\mathbf{x}_0)^2/\sigma_0^2}, \tag{8}$$

where σ_0= $1/16$ of a side of the numerical box and $\mathbf{x}_0$ lies at the center of the computational box. The value of σ_0 ensures that the scalar is injected in the inertial range of turbulence. The scalar field follows the continuity equation: $\partial\psi/\partial t + \nabla \cdot (\psi\mathbf{v}) = 0$.

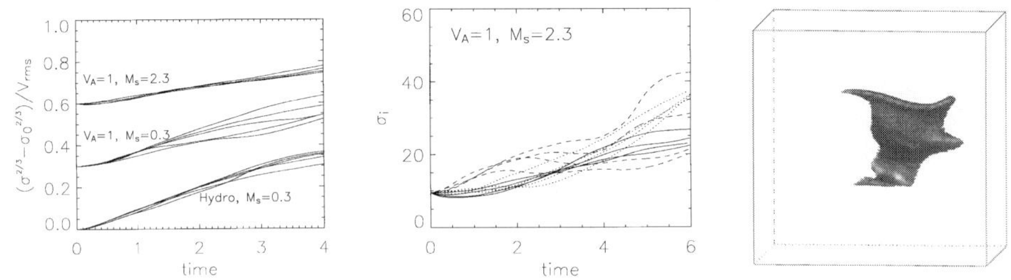

Figure 2. Left: $(\sigma^{2/3} - \sigma_0^{2/3})/V_L$ vs. time. σ is the standard deviation of the passive scalar field. Y-axis is in the unit of box-size and Y-values are shifted by 0.3 units for convenience. Note that the slope does not strongly depend on the mean field B_0 or sonic Mach number M_s. *Middle:* σ_i (i=x, y, and z) vs. time. Solid lines=parallel to $\mathbf{B}_0$; dashed and dotted lines=perpendicular to $\mathbf{B}_0$. *Right:* Snapshot of the passive scalar field for $V_A(\equiv B_0/\sqrt{4\pi\rho}) \sim V_L$ and $M_s = 0.3$. $\mathbf{B}_0$ is parallel to the dashed line (i.e. vertical). Turbulent motions provide efficient mixing of the passive scalar. *Left* and *Middle* panels suggest that magnetic field does not suppress this turbulent mixing process. Obviously, when V_A is lower than V_L, the efficiency of turbulent mixing is no less than that in the $V_A \sim V_L$ case. Results are obtained from using 192^3 and 216^3 grid points. From Cho et al. (2003, ApJ, 589, L77).

In Figure 2, we compare time evolution of σ in hydrodynamic case and in MHD cases. The results are very close and this enables us to suggest that the coefficient of thermal diffusion is

$$\kappa_{dynamic} = C_{dyn} L V_L, \tag{9}$$

where C_{dyn} is a constant of order unity, V_L is the amplitude of the r.m.s. turbulent velocity, and L is the scale of the turbulent motions. This is the effective diffusion by turbulent motions suitable for scales larger than injection scale L. The value of C_{dyn} remains almost constant for V_A's of up to the equipartition value $V_A \sim V_L$. The exact value of C_{dyn} is uncertain. In hydrodynamic cases, C_{dyn} is of order of ~ 0.3 (see Lesieur, 1990 chapter VIII and references therein).

5. Astrophysical Implications

In the sections above we discussed two issues: magnetic stochastic reconnection and turbulent transport of heat through mixing motions. We believe that the two processes are intrinsically connected and fast reconnection implies efficient turbulent transport. Nevertheless, while testing of stochastic reconnection is still a challenging problem, efficient mixing and heat transport has been already tested numerically.

The astrophysical implications of stochastic reconnection were discussed in LV99 where it was shown that it naturally accounts for Solar flares, for operation of galactic dynamo, etc. In Lazarian et al. (2003), its implications for gamma ray busts were analyzed, however. This does not limit the list of the possible applications. For instance, it is easy to see that stochastic reconnection can provide efficient removal of magnetic flux from starforming clouds. A proper discussion of these implications requires a more extended review. Here we briefly consider implications of fast turbulent transport.

Clusters of Galaxies

It is widely accepted that ubiquitous X-ray emission due to hot gas in clusters of galaxies should cool significant amounts of intracluster medium and this must result in cooling flows (Fabian, 1994). However, observations do not support the evidence for the cool gas (see Fabian et al., 2001) which is suggestive of the existence of heating that replenishes the energy lost via X-ray emission. Heat transfer from the outer hot regions can do the job, provided that the heat transfer is sufficiently efficient.

Gas in the clusters of galaxies is magnetized and the conventional wisdom maintains that the magnetic fields strongly suppress thermal conduction perpendicular to their direction. Realistic magnetic fields are turbulent and the issue of the thermal conduction in such a situation has been long debated. A recent paper by Narayan and Medvedev (2001) obtained the estimates for the thermal conductivity of the turbulent magnetic fields.

However, Narayan and Medvedev (2001) treated the turbulent magnetic fields as static. In hydrodynamical turbulence it is possible to neglect plasma turbulent motions only when the diffusion of electrons which is the product of the electron thermal velocity v_{elect} and the electron mean free path in plasma l_{mfp}, i.e. $v_{elect}l_{mfp}$, is greater than the turbulent velocity V_L times the turbulent injection scale L, i.e. LV_L. If such scaling estimates are applicable to heat transport in magnetized plasma, the turbulent heat transport should be accounted for heat transfer within clusters of galaxies. Indeed, data for l_{mfp} $v_{elect}l_{mfp}$ given in Zakamska and Narayan (Narayan and Medvedev, 2001) provide the diffusion coefficient $\kappa_{Sp} \equiv v_{elect}l_{mfp} \sim 6.22 \times 10^{30}$ cm^2 sec^{-1} (for Hydra A). If turbulence in the cluster of galaxies is of the order of the velocity dispersion of galaxies, while the injection scale L is of the order of 20 kpc, the turbulent diffusion coefficient is $\kappa_{dynamic} \sim LV_L \sim 6 \times 10^{30}$ cm^2 sec^{-1}, where we take $V_L \sim 1000$ km/sec.

Although the estimates for the electron thermal diffusivity and our estimate of the turbulent diffusivity are of the same order the applicability of Narayan and Medvedev's model is a bit restricted – their model requires strong (i.e. $B_0 \sim \delta B$) mean magnetic field. While there are strong mean magnetic fields in the Galaxy, this is unlikely for the intracluster medium. When the mean field is weak, turbulence at the scales smaller than the characteristic magnetic field scale ($\equiv l_B$) may follow the GS95 model. However, this requires further studies. Our turbulent mixing model gives the same $\kappa_{dynamic}$ regardless of magnetic field geometry.

Local Bubble and SNRs

Now let us compare the results for thermal diffusivity for other important astrophysical situations. The Local Bubble (LB) is a hot ($T \sim 10^6 K; kT \sim 100$ eV), tenuous ($n \sim 0.008/cm^3$) cavity immersed in the interstellar medium (Berghofer et al., 1998; Smith and Cox, 2001). Turbulence parameters are uncertain. We take typical interstellar medium values: $L \sim 10$ pc and $V_L \sim 5$ km/sec. For these parameters, the ratios of $\kappa_{dynamic}$ to κ_{Sp} are

$$f_{in} = \kappa_{dynamic}/\kappa_{Sp} \sim 0.05, \quad \text{(inside LB)} \tag{10}$$

$$f_{mix} = \kappa_{dynamic}/\kappa_{Sp} \sim 100, \quad \text{(in mixing layer)}, \tag{11}$$

where we take $\bar{T} \sim \sqrt{T_c T_h} \sim 10^5 K$, $\bar{n} \sim \sqrt{n_c n_h} \sim 0.1/cm^3$ (Begelman and Fabian 1990), $T_c \sim 10^4 K$, $n_c \sim 1/cm^3$, $T_h \sim 10^6 K$, and $n_h \sim 0.008/cm^3$. Subscripts 'h' and 'c' stand for 'hot' and 'cold', respectively. This implies that thermal conduction along static magnetic field lines is suppressed when electrons gyrate through the 'cold' and 'dense' mixing layer. Note that turbulent diffusion is still effective in the mixing layer. We expect similar results for supernova remnants since parameters are similar.

6. Summary

In the article above we briefly reviewed a number of existing ideas about reconnection. Analysis of the solar and other reconnection-related data provides a number of requirements to the successful candidate. It should be robust to act in the situation when the boundary conditions constantly change due to turbulence. It should typically provide reconnection rates of the order of one tenth V_A, but sometimes it should be much slower to allow the accumulation of the magnetic flux. Our analysis shows that the stochastic reconnection satisfies naturally to all these criteria.

Stochastic reconnection is different from the 'turbulent magnetic diffusivity' idea. It does not require fine mixing of the magnetic fields of opposite polarity. However, we show that the turbulent transport of heat happens in the presence of magnetic field very similarly to how it happens in pure hydrodynamics. In particular, we show that turbulent transport of heat in the intracluster matter is at least as fast as the earlier estimates based on the electron thermal conductivity suggested.

Acknowledgements

A.L. acknowledges the support of NSF Grant AST-0125544. The study of turbulent diffusion was done in collaboration with P. Moin and A. Honein at the Center for Turbulence Research in Stanford University as a part of Visiting Scholar Program in Summer 2002.

References

Balbus, S.A. and Hawley, J.F.: 1998, *Rev. Mod. Phys.* **70**, 1.
Begelman, M. and Fabian, A.: 1990, *MNRAS* **244**, 26.
Berghöfer, T.W., Bowyer, S., Lieu, R. and Knude, J.: 1998, *Astrophys. J.* **500**, 838.
Biskamp, D.: 1984, *Phys. Lett. A* **105**, 124.
Biskamp, D.: 1986, *Phys. Fluids* **29**, 1520.
Biskamp, D.: 1996, *Astrophys. & Sp. Sci.* **242**, 165.
Chandran, B. and Cowley, S.: 1998, *Phys. Rev. Lett.* **80**, 3077.
Cho, J. and Lazarian, A.: 2002, *Phys. Rev. Lett.* **88**(24), 245001.
Cho, J. and Lazarian, A.: 2003, astro-ph/0301462.
Cho, J., Lazarian, A.: and Vishniac, E.T.: 2002, *Astrophys. J.* **564**, 291.
Fabian, A.C.: 1994, *Ann. Rev. Astro. & Astrophys.* **32**, 277.
Fabian, A.C., Mushotzky, R.F., Nulsen, P.E.J. and Peterson, J.R.: 2001, *MNRAS* **321**, L20.
Furth, H.P., Killeen, J. and Rosenbluth, M.N.: 1963, *Phys. Fluids* **6**, 459.
Goldreich, P. and Sridhar, S.: 1995, *Astrophys. J.* **438**, 763 (GS95).
Hameiri, E. and Bhattacharjee, A.: 1987, *Phys. Fluids* **30**, 1743.
Kim, E.-J. and Diamond, P.H.: 2001, *Astrophys. J.* **556**, 1052.
Lazarian, A., Petrosian, V., Yan, H. and Cho, J.: 2003, astro-ph/0301181.
Lazarian, A. and Vishniac, E.T.: 1999, *Astrophys. J.* **517**, 700 (LV99).
Lazarian, A. and Vishniac, E.T.: 2000, *Revista Mexicana de Astronomia y Astrofisica* **9**, 55.
Lazarian, A., Vishniac, E.T. and Cho, J.: 2003, *Astrophys. J.*, submitted.
Lesieur, M.: 1990, *Turbulence in Fluids: Stochastic and Numerical Modelling*, 2nd. rev. ed., Kluwer Academic Publishers, Dordrecht.
Ma, Z.W. and Bhattacharjee, A.: 1996, *J. Geophys. Res.* **101**, 2641.
Matthaeus, W.H. and Lamkin, S.L.: 1985, *Phys. Fluids* **28**, 303.
Narayan, R. and Medvedev, M.V.: 2001, *Astrophys. J.* **562**, L129.
Parker, E.N.: 1957, *J. Geophys. Res.* **62**, 509.
Parker, E.N.: 1979, *Cosmical Magnetic Fields*, Clarendon Press, Oxford.
Parker, E.N.: 1992, *Astrophys. J.* **401**, 137.
Petschek, H.E.: 1964, Magnetic field annihilation, in: W.H. Hess (ed.), *The Physics of Solar Flares*, NASA Special Pub. **50**, Washington, DC, pp. 425–439.
Priest, E. and Forbes, T.: 2000, *Magnetic Reconnection: MHD Theory and Applications*, Cambridge University Press, Cambridge.
Shay, M. and Drake, J.: 1998, *Geophys. Res. Lett.* **25**(20), 3759.
Shay, M., Drake, J., Denton, R. and Biskamp, D.: 1998, *J. Geophys. Res.* **103**, 9165.
Shay, M., Drake, J., Rogers, B. and Denton, R.: 1999, *Geophys. Res. Lett.* **26**(14), 2163.
Slavin, J.D., Shull, J.M. and Begelman, M.C.: 1993, *Astrophys. J.* **407**, 83.
Smith, R. and Cox, D.: 2001, *Astrophys. J. Supp.* **134**, 283.
Strauss, H.R.: 1988, *Astrophys. J.* **326**, 412.
Sweet, P.A.: 1958, The neutral point theory of solar flares, in: B. Lehnert (ed.), *IAU Symp. 6, Electromagnetic Phenomena in Cosmical Plasma*, pp. 123–134, Cambridge Univ. Press, NY.
Vishniac, E.T. and Lazarian, A.: 1999, *Astrophys. J.* **511**, 193.
Vishniac, E.T., Lazarian, A. and Cho, J.: 2003, in: T. Passot and E. Falgarone (eds.), *Simulations of Magnetohydrodynamic Turbulence in Astrophysics*, Springer-Verlag, Lecture Notes in Physics, astro-ph/0205286.
Wang, X., Ma, Z.W. and Bhattacharjee, A.: 1996, *Phys. Plasmas* **3**(5), 2129.
Yamada, M., Ji, H., Hsu, S., Carter, T., Kulsrud, R. and Trintchouk, F.: 2000, *Phys. Plasmas* **7**(5), 1781.
Zakamska, N.L. and Narayan, R.: 2002, astro-ph/0207127.
Zweibel, E.: 1989, *Astrophys. J.* **343**, 994.

THE EFFECTS OF SPIRAL ARMS ON THE MULTI-PHASE ISM

ANVAR SHUKUROV and GRAEME R. SARSON
School of Mathematics & Statistics, University of Newcastle, Newcastle NE1 7RU, U.K.

ÅKE NORDLUND
NBIfAFG & TAC, Juliane Maries Vej 30, DK-2100 Copenhagen Ø, Denmark

BORIS GUDIKSEN
Inst. Solar Physics, Royal Swedish Acad. of Sciences, SE-106 91 Stockholm, Sweden

AXEL BRANDENBURG
NORDITA, Blegdamsvej 17, DK-2100 Copenhagen Ø, Denmark

Abstract. Statistical parameters of the ISM driven by thermal energy injections from supernova explosions have been obtained from 3D, nonlinear, magnetohydrodynamic, shearing-box simulations for spiral arm and interarm regions. The density scale height obtained for the interarm regions is 50% larger than for the spiral arms because of the higher gas temperature. The filling factor of the hot gas is also significantly larger between the arms and depends sensitively on magnetic field strength.

Keywords: ISM, turbulence, galactic spiral structure

Detailed observations of the multi-phase interstellar medium (ISM) have recently been supplemented by extensive numerical simulations that capture much of the physics involved, including driving by supernovae (SNe) and radiative cooling. Several of the models are in 3D (e.g., de Avillez, 2000; de Avillez and Berry, 2001; de Avillez and MacLow, 2001), but only that of Korpi et al. (1999a,b) includes rotation and so is suitable for realistic modelling of magnetic fields. Here we discuss a further development of this model, where we concentrate on the effects of the total gas column density and magnetic field strength on the properties of the ISM.

Our models for regions inside and outside the spiral arms differ in the gas column density; the midplane density ρ_0 is given in Table I. The three models with low, intermediate and high densities are referred to as Interarm, Average and Arm. The column density does not vary in time in our simulations. Thus, our model is applicable not far from the corotation radius of the spiral pattern, and so is complementary to the model of Gómez and Cox (2002) which focuses on dynamic effects due to the passage of a density wave through the ISM.

We solve numerically the 3D, non-ideal MHD equations with rotation, density stratification in external gravity, heat sources and radiative cooling (Korpi et al., 1999a). The simulations start with an imposed azimuthal magnetic field of variable strength B_0, with vertical profile $B_0 \cosh^{-2}(z/0.3\,\mathrm{kpc})$. The system is driven by localised thermal energy injections, modelling SN explosions. Type II SNe are initiated at those locations where the gas density ρ exceeds $\rho_c = 10^{-24}\,\mathrm{g\,cm^{-3}}$ and temperature T is lower than $T_c = 4000\,\mathrm{K}$; the probability of an SN explosion at

Astrophysics and Space Science **289:** 319–322, 2004.

TABLE I

Three models with varying initial mid-plane density, ρ_0. The mean temperature, pressures, root mean square velocity and filling factor are all calculated within $|z| < 0.2$ kpc. The filling factors are for $T > 10^5$ K; the initial magnetic field at the midplane is $B_0 = 6\,\mu$G in all models.

	Unit	Interarm	Average	Arm
Initial midplane density, ρ_0	10^{-24} g cm^{-3}	0.7	1.4	2.9
Density Gaussian scale height	kpc	0.23	0.20	0.16
Mean temperature	10^4 K	35	7.8	3.4
Mean thermal pressure	10^{-14} dyn cm^{-2}	42	68	120
Rms vertical velocity	km s^{-1}	23	20	20
Mean turbulent pressure	10^{-14} dyn cm^{-2}	39	63	110
Hot gas filling factor		0.12	0.07	0.04
SN II rate	kpc^{-2} Myr^{-1}	11	38	111
SN II Gaussian scale height	kpc	0.30	0.16	0.14

a given position is proportional to the local gas density within those regions. The SN II rate, $\nu_{\rm SN\,II}$, depends on the local ISM parameters via (Gudiksen, 1999)

$$\nu_{\rm SN\,II} = \frac{M_{\rm c} X_{\rm SN} X_*}{M_{\rm SN} \tau_{\rm c}} ,$$

where $M_{\rm c}$ is the mass of gas with $\rho > \rho_{\rm c}$ and $T < T_{\rm c}$, $X_{\rm SN} = 0.1$ is the fraction of the stellar mass in SN progenitors obtained from a suitable initial mass function, $X_* = 0.02$ is the gas mass fraction converted into stars, $M_{\rm SN} = 10\,M_\odot$ is the SN progenitor mass, and $\tau_{\rm c} = 20$ Myr is the gas recycling time. The latter has been adjusted to have $\nu_{\rm SN\,II} \simeq 40\,{\rm kpc}^{-2}\,{\rm Myr}^{-1}$ (1/33 yr for the whole Galaxy) for the *average* gas density; we stress that both the rate and spatial distribution of Type II SNe are not prescribed but rather follow from the simulations. The resulting dependence of the SN II rate on the mean gas density, shown in Figure 1, is consistent with the observed scaling of the star formation rate with gas surface density, SFR $\propto \Sigma^\kappa$ with $\kappa = 0.9$–1.7 (Kennicutt, 1998). The SN II rates and scale heights are given in Table I. Our model also includes Type I SNe implemented as in Korpi et al. (1999a).

Our results, as reported here, are preliminary because we use a relatively small computational domain with the horizontal and vertical (z) dimensions of $0.25 \times 0.25 \times 1$ kpc (with the midplane in the centre), modest spatial resolution of about 4 pc, and closed boundary conditions in z. Furthermore, the cooling function was truncated at 500 K to avoid high densities. All these restrictions are of a technical nature and can readily be relaxed.

An unexpected result of our simulations is that the density scale height is significantly larger in the Interarm model, although both thermal and turbulent pressures

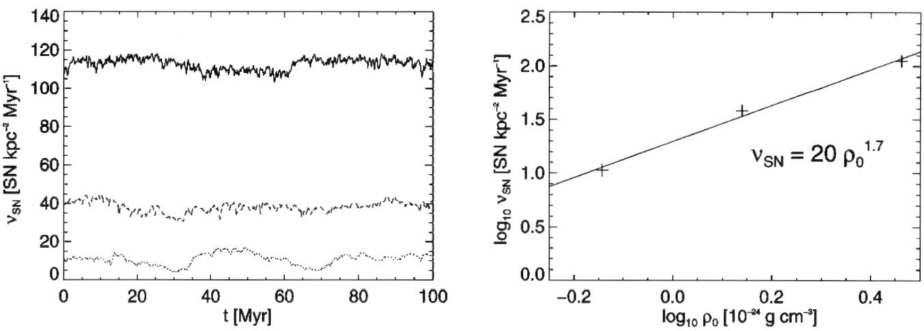

Figure 1. The SN II rate versus time (left), for the Arm (solid), Average (dashed) and Interarm (dotted) models. The scaling of the mean SN II rate with mid-plane density, ρ_0 (right).

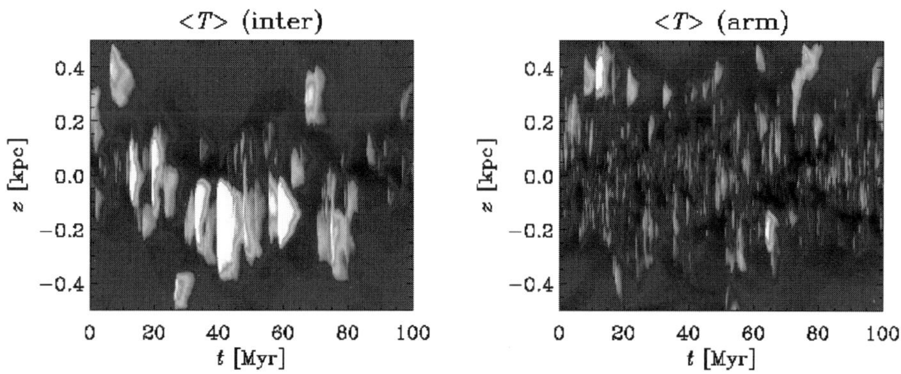

Figure 2. Horizontally averaged temperature as a function of height and time, for the Interarm (left panel) and Arm (right panel) solutions. Lighter shades of grey indicate higher temperatures.

are a factor of about 3 larger in the Arm model. The reason is that the gas temperatures are higher in the Interarm case (Figure 2), even though the SN rate is lower. The apparent reason is that the cooling rate has a stronger net dependence on gas density than the SN energy injection rate.

Another surprising feature of the results presented in Table I is that the filling factor of the hot gas is lower than expected by a factor of 2–3. This can be attributed to the geometry of the magnetic field in our models; it is uniform initially, and therefore effective in confining expanding bubbles of hot gas. The initial midplane field strength, $6\,\mu$G, is close to that of the *total* field in the Solar vicinity, but the field is implausibly well ordered. A more realistic simulation would initialize the model with a ratio of turbulent to ordered magnetic energy of about 3. We postpone such calculations for future work, but here we explore the dependence of the results on the strength of the initially uniform magnetic field; the results are shown in Table II. The filling factor of the hot gas is sensitive to the field strength and increases to 0.2 as the field becomes weaker. The density scale height marginally

TABLE II

The effect of magnetic field on the ISM illustrated by two runs with varying initial magnetic field, B_0. All variables are as defined in Table I. All runs have $\rho_0 = 0.7 \times 10^{-24}\ \mathrm{g\,cm^{-3}}$ (the Interarm model).

Initial magnetic field strength, B_0	μG	0	6
Density scale height	kpc	0.20	0.23
Mean thermal pressure	$10^{-14}\ \mathrm{dyn\,cm^{-2}}$	50	42
Rms vertical velocity	$\mathrm{km\,s^{-1}}$	43	23
Mean turbulent pressure	$10^{-14}\ \mathrm{dyn\,cm^{-2}}$	54	39
Hot gas filling factor		0.19	0.12

increases with magnetic field strength, but this effect is much less pronounced than the suppression of the hot phase; the field strongly suppresses turbulence in the hot gas. Thus, the magnetic field can affect the disc-halo connection and the global structure of the ISM in crucial, diverse and unexpected ways. This aspect of the ISM dynamics has not yet been fully explored.

Acknowledgements

We are grateful to M.-M. MacLow for helpful comments on the dependence of the filling factor of the hot gas on magnetic field. Use of the Copenhagen branch of the Danish Center for Scientific Computing and the PPARC supported UKAFF computer facility at Leicester is acknowledged. This work was supported by the PPARC Grant PPA/G/S/2000/00528.

References

de Avillez, M.A.: 2000, *MNRAS* **315**, 479.
de Avillez, M.A. and Berry, D.L.: 2001, *MNRAS* **328**, 708.
de Avillez, M.A. and MacLow, M.-M.: 2001, *ApJ* **551**, L57.
Gómez, G.C. and Cox, D.P.: 2002, *ApJ* **580**, 235.
Gudiksen, B.V.: 1999, Master Thesis, Astron. Obs., Univ. Copenhagen.
Kennicutt, R.C.: 1998, *ARAA* **36**, 189.
Korpi, M.J., Brandenburg, A., Shukurov, A., Tuominen, I. and Nordlund, Å.: 1999a, *ApJ* **514**, L99.
Korpi, M.J., Brandenburg, A., Shukurov, A. and Tuominen, I.: 1999b, *A&A* **350**, 230.

TURBULENCE IN THE INTERSTELLAR MEDIUM

Energetics and Driving Mechanisms

MORDECAI-MARK MAC LOW
Department of Astrophysics, American Museum of Natural History
New York, NY 10024-5192, USA; E-mail: mordecai@amnh.org

Abstract. Interstellar turbulence is expected to dissipate quickly in the absence of continuous energy input. I examine the energy available for driving the turbulence from six likely mechanisms: magnetorotational instability, gravitational instability, protostellar outflows, H II region expansion, stellar winds, and supernovae. I conclude that supernovae contribute far more energy than the other mechanisms, and so form the most likely driving mechanism.

Keywords: turbulence, interstellar medium

1. Introduction

Turbulent flows appear more and more central to our understanding of the interstellar medium. The distribution of pressures and densities are probably determined as much by turbulent ram pressures as by thermal phase transitions (Vázquez-Semadeni, Gazol and Scalo, 2000; Mac Low et al., 2002). Furthermore, compression by large-scale turbulent flows may form molecular clouds (Ballesteros-Paredes, Hartmann and Vázquez-Semadeni, 1999), and drive turbulence within those clouds (Ossenkopf and Mac Low, 2002) to support them against gravitational collapse (Klessen, Heitsch and Mac Low, 2000).

Both support against gravity and maintenance of observed motions appear to depend on continued driving of the turbulence, which has kinetic energy density $e = (1/2)\rho v_{\rm rms}^2$. Mac Low (1999, 2002) estimates that the dissipation rate for isothermal, supersonic turbulence is

$$\begin{aligned}\dot{e} &\simeq -(1/2)\rho v_{\rm rms}^3/L_d \\ &= -(3\times10^{-27}\ {\rm erg\ cm^{-3}\ s^{-1}})n\left(\frac{v_{\rm rms}}{10\ {\rm km\ s^{-1}}}\right)^3\left(\frac{L_d}{100\ {\rm pc}}\right)^{-1}, \end{aligned} \quad (1)$$

where n is the number density in units of cm^{-3}, L_d is the driving scale, which we have somewhat arbitrarily taken to be 100 pc (though it could well be smaller), and we have assumed a mean mass per particle $\mu = 2.11\times10^{-24}$ g. The dissipation time for turbulent kinetic energy

$$\tau_d = e/\dot{e} \simeq L/v_{\rm rms} = (9.8\ {\rm Myr})\left(\frac{L_d}{100\ {\rm pc}}\right)\left(\frac{v_{\rm rms}}{10\ {\rm km\ s^{-1}}}\right)^{-1}, \quad (2)$$

Astrophysics and Space Science **289:** 323–331, 2004.

which is just the crossing time for the turbulent flow across the driving scale (Elmegreen, 2000b). What then is the energy source for this driving? We here review the energy input rates for a number of possible mechanisms.

2. Magnetorotational Instabilities

One energy source for interstellar turbulence that has long been considered is shear from galactic rotation (Fleck, 1981). However, the question of how to couple from the large scales of galactic rotation to smaller scales remained open. Work by Sellwood and Balbus (1999) has shown that the magnetorotational instability (Balbus and Hawley, 1991, 1998) could couple the large-scale motions to small scales efficiently. The instability generates Maxwell stresses (a positive correlation between radial B_R and azimuthal B_Φ magnetic field components) that transfer energy from shear into turbulent motions at a rate $\dot{e} = T_{R\Phi}\Omega$ (Sellwood and Balbus, 1999). Numerical models suggest that the Maxwell stress tensor $T_{R\Phi} \simeq 0.6B^2/(8\pi)$ (Hawley, Gammie and Balbus, 1995). For the Milky Way, the value of the rotation rate recommended by the IAU is $\Omega = (220\ \mathrm{Myr})^{-1} = 1.4 \times 10^{-16}\ \mathrm{rad\ s}^{-1}$, though this may be as much as 15% below the true value (Olling and Merrifield 1998, 2000). The magnetorotational instability may thus contribute energy at a rate

$$\dot{e} = (3 \times 10^{-29}\ \mathrm{erg\ cm}^{-3}\ \mathrm{s}^{-1}) \left(\frac{B}{3\mu\mathrm{G}}\right)^2 \left(\frac{\Omega}{(220\ \mathrm{Myr})^{-1}}\right). \tag{3}$$

For parameters appropriate to the H I disk of a sample small galaxy, NGC 1058, Sellwood and Balbus (1999) find that the magnetic field required to produce the observed velocity dispersion of 6 km s^{-1} is roughly 3 μG, a reasonable value for such a galaxy. This instability may provide a base value for the velocity dispersion below which no galaxy will fall. If that is sufficient to prevent collapse, little or no star formation will occur, producing something like a low surface brightness galaxy with large amounts of H I and few stars.

3. Gravitational Instabilities

Motions coming from gravitational collapse have often been suggested as a local driving mechanism in molecular clouds, but fail due to the quick decay of the turbulence (Klessen, Burkert and Bate, 1998). If the turbulence decays in less than a free-fall time, as suggested by equation 1, then it cannot delay collapse for substantially longer than a free-fall time.

On the galactic scale, spiral structure can drive turbulence in gas disks. Roberts (1969) first demonstrated that shocks would form in gas flowing through spiral arms formed by gravitational instabilities in the stellar disk (Lin and Shu, 1964; Lin, Yuan and Shu, 1969). These shocks were studied in thin disks by Tubbs (1980) and Soukup and Yuan (1981), who found few vertical motions. More recently, it has been realized that in a more realistic thick disk, the spiral shock will take on some properties of a hydraulic bore, with gas passing through a sudden vertical jump at the position of the shock (Martos and Cox, 1998; Gómez and Cox, 2002). Behind the shock, downward flows of as much as 20 km s^{-1} appear (Gómez and Cox, 2002). Some portion of this flow will contribute to interstellar turbulence. However, the observed presence of interstellar turbulence in irregular galaxies without spiral arms, as well as in the outer regions of spiral galaxies beyond the regions where the arms extend suggest that this cannot be the only mechanism driving turbulence. A more quantitative estimate of the energy density contributed by spiral arm driving has not yet been done.

The interaction between rotational shear and gravitation can, at least briefly, drive turbulence in a galactic disk, even in the absence of spiral arms. This process has been numerically modeled at high resolution (sub-parsec zones) in two dimensions in a series of papers by Wada and Norman (1999, 2001), Wada, Spaans and Kim (2000), and Wada, Meurer and Norman (2002). However, these models all share two limitations: they do not include the dominant stellar component, and gravitational collapse cannot occur beneath the grid scale. The computed filaments of dense gas are thus artificially supported, and would actually continue to collapse to form stars, rather than driving turbulence in dense disks (see Sánchez-Salcedo [2001] for a detailed critique). In very low density disks, where even the dense filaments remained Toomre stable, this mechanism might operate, however.

Wada et al. (2002) estimated the energy input from this mechanism following the lead of Sellwood and Balbus (1999), but substituting Newton stresses (Lynden-Bell and Kalnajs 1972) for Maxwell stresses. The Newton stresses will only add energy if a positive correlation between radial and azimuthal gravitational forces exists, however, which is not demonstrated by Wada et al. (2002). Nevertheless, they estimate the order of magnitude of the energy input from Newton stresses as

$$\begin{aligned} \dot{e} &\simeq G(\Sigma_g/H)\lambda^2\Omega \\ &\simeq (4 \times 10^{-29}\ \mathrm{erg\ cm^{-3}\ s^{-1}}) \left(\frac{\Sigma_g}{10\ \mathrm{M_\odot\ pc^{-2}}}\right)^2 \times \\ &\times \left(\frac{H}{100\ \mathrm{pc}}\right)^{-2} \left(\frac{\lambda}{100\ \mathrm{pc}}\right)^2 \left(\frac{\Omega}{(220\ \mathrm{Myr})^{-1}}\right), \end{aligned} \quad (4)$$

where G is the gravitational constant, Σ_g the density of gas, H, the scale height of the gas, λ a length scale of turbulence, and Ω the angular velocity of the disk. Values chosen are appropriate for the Milky Way. This is two orders of magnitude below the value required to maintain interstellar turbulence (eq. [1]).

4. Protostellar Outflows

Protostellar jets and outflows are a popular suspect for the energy source of the observed turbulence. We can estimate their average energy input rate, following McKee (1989), by assuming that some fraction f_w of the mass accreted onto a star during its formation is expelled in a wind travelling at roughly the escape velocity. Shu et al. (1988) argue that $f_w \simeq 0.4$, and that most of the mass is ejected from close to the stellar surface, where the escape velocity

$$v_{\rm esc} = (2GM/R)^{1/2} = (200\ {\rm km\ s^{-1}})(M/\ {\rm M_\odot})^{1/2}(R/10\ {\rm R_\odot})^{-1/2}, \tag{5}$$

where the scaling is appropriate for a protostar with mass $M = 1\ {\rm M_\odot}$ and radius $R = 10\ {\rm R_\odot}$. Observations of neutral atomic winds from protostars suggest outflow velocities of roughly this value (Lizano et al., 1988; Giovanardi et al., 2000).

The total energy input from protostellar winds will substantially exceed the amount that can be transferred to the turbulence due to radiative cooling at the wind termination shock. We represent the fraction of energy lost there by η_w. A reasonable upper limit to the energy loss is offered by assuming fully effective radiation and momentum conservation, so that

$$\eta_w < \frac{v_{\rm rms}}{v_w} = 0.05 \left(\frac{v_{\rm rms}}{10\ {\rm km\ s^{-1}}}\right)\left(\frac{200\ {\rm km\ s^{-1}}}{v_w}\right), \tag{6}$$

where $v_{\rm rms}$ is the rms velocity of the turbulence, and we have assumed that the flow is coupled to the turbulence at typical velocities for the diffuse ISM. If we assumed that most of the energy went into driving dense gas, the efficiency would be lower, as typical rms velocities for CO outflows are 1–2 km s^{-1}. The energy injection rate

$$\begin{aligned}\dot{e} &= \frac{1}{2} f_w \eta_w \frac{\dot{\Sigma}_*}{H} v_w^2 \\ &\simeq (2 \times 10^{-28}\ {\rm erg\ cm^{-3}\ s^{-1}}) \left(\frac{H}{200\ {\rm pc}}\right)^{-1} \left(\frac{f_w}{0.4}\right) \left(\frac{v_{\rm rms}}{10\ {\rm km\ s^{-1}}}\right) \times \\ &\times \left(\frac{v_w}{200\ {\rm km\ s^{-1}}}\right) \left(\frac{v_{\rm rms}}{10\ {\rm km\ s^{-1}}}\right) \left(\frac{\dot{\Sigma}_*}{4.5 \times 10^{-9}\ {\rm M_\odot\ pc^{-2}\ yr^{-1}}}\right),\end{aligned} \tag{7}$$

where $\dot{\Sigma}_*$ is the surface density of star formation, and H is the scale height of the star-forming disk. The scaling value used for $\dot{\Sigma}_*$ is the solar neighborhood value (McKee, 1989).

Although protostellar jets and winds are indeed quite energetic, they deposit most of their energy into low density gas (Henning, 1989), as is shown by the observation of multi-parsec long jets extending completely out of molecular clouds (Bally and Devine, 1994). Furthermore, observed motions of molecular gas show increasing power on scales all the way up to and perhaps beyond the largest scale of molecular cloud complexes (Ossenkopf and Mac Low, 2002). It is hard to see how such large scales could be driven by protostars embedded in the clouds.

5. Massive Stars

In active star-forming galaxies, however, massive stars appear likely to dominate the driving. They do so through ionizing radiation and stellar winds from O stars, and clustered and field supernova explosions, predominantly from B stars no longer associated with their parent gas. The supernovae appear likely to dominate, as we now show.

5.1. STELLAR WINDS

First, we consider stellar winds. The total energy input from a line-driven stellar wind over the main-sequence lifetime of an early O star can equal the energy from its supernova explosion, and the Wolf-Rayet wind can be even more powerful. However, the mass-loss rate from stellar winds drops as roughly the sixth power of the star's luminosity if we take into account that stellar luminosity varies as the fourth power of stellar mass (Vink, de Koter and Lamers, 2000), and the powerful Wolf-Rayet winds (Nugis and Lamers, 2000) last only 10^5 years or so, so only the very most massive stars contribute substantial energy from stellar winds. The energy from supernova explosions, on the other hand, remains nearly constant down to the least massive star that can explode. As there are far more lower-mass stars than massive stars, with a Salpeter (1955) IMF giving a power-law in mass of $\alpha = -2.35$, supernova explosions will inevitably dominate over stellar winds after the first few million years of the lifetime of an OB association, until the lifetime of the least massive star to explode of around 40–50 Myr.

5.2. HII REGION EXPANSION

Next, we consider ionizing radiation from OB stars. The total amount of energy contained in ionizing radiation is vast. Abbott (1982) estimates the total luminosity of ionizing radiation in the disk to be

$$\dot{e} = 1.5 \times 10^{-24} \text{ erg s}^{-1} \text{ cm}^{-3}. \tag{8}$$

However, only a very small fraction of this total energy goes to driving interstellar motions, as we now show.

Ionizing radiation primarily contributes to interstellar turbulence by ionizing HII regions, heating them to 7000–10,000 K, and raising their pressures above that of surrounding neutral gas, so that they expand supersonically. Matzner (2002) computes the momentum input from the expansion of an individual HII region into a surrounding molecular cloud, as a function of the cloud mass and the ionizing luminosity of the central OB association. By integrating over the HII region luminosity function derived by McKee and Williams (1997), he finds that the average momentum input from a Galactic region is

$$\langle \delta p \rangle \simeq (260 \text{ km s}^{-1}) \left(\frac{N_H}{1.5 \times 10^{22} \text{ cm}^{-2}} \right)^{-3/14} \left(\frac{M_{cl}}{10^6 \text{ M}_\odot} \right)^{1/14} \langle M_* \rangle. \tag{9}$$

The column density N_H is scaled to the mean value for Galactic molecular clouds (Solomon et al., 1987), which varies little as cloud mass M_{cl} varies. The mean stellar mass per cluster in the Galaxy $\langle M_* \rangle = 440\ \mathrm{M}_\odot$ (Matzner, 2002).

The number of OB associations contributing substantial amounts of energy can be drawn from the McKee and Williams (1997) cluster luminosity function

$$\mathcal{N}(> S_{49}) = 6.1 \left(\frac{108}{S_{49}} - 1 \right), \tag{10}$$

where $\mathcal{N}$ is the number of associations with ionizing photon luminosity exceeding $S_{49} = S/(10^{49}\ \mathrm{s}^{-1})$. The luminosity function is rather flat below $S_{49} = 2.4$, the luminosity of a single star of $120\ \mathrm{M}_\odot$, which was the highest mass star considered, so taking its value at $S_{49} = 1$ is about right, giving $\mathcal{N}(> 1) = 650$ clusters.

To derive an energy input rate per unit volume $\dot{e}$ from the mean momentum input per cluster $\langle \delta p \rangle$, we need to estimate the average velocity of momentum input v_i, the time over which it occurs t_i, and the volume V under consideration. Typically expansion will not occur supersonically with respect to the interior, so $v_i < c_{s,i}$, where $c_{s,i} \simeq 10\ \mathrm{km\ s}^{-1}$ is the sound speed of the ionized gas. McKee and Williams (1997) argue that clusters typically last for about five generations of massive star formation, where each generation lasts $\langle t_* \rangle = 3.7$ Myr. The scale height for massive clusters is $H_c \sim 100$ pc (e.g. Bronfman et al., 2000), and the radius of the star-forming disk is roughly $R_{sf} \sim 15$ kpc, so the relevant volume $V = 2\pi R_{sf}^2 H_c$. The energy input rate from HII regions is then

$$\begin{aligned} \dot{e} &= \frac{\langle \delta p \rangle \mathcal{N}(> 1) v_i}{V t_i} \\ &= (3 \times 10^{-30}\ \mathrm{erg\ s^{-1}\ cm^{-3}}) \left(\frac{N_H}{1.5 \times 10^{22}\ \mathrm{cm}^{-2}} \right)^{-3/14} \times \\ &\times \left(\frac{M_{cl}}{10^6\ \mathrm{M}_\odot} \right)^{1/14} \left(\frac{\langle M_* \rangle}{440\ \mathrm{M}_\odot} \right) \left(\frac{\mathcal{N}(> 1)}{650} \right) \times \\ &\times \left(\frac{v_i}{10\ \mathrm{km\ s}^{-1}} \right) \left(\frac{H_c}{100\ \mathrm{pc}} \right)^{-1} \left(\frac{R_{sf}}{15\ \mathrm{kpc}} \right)^{-2} \left(\frac{t_i}{18.5\ \mathrm{Myr}} \right)^{-1}, \end{aligned} \tag{11}$$

where all the scalings are appropriate for the Milky Way as discussed above. Nearly all of the energy in ionizing radiation goes towards maintaining the ionization of the diffuse ionized medium, and hardly any towards driving turbulence. Flows of ionized gas may be important very close to young clusters, but do not appear to contribute significantly on a global scale.

5.3. SUPERNOVAE

The largest contribution from massive stars to interstellar turbulence comes from supernova explosions. To estimate their energy input rate, we begin by finding the supernova rate in the Galaxy σ_{SN}. Cappellaro et al. (1999) estimate the total supernova rate in supernova units to be 0.72 ± 0.21 SNu for galaxies of type S0a-b and 1.21 ± 0.37 SNu for galaxies of type Sbc-d, where 1 SNu = 1 SN $(100\ \mathrm{yr})^{-1}(10^{10} L_B/\mathrm{L}_\odot)^{-1}$, and L_B is the blue luminosity of the galaxy. Taking the Milky Way as lying between Sb and Sbc, we estimate $\sigma_{SN} = 1$ SNu. Using a Galactic luminosity of $L_B = 2 \times 10^{10}\ \mathrm{L}_\odot$, we find a supernova rate of $(50\ \mathrm{yr})^{-1}$, which agrees well with the estimate in equation (A4) of McKee (1989). If we use the same scale height H_c and star-forming radius R_{sf} as above, we can compute the energy input rate from supernova explosions with energy $E_{SN} = 10^{51}$ erg to be

$$\begin{aligned} \dot{e} &= \frac{\sigma_{SN}\eta_{SN}E_{SN}}{\pi R_{sf}^2 H_c} \\ &= (3 \times 10^{-26}\ \mathrm{erg\ s^{-1}\ cm^{-3}}) \left(\frac{\eta_{SN}}{0.1}\right)\left(\frac{\sigma_{SN}}{1\ \mathrm{SNu}}\right) \times \\ &\times \left(\frac{H_c}{100\ \mathrm{pc}}\right)^{-1}\left(\frac{R_{sf}}{15\ \mathrm{kpc}}\right)^{-2}\left(\frac{E_{SN}}{10^{51}\ \mathrm{erg}}\right). \end{aligned} \tag{12}$$

The efficiency of energy transfer from supernova blast waves to the interstellar gas η_{SN} depends on the strength of radiative cooling in the initial shock, which will be much stronger in the absence of a surrounding superbubble (e.g. Heiles, 1990). Substantial amounts of energy can escape in the vertical direction in superbubbles as well, however. (Norman and Ferrara [1996] make an analytic estimate of the effectiveness of driving by SN remnants and superbubbles.) The scaling factor $\eta_{SN} \simeq 0.1$ used here was derived by Thornton et al. (1998) from one-dimensional numerical simulations of SNe expanding in a uniform ISM, or can alternatively be drawn from momentum conservation arguments comparing a typical expansion velocity of 100 km s^{-1} to typical interstellar turbulence velocity of 10 km s^{-1}. Detailed multi-dimensional modeling of the interactions of multiple SN remnants (e.g. Avillez, 2000) will be required to better determine it.

Supernova driving appears to be powerful enough to maintain the turbulence even with the dissipation rates estimated in equation (1). It provides a large-scale self-regulation mechanism for star formation in disks with sufficient gas density to collapse despite the velocity dispersion produced by the magnetorotational instability. As star formation increases in such galaxies, the number of OB stars increases, ultimately increasing the supernova rate and thus the velocity dispersion, which restrains further star formation.

Supernova driving not only determines the velocity dispersion, but may actually form molecular clouds by sweeping gas up in a turbulent flow. Clouds that are turbulently supported will experience inefficient, low-rate star formation, while clouds that are too massive to be supported will collapse (e.g. Kim and Ostriker, 2001), undergoing efficient star formation to form OB associations or even starburst knots.

6. Conclusions

Interstellar gas is only quiescent on the very smallest scales (< 0.1 pc) in dense cores. The transsonic and supersonic flows observed elsewhere must be driven, as turbulent motions otherwise decay away on timescales of order the crossing time of their outer scale. The largest energy source in star-forming regions of galaxies is supernova explosions. Even turbulence within dense molecular clouds may derive primarily from field supernovae (not the few supernovae that may be associated with the cloud itself before its disappearance, but the background explosions from B stars formed within the previous 50 Myr or so.) Outside of star-forming parts of galactic disks, turbulence may be driven by magnetorotational instabilities, or even interaction between gravitation and rotation.

Acknowledgements

I thank the organizers of the conference for their invitation and partial support of my attendance. This talk was adapted from a section of Mac Low and Klessen (2004), in press. This research was partially supported by US NSF CAREER grant AST99-85392.

References

Abbott, D.C.: 1982, *Astrophys. J.* **263**, 723–735.
Avillez, M.A.: 2000, *Mon. Not. R. Astron. Soc.* **315**, 479–497.
Balbus, S.A. and J.F. Hawley: 1991, *Astrophys. J.* **376**, 214–233.
Balbus, S.A. and J.F. Hawley: 1998, *Rev. Mod. Phys.* **70**, 1–53.
Ballesteros-Paredes, J., Hartmann, L. and Vázquez-Semadeni, E.: 1999, *Astrophys. J.* **527**, 285–297.
Bally, J. and Devine, D.: 1994, *Astrophys. J.* **428**, L65–L68.
Bronfman, L., Casassus, S., May, J. and Nyman, L.-Å.: 2000, *Astron. Astrophys.* **358**, 521–534.
Cappellaro, E., Evans, R. and Turatto, M.: 1999, *Astron. Astrophys.* **351**, 459–466.
Elmegreen, B.G.: 2000b, *Astrophys. J.* **530**, 277–281.
Fleck, R.C.: 1981, *Astrophys. J.* **246**, L151–L154.
Giovanardi, C., Rodríguez, L.F., Lizano, S. and Cantó, J.: 2000, *Astrophys. J.* **538**, 728–737.
Gómez, G.C. and Cox, D.P.: 2002, *Astrophys. J.* **580**, 235–252.
Hawley, J.F., Gammie, C.F. and Balbus, S.A.: 1996, apj **464**, 690–703.
Heiles, C.: 1990, *Astrophys. J.* **354**, 483–491.
Kim, W.-T. and Ostriker, E.C.: 2001, *Astrophys. J.* **559**, 70–95.
Klessen, R.S., Burkert, A. and Bate, M.R.: 1998, *Astrophys. J.* **501**, L205–L208.
Klessen, R.S., Heitsch, F. and Mac Low, M.-M.: 2000, *Astrophys. J.* **535**, 887–906.
Lin, C.C. and Shu, F.H.: 1964, *Astrophys. J.* **140**, 646–655.
Lin, C.C., Yuan, C. and Shu, F.H.: 1969, *Astrophys. J.* **155**, 721–746.
Lizano, S., Heiles, C., Rodriguez, L.F., Koo, B.-F., Shu, F.H., Hasegawa, T., Hayashi, S. and Mirabel, I.F.: 1988, *Astrophys. J.* **328**, 763–776.
Lynden-Bell, D. and Kalnajs, A.J.: 1972, *Mon. Not. R. Astron. Soc.* **157**, 1–30.
Mac Low, M.-M.: 1999, *Astrophys. J.* **524**, 169–178.

Mac Low, M.-M.: 2002, in: T. Passot and E. Falgarone (eds.), *Simulations of Magnetohydrodynamic Turbulence in Astrophysics*, Springer, Heidelberg, 182–207.
Mac Low, M.-M. and Klessen, R.S.: 2003, *Rev. Mod. Phys.*, in preparation.
Mac Low, M.-M., Balsara, D.S., Avillez, M.A. and Kim, J.: 2002, *Astrophys. J.*, in revision (astro-ph/0106509).
Martos, M.A. and Cox, D.P.: 1998, *Astrophys. J.* **509**, 703–716.
Matzner, C.D.: 2002, *Astrophys. J.* **566**, 302–314.
McKee, C.F.: 1989, *Astrophys. J.* **345**, 782–801.
McKee, C.F. and Williams, J.P.: 1997, *Astrophys. J.* **476**, 144–165.
Norman, C.A. and Ferrara, A.: 1996, *Astrophys. J.* **467**, 280–291.
Nugis, T. and Lamers, H.J.G.L.M.: 2000, *Astron. Astrophys.* **360**, 227–244.
Olling, R.P. and Merrifield, M.R.: 1998, *Mon. Not. R. Astron. Soc.* **297**, 943–952.
Olling, R.P. and Merrifield, M.R.: 2000, *Mon. Not. R. Astron. Soc.* **311**, 361–369.
Ossenkopf V. and Mac Low, M.-M.: 2002, *Astron. Astrophys.* **390**, 307–326.
Roberts, W.W.: 1969, *Astrophys. J.* **158**, 123–143.
Salpeter, E.E.: 1955, *Astrophys. J.* **121**, 161–167.
Sánchez-Salcedo, F.J.: 2001, *Astrophys. J.* **563**, 867–874.
Sellwood, J.A. and Balbus, S.A.: 1999, *Astrophys. J.* **511**, 660–665.
Shu, F.H., Lizano, S., Ruden, S.P. and Najita, J.: 1998, *Astrophys. J. (Letters)* **328**, L19–L23.
Solomon, P.M., Rivolo, A.R., Barrett, J. and Yahil, A.: 1987, *Astrophys. J.* **319**, 730–741.
Soukup, J.E. and Yuan, C.: *Astrophys. J.* **246**, 376–385.
Thornton, K., Gaudlitz, M., Janka, H.-Th. and Steinmetz, M.: 1998, *Astrophys. J.* **500**, 95–119.
Tubbs, A.D.: 1980, *Astrophys. J.* **239**, 882–892.
Vázquez-Semadeni, E., Gazol, A. and Scalo, J.: 2000, *Astrophys. J.* **540**, 271–285.
Vink, J.S., de Koter, A. and Lamers, H.J.G.L.M.: 2000, *Astron. Astrophys.* **362**, 295–309.
Wada, K., Meurer, G. and Norman, C.A.: 2002, *Astrophys. J.* **577**, 197–205.
Wada, K. and Norman, C.A.: 1999, *Astrophys. J. (Letters)* **516**, L13–L16.
Wada, K. and Norman, C.A.: 2001, *Astrophys. J.* **547**, 172–186.
Wada, K., Spaans, M. and Kim, S.: 2000, *Astrophys. J.* **540**, 797–807.

MOLECULE DESTRUCTION AND FORMATION IN MOLECULAR CLOUDS

MICHAEL D. SMITH[1], GEORGI PAVLOVSKI[1], MORDECAI-MARK MACLOW[2], ALEXANDER ROSEN[1], TIGRAN KHANZADYAN[1], ROLAND GREDEL[3] and THOMAS STANKE[4]

[1]*Armagh Observatory, Armagh, U.K.*
[2]*American Museum of Natural History, New York, U.S.A.*
[3]*Max-Planck-Institut für Astronomie, Heidelberg, Germany*
[4]*Max-Planck-Institut für Radioastronomie, Bonn, Germany*

Abstract. We show that supersonic turbulence accelerates the transition of an atomic cloud into a molecular cloud, quantified here through a demo model and 3D numerical simulations which explicitly include atomic-molecular chemistry. Specific sites where amplified formation may be detectable are suggested.

1. Introduction

The means of production of a molecular interstellar phase is a central issue for both star and galaxy formation theories. *Clouds* of molecules could form by coagulation (collisional agglomeration) of smaller pre-existing clouds, gravi-thermal instability, condensation out of large atomic clouds and by shock-sweeping. Individual molecules form by association on dust grains or, if dust is absent, via H collisions with available H^-. If the molecules in the interstellar medium form at a typical density of 30 cm^{-3}, then a uniform atomic cloud could form on a timescale (Hollenbach and McKee, 1979):

$$t_R \sim 10^8\ g\ (n/30\ \mathrm{cm}^{-3})^{-1}\mathrm{yr}, \tag{1}$$

with a factor g to account for correction terms, uncertainties (e.g. Biham et al., 2001; Hartmann et al., 2001) and the dust properties relative to standard interstellar values. This, with g close to unity, was considered as roughly consistent with estimates of cloud ages, until recently.

There is now ample evidence that clouds of various types are generally young: their lifetimes are comparable to their dynamical time scales (e.g. Ballesteros-Paredes et al., 1999). As a consequence, molecular material is recycled through the ISM quite rapidly with the aid of supersonic turbulence and shock sweeping. Moreover, supersonic turbulence clearly dominates the motions in clouds. Since turbulence is expected to decay on very short time scales (MacLow, 1999), there is no force to balance the clouds against gravitational collapse and star formation occurs globally on a few free-fall times (Elmegreen, 2000).

Astrophysics and Space Science **289:** 333–336, 2004.

This implies that molecule formation is also somehow accelerated. We explore here the means of achieving this through turbulence. Supersonic turbulence leads to non-uniformity: at any instant, a fraction of the gas is highly compressed and will be forming molecules locally at a high rate. Can this proceed fast enough *and* be sufficiently global to generate a molecular cloud?

To answer this we require a 3D model for supersonic turbulence which follows molecule formation and destruction. Relevant numerical models have so far been limited to simple single-component equations of state. By assuming an atomic-molecular transition to take place at a density of, say, 30 cm^{-3}, distributions of molecular clouds are then predicted.

Here, we summarise the major findings of the first simulations of decaying supersonic turbulence with molecular chemistry and cooling. We begin by demonstrating that supersonic turbulence should accelerate molecule formation. We confirm this through the simulations. Finally, we discuss locations where the signals of cloud and molecule formation may be detectable.

1.1. A DEMONSTRATION MODEL

Assume a cloud of atomic gas of uniform density, ρ_o and volume V. The molecule formation rate is $R\rho_o^2$ V. Now, mix in the turbulence characterised by a Mach number, M. This 'Mach turbulence' provides density contrasts of order M^2 and we represent the cloud now by two components. A low density component of density ρ occupies a volume fraction $M^2/(M^2+1)$ and a high density component of density $(1+M^2)\rho$ occupies a volume $1/(M^2+1)$. Mass conservation yields ρ, and the molecule formation rate then becomes

$$\int R\,\rho\,dV = R\,\rho_o^2\,V \times \frac{(M^2+1)(M^4+3M^2+1)}{(2M^2+1)^2}. \tag{2}$$

Hence, at high M, turbulence accelerates H_2 formation by a factor $M^2/4$. Even for moderate M, the speed up is considerable: for M = 4, the rate is 4.8 $R\rho_o^2$ V.

1.2. SIMULATIONS OF TURBULENCE: MOLECULE REFORMATION

3D simulations of decaying supersonic turbulence with the ZEUS-3D code have been executed (Pavlovski et al., 2003). The code was modified by physics and chemistry pertaining to H, H_2, C, CO, O, OH and H_2O, appropriate for high density regions, assumed to be shielded from ionising, dissociating or heating radiation. Many of the findings, however, have broader implications. Beginning with a uniform density of 10^6 cm^{-3} and a 60 km s^{-1} root mean square turbulence, collisional dissociation reduced the 10^{16} cm cloud to an atomic form within 40 years. Then, within 300 years over half the molecules had reformed and at 600 yr, 80% had reformed, as illustrated in Figure 1. In an undisturbed medium, however, the H_2 would take 3000 yr to reform according to the above formula.

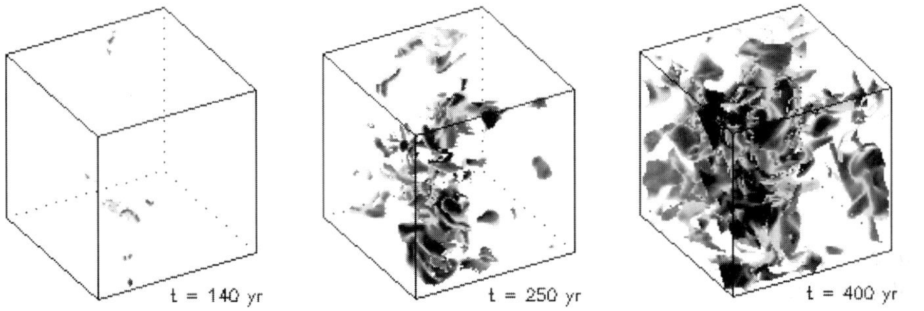

Figure 1. The molecular fraction during decaying turbulence, illustrating the distributed reformation. See Pavlovski et al. (2003) for details.

Moreover, the distribution of molecules closely follows the distribution of total density. Hence, molecules are present throughout the cloud, not confined to the high density regions where molecule formation is presently taking place. We thus conclude that the generated shocks sweep through the entire cloud, leaving almost no region unturned.

In decaying turbulence, the speed up in reformation mainly takes place during the short time that the turbulence is strong, immediately following the decay (if necessary) of shocks to non-dissociative speeds. Pavlovski et al. (2003) found that the energy in molecular turbulence decays steeper than isothermal turbulence. Hence the brief time in which high Mach number turbulence is present is responsible for the majority of reformed molecules.

1.3. Observing Molecule Formation

To form clouds and molecules in 1 Myr, a density of 3000 cm^{-3} is required in a uniform medium. This is reduced to an average density of just $\sim$ 500 $(M/5)^2$ cm^{-3} in a turbulent medium with an effective RMS Mach number of 5. Compressed layers possess a density

$$\rho = \frac{(M^2+1)^2}{2M^2+1}\rho_o, \tag{3}$$

i.e. a number density of 6600 cm^{-3} for our example. Such a clump could just be observed, as now demonstrated.

The question remains: at what density will the formation of molecules become detectable in the infrared? If we assume that each H_2 formation generates a fraction of ξ photons in transitions from the first to the ground vibrational level, 11% per cent of these are in the 2.12μm S(1) line and this line represents 19% of the K-band emission (the percentages being taken from the cascade calculation following UV fluorescence), then the total K-band surface brightness is
$3.4 \times 10^{-7}\xi\, f_H\, T^{0.5}\, L_{pc}\, n_4^2\, \mathrm{W\,m^{-2}\,s^{-1}\,sr^{-1}}$ where f_H is the atomic hydrogen fraction, $n_4 = n/(10^4\, cm^{-3})$ and $L_{pc} = L/(1\, pc)$. In observer's mumbo-jumbo, this

converts to a K-band flux of
2.0×10^{-18} W m^{-2} arcsec^{-2} and a S(1) flux of 3.8×10^{-19} W m^{-2} arcsec^{-2} taking $\xi = 0.5$, $f_H = 0.5$ and the other fiducial values. These values are indeed just on the verge of detectability with modern instruments. The flux derived here is over 1,000 times that predicted by Le Bourlot et al. (1995) for a dense cloud under equilibrium conditions.

We believe that certain Bok globules may provide the isolated dense conditions corresponding to the above estimates of physical parameters. Large Bok globules such as CB 34 (Khanzadyan et al., 2002) and B 335 (Hodapp, 1998) may be in the process of condensing out of atomic halos within an interstellar medium recently shaken by strong turbulence. A search for atomic hydrogen halos, other indicators for atomic gas or direct narrow-band spectroscopy may help determine if the broad-band emission is being produced by H_2 ro-vibrational transitions or scattering from internal or external radiation sources.

MML is partially supported by NSF grant AST99-85392, simulations on FORGE & UKAFF (PPARC), and support for AR from PPARC.

References

Ballesteros-Paredes, J., Hartmann, L. and Vázquez-Semadeni, E.: 1999, *ApJ* **527**, 285.
Biham, O., Furman, I., Pirronello, V. and Vidali, G.: 2001, *ApJ* **553**, 595.
Elmegreen, B.G.: 2000, *ApJ* **530**, 277.
Hartmann, L., Ballesteros-Paredes, J. and Bergin, E.A.: 2001, *ApJ* **562**, 852.
Hodapp, K.: 1998, *ApJ* **500**, L183.
Hollenbach, D. and McKee, C.F.: 1979, *ApJS* **41**, 555.
Khanzadyan, T., Smith, M.D., Gredel, R., Stanke, T. and Davis, C.J.: 2002, *A&A* **383**, 502.
Le Bourlot, J., Pineau des Forets, G., Roueff, E., Dalgarno, A. and Gredel, R.: 1995, *ApJ* **449**, 178.
MacLow, M.: 1999, *ApJ* **524**, 169.
Pavlovski G., Smith, M.D., MacLow, M.-M. and Rosen, A.: 2003, *MNRAS*, in press.

COSMIC RAYS AND GALACTIC WINDS

ERNST A. DORFI
Institut für Astronomie der Universität Wien, Türkenschanzstr. 17, A-1180 Wien, Austria

Abstract. Repeated SN-explosion provide large amounts of thermal energy as well as energetic particles through a 1. order Fermi-process. Both effects together with the generation of Alfvén-waves are considered to drive a large scale outflow from a galaxy. These so-called galactic winds transport stellar material enriched by heavy elements into the intergalactic space explaining also the large amount of metals found in the intergalactic gas. The present contribution is focused on time-dependent effects which originate from galactic winds driven by a star burst activity. Shock waves travelling through the galactic wind and radiative cooling within the expanding plasma lead to complex flow structures. Depending e.g. on the SFR of the galaxy galactic winds can remove almost all ISM into the galactic halo and therefore cease a subsequent star formation.

Keywords: cosmic rays, galactic winds, SN-explosions, starburst

1. Introduction

Comparing the energy densities in our galaxy it becomes evident that cosmic rays have to play a vital role in the overall dynamics of the interstellar medium. The heating and ionization of interstellar clouds as well as the production of gamma-rays (above 100 MeV) through collisions and subsequent π^0-decay reveal such a close interaction. In discussing the production of galactic cosmic rays we assume evolving SNR's to be the main source of the observed energetic particles (e.g. Blandford and Ostriker, 1980). In the context of SN-rates occurring within a galaxy we can question how much of the SN-explosion energy has to be converted into cosmic rays to explain their average properties. From gyration radii as well as radiactive ^{10}Be one can estimate the typical galactic confinement time $\tau_{\rm CR} \simeq 2 \cdot 10^7$ yr (e.g. Simpson and Garcia-Munoz, 1988). Taking e.g a mean storage volume of the hot interstellar gas to be equal to the volume occupied by cosmic rays $V_{\rm gal,CR} \simeq 10^{67} {\rm cm}^{-3}$ a galactic CR luminosity $\bar{L}_{\rm gal,CR}$ is calculated from a typical CR energy density of $u_{\rm CR} \simeq 1\,{\rm eV\,cm}^{-3}$, i.e.

$$\bar{L}_{\rm gal,CR} \simeq \frac{u_{\rm CR}}{\tau_{\rm CR}} V_{\rm gal,CR} \simeq 4 \cdot 10^{40}\,{\rm erg\,s}^{-1} \simeq 0.1\,\tau_{\rm SN}\,E_{\rm SN}\,, \qquad (1)$$

where the last expression is based on a SN-rate of $\tau_{\rm SN} = 1/30\,{\rm yr}^{-1}$ with an average explosion energy of $E_{\rm SN} = 10^{51}$ erg. More elaborate estimates which depend on the energy-dependent galactic scale time and the details of the source spectrum can be found in Drury et al. (1989).

Astrophysics and Space Science **289:** 337–346, 2004.

According to the standard picture of particle acceleration by a 1.-order Fermi process occurring at shock waves the particles gain energy by multiple reflections between the up-stream and down-stream regions (Axford et al., 1977; Krymsky, 1977; Bell, 1978a,b; Blandford and Ostriker, 1978). However, the size as well as the lifetime of the shock waves limit the maximum energy available to these cosmic rays. Assuming a typical magnetic field of 5 μG the current estimates based on SNR models lead to particles energies less than 10^{15} eV/nuc (e.g. Axford, 1981), i.e. the population of the so-called galactic cosmic rays can be explained by this process. However, up to now the observational evidence for shock acceleration is marginal and only in the case of SN 1006 the X-ray data (Koyama et al., 1995) allow such interpretations.

In a number of observed edge-on galaxies the increasing power of modern telescopes reveals the existence of large amounts of material above the galactic plane. In the case of NGC 1482, a bright SA0 IR-galaxy, the observations of ionized Nitrogen show an outflow up to distances of at least 1.5 kpc with velocities around 250 km/s determined by spectroscopic studies (Veilleux and Rupke, 2002). The kinetic energy contained in this outflow can be estimated to be at least $E_{\rm kin} \simeq 2 \cdot 10^{53}$ erg, which can be attributed to the collective effect of several SN-explosions. Such large scale outflows, also called galactic winds, seem to be a common phenomenon for a large variety of normal spiral galaxies. Already at early epochs of galactic evolution at a redshift of $z = 5.19$ the observed Lyα-profiles (Dawson et al., 2002) can be interpreted by an outflow of hot gas with velocities around 300 km/s. The line profile can be fitted by the leaving and ascending parts of a large scale wind.

Summarizing these few examples as well as e.g. recent X-ray (e.g. Wang et al., 2001; Pietsch et al., 2001) observations of nearby edge-on galaxies we emphasize that normal galaxies can undergo phases where the combined effects of several SNe generate not only superbubbles but these superbubbles break out into the intergalactic space. Another piece of evidence comes from the observations of metal-enriched intercluster gas where such galactic winds provide a natural explanation for the deposition of heavy elements from the exploded SNe.

2. Cosmic Rays and Galactic Winds

The results presented in the next sections are obtained from numerical simulations adopting a simple 1-dimensional flux-tube geometry. The area of the flux tube has to be prescribed, e.g.

$$A(z) = A_0 \left[1 + \left(\frac{z}{z_0} \right)^2 \right], \tag{2}$$

where $z_0 \simeq 15$ kpc defines a typical scale above which the flux tube opens towards a more spherical shape. For small distances $z \ll z_0$ a planar flow is established.

The other physical variables depend only on distance z from the galactic plane and on time t. The physical properties of the models are described in detail elsewhere (Breitschwerdt et al., 1991; Dorfi and Breitschwerdt, 2003). Since the energetic particles are accelerated at shock waves and can diffuse through the interstellar space the energy equation of cosmic rays contains a mean diffusion coefficient $\bar{\kappa}$ averaged over the particle distribution function (e.g. Drury, 1983). According to cosmic ray propagation models (e.g. Ginzburg and Ptuskin, 1985) this diffusion coefficient is estimated to be of the order of $10^{29}\,\mathrm{cm\,s^{-1}}$. For galactic winds the inner boundary condition located at 1 kpc is modified to yield the total outward directed flux

$$F_{\mathrm{c},0} = \frac{\gamma_\mathrm{c}}{\gamma_\mathrm{c}-1}(u_0 + v_\mathrm{A})P_{\mathrm{c},0} - \frac{\bar{\kappa}}{\gamma_\mathrm{c}-1}\left.\frac{\partial P_\mathrm{c}}{\partial z}\right|_0 , \tag{3}$$

where u_0 and v_A define the gas velocity and the Alfvén velocity at the inner boundary, respectively, $\gamma_\mathrm{c} = 4/3$ denotes the adiabatic index of the cosmic rays. Since we consider only outgoing magnetic waves with Alfvén velocity v_A the advection of the cosmic rays is linked to the combined speed $u_0 + v_\mathrm{A}$.

Figure 1 shows a typical flow pattern of a galactic wind generated by the combined effect of thermal pressure, cosmic ray pressure and wave pressure. The galactic gravitational potential is calculated for a flux tube situated at a galactocentric distance of $R_0 = 10\,\mathrm{kpc}$. The cosmic ray flux at the inner boundary $F_{\mathrm{c},0}$ is kept constant. According to eq. (3) an increasing cosmic ray diffusion coefficient $\bar{\kappa}$ reduces the innermost gas outflow velocity u_0 and thereby also the mass flux through the surface area A_0 given by $\dot{M} = A_0\rho_0 u_0$, where $\rho_0 = 1.67\cdot10^{-27}\,\mathrm{g\,cm^{-3}}$ denotes the fixed gas density at this reference level.

As seen in Figure 1 the increase of the cosmic ray diffusion coefficient $\bar{\kappa}$ results in higher final wind velocities. The cosmic rays can flow towards the extragalactic space without being so tightly coupled to the thermal plasma. As depicted in Figure 1d the diffusive effect changes the ratio of the gas pressure to the cosmic ray pressure at large distances. Since the basic scale is given by the diffusion coefficient and a typical flow velocity U through $L \simeq \bar{\kappa}/U$, this spatial scale is also increasing which leads to a growing overall acceleration region for the whole galactic wind. The final wind velocity changes from 320 km/s for the case without cosmic ray diffusion up to 1030 km/s in the case of $\bar{\kappa} = 10^{30}\,\mathrm{cm^2\,s^{-1}}$. Correspondingly, the velocity at the inner boundary decreases from 9.7 km/s to 2.5 km/s reducing the mass loss by a factor of 4. The critical point, where the flow becomes supersonic, decreases from 36 kpc to 27 kpc and typical flow times up to a distance of 1 Mpc are in the range of 10^9 years. In general we can state that diffusion of cosmic rays leads to faster but thinner galactic winds.

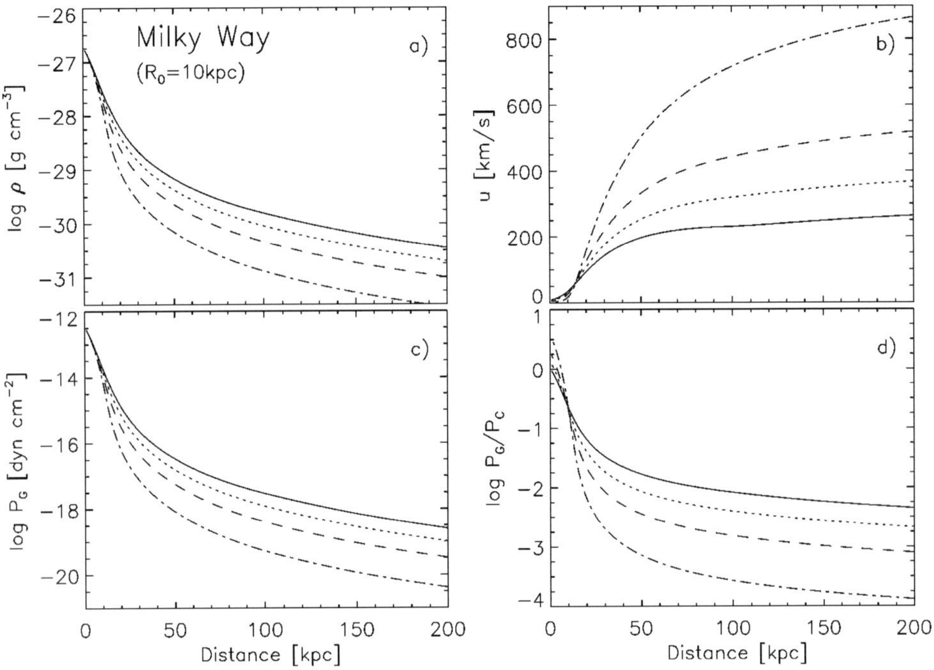

Figure 1. A time asymptotic galactic wind solution within a flux tube for parameters corresponding to our milky way. The radial structure plotted for the innermost 200 kpc is parameterized by different cosmic ray diffusion coefficients $\bar{\kappa}$. The case $\bar{\kappa} = 10^{27}\,\mathrm{cm}^2\,\mathrm{s}^{-1}$ (full line) is almost identical to $\bar{\kappa} = 0$, dotted line: $\bar{\kappa} = 10^{29}\,\mathrm{cm}^2\,\mathrm{s}^{-1}$, dashed line: $\bar{\kappa} = 3 \cdot 10^{29}\,\mathrm{cm}^2\,\mathrm{s}^{-1}$, dashed-dotted line: $\bar{\kappa} = 10^{30}\,\mathrm{cm}^2\,\mathrm{s}^{-1}$. a) gas density, b) gas velocity, c) thermal gas pressure, d) ratio of gas pressure to cosmic ray pressure.

3. Galactic Winds from Starburst Galaxies

3.1. Simple Flow Structures

In the following section we discuss a galactic wind originating from a star burst region. As a particular example we take typical parameters corresponding to the innermost 700 pc of the starburst region of M82 (e.g. Muxlow et al., 1994; Lehnert et al., 1999). The inner boundary is located 100 pc above the galactic plane. The large amount of energy supplied by frequent SN-explosions allows the development of galactic outflows driven mainly by the thermal pressure gradients. Therefore cosmic rays play a minor role for the overall properties of these winds.

A number of nonlinear waves is initiated by increasing the pressures at the inner boundary by a factor of 10 on a time scale of 10^6 years. The reaction of a galactic wind is plotted in Figure 2 where a flow pattern similar to early SN-explosions develops, namely two shock waves, separated by a contact discontinuity, are traveling outwards. The velocities of these waves are given by 1700 km/s, 1900 km/s and 2100 km/s, respectively and lead to a growing separation between the forward and

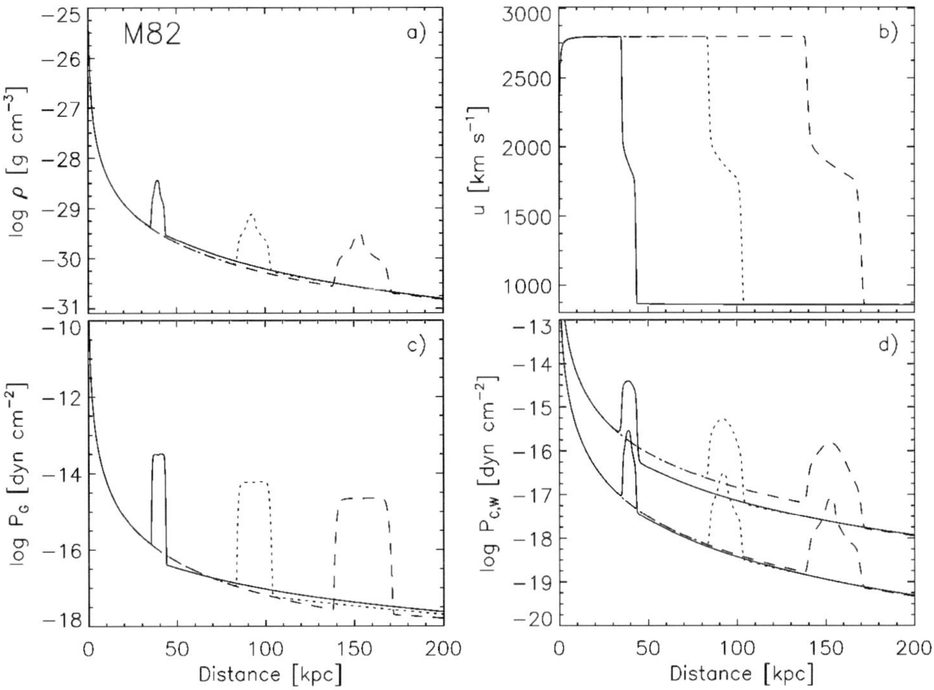

Figure 2. A time-dependent galactic wind where at the inner boundary the pressure has been increase by a factor of 10 within 10^6 years. The three groups of curves correspond to an evolution time of $2 \cdot 10^7$ years (full line), $4.8 \cdot 10^7$ years (dotted line) and $8 \cdot 10^7$ years (dashed line). The distance of the waves grows with time almost linearly.

reverse shock. The diffusion of cosmic rays lead to smoother variations of the flow quantities at larger radii compared to computations without particle acceleration. As aforementioned the dynamical influence is rather minute but these discontinuities can be used to reaccelerate a preexisting population of galactic cosmic rays as discussed in Sect. 4. From such simple simulations we conclude that variations of the inner boundary can be traced up to 200 kpc over time scale of 10^8 years.

3.2. Starburst simulations

More realistic models of galactic outflows can be obtained by adopting a STARBURST simulation (Leitherer el al., 1999) for the inner boundary. These simulations provide for a specified initial mass function a synthetic stellar evolution including the overall stellar mass loss $\dot{M}_*(t)$ as well as the energy input $\dot{E}_{\rm SN}$ from the various initial SN masses. Our innermost computational cell is specified by a simple spatially averaged model for the ongoing starburst activities, e.g. the time-dependent mass balance of the ISM is given by

$$M_{\rm ISM}(t) = M_{\rm ini} + \int_0^t \dot{M}_*(t)\,dt - \int_0^t \dot{M}_{\rm GW}(t)\,dt\,, \tag{4}$$

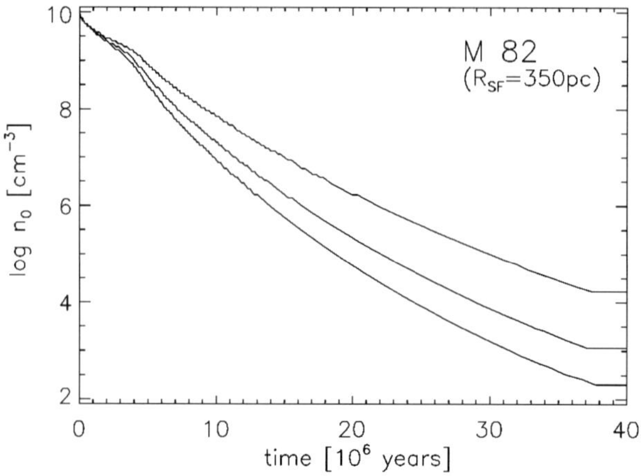

Figure 3. The time-dependent decrease of the ISM number density n_0 within a starburst region containing $9 \cdot 10^8\ M_\odot$. The first SN in this STARBURST simulation occurs at $4 \cdot 10^6$ years, the last one at $37 \cdot 10^6$ years. The curves correspond to different amounts of energy converted into the galactic wind (see text for details).

which is related to the density $\rho_0 = M_{\rm ISM}/V_{\rm SF}$ at the inner boundary through the volume of the starburst region $V_{\rm SF}$. The mass loss rate of the generated galactic wind is denoted by $\dot{M}_{\rm GW}$. The total mass of the starburst has been fixed to $M_{\rm SB} = 10^9\ M_\odot$ and we assume that during the starburst 10% of this mass can be transformed into stars, i.e. $10^8\ M_\odot$. Due to these values we have $M_{\rm ini} = 9 \cdot 10^8\ M_\odot$ and all further parameters have been chosen according to the standard values of STARBURST (see Leitherer el al., 1999 for more details).

Such a time-dependent evolution of the number density n_0 is shown in Figure 3. Since the amount of energy available to accelerate the galactic wind depends on the density of the interstellar medium (e.g. Dorfi 2000) the three curves correspond to different conversion factors. In the upper curve only $0.1\ \dot{E}_{\rm SN}$ has been used to drive the galactic wind, $0.2\ \dot{E}_{\rm SN}$ labels the middle curve and $0.3\ \dot{E}_{\rm SN}$ for the lower curve. The more SN energy is pumped into the starburst region the more effective the galactic wind removes the ISM form the starburst region. The small bump at early times ($t < 5 \cdot 10^6$ years) is caused by the mass return into the ISM due to stellar winds, i.e. the second RHS-term in eq. (4).

The cumulative effect of this mass loss is plotted in Figure 4 where in the case of $0.3\ \dot{E}_{\rm SN}$ (upper curve) almost $6 \cdot 10^8\ M_\odot$ of the ISM are ejected into the intergalactic space within $37 \cdot 10^6$ years. At the beginning of the SN activity the mass loss rate can reach up 30 $M_\odot$/year.

Equating the energy sources and sinks within the starburst region we can also calculate the mean temperature $T_{\rm SF}$ of the ISM during the ongoing starburst which drives the galactic wind. As seen in Figure 5 the mean temperature remains almost constant after the initial SNe have been deposited their energy to establish the outflow. Adopting the parameters for an initial starburst of $10^8\ M_\odot$ we end up with

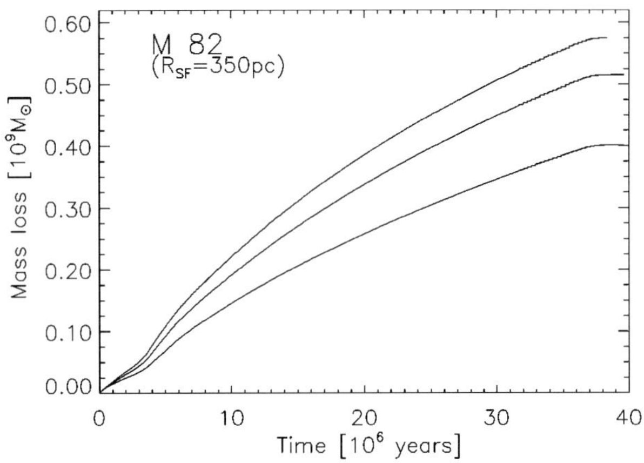

Figure 4. The integrated mass loss due to galactic winds for the starburst region in M82. The curves correspond to different SN-energy efficiencies of $0.3\,\dot{E}_{\mathrm{SN}}$, $0.2\,\dot{E}_{\mathrm{SN}}$ and $0.1\,\dot{E}_{\mathrm{SN}}$ (from top to down) available for driving the galactic wind.

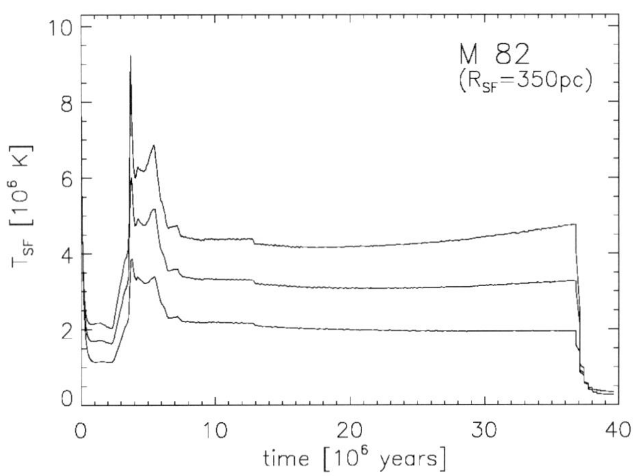

Figure 5. Evolution of the mean temperature at the bottom of the flux tube in the case of a M82 model. After the initial disturbances due to the first SN occurring in the starburst a rather smooth outflow is generated, which leads to almost constant gas temperatures until the SN-activity ends at $37 \cdot 10^6$ years.

temperatures between $2 \cdot 10^6$ K and $5 \cdot 10^6$ K over a period of about $30 \cdot 10^6$ years. With these STARBURST values the SN activity ceases after $37 \cdot 10^6$ years and the galactic winds breaks down. Clearly, ongoing SN explosions due to a more continuous stellar birth rate would result in a long lasting galactic outflow.

4. Cosmic Rays from Galactic Winds

Variations at the bottom of the flux tube introduce disturbances which propagate outwards and due to the large gradients quickly develop into shock waves. These

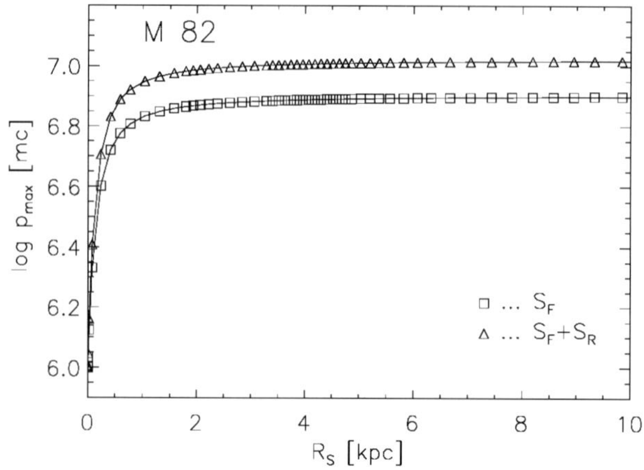

Figure 6. The particle acceleration in galactic winds boosts the cosmic rays to higher energies. The particle momentum (in units of mc) increases rapidly already within a few kpc above the galactic plane. The upper curve (open triangles) includes also particle acceleration at the reverse shock (S_R).

time-dependent waves can reaccelerate the cosmic rays advected by the flow to higher energies relative to maximum of 10^{15}eV obtainable from SNRs.

In Figure 6 the maximum momentum p_{max} has been integrated in the so-called test particle limit according to

$$\frac{dp_{max}}{dt} = \frac{p_{max}}{\tau_{acc}}, \qquad \tau_{acc} = \frac{3}{u_1 - u_2}\left(\frac{\kappa_1}{u_1} + \frac{\kappa_2}{u_2}\right), \tag{5}$$

where the diffusive acceleration time τ_{acc} at a shock wave is given the up-stream u_1, κ_1 and down-stream values u_2, κ_2. The diffusion coefficients $\kappa = \kappa(p)$ depend on the particle momentum p and can be estimated from the Bohm-limit (e.g. Drury, 1983). For the shock waves plotted in Figure 2 the integration of eq. (5) leads to a simple asymptotic expression with a constant A

$$p_{max} = p_0 + A\left(\frac{1}{t_0} - \frac{1}{t}\right), \tag{6}$$

which depends basically only at initial momentum $p(t_0) = p_0$ at some time t_0. More details can be found in Dorfi and Breitschwerdt (2003).

The typical shock speeds are around $u_s \simeq 2000\,\mathrm{km/s}$ and with a diffusion coefficient of $\bar{\kappa} \simeq 10^{30}\,\mathrm{cm^2 s^{-1}}$ the acceleration length scale can be estimated, i.e. $\bar{\kappa}/u_s \simeq 1.6\,\mathrm{kpc}$. The corresponding acceleration time is determined through $\tau_{acc} \simeq \bar{\kappa}/u_s^2 \simeq 8 \cdot 10^5$ years. Hence, after a time of about 10^6 years and within a distance of less than about 3 kpc the particles have reached their final energies and can diffuse also back into the galactic plane. Both particle populations, the galactic cosmic rays below 10^{15}eV accelerated directly at the SNR shock waves as well as the ones energized by the shock waves travelling through the galactic winds are produced by the same physical mechanism, namely SN-explosions.

5. Discussion and Conclusions

The data from STARBURST simulations, i.e. the stellar mass loss and energy input from stellar winds and SN explosions, are used as an inner boundary for the galactic wind models. By such computations the star forming properties of the galaxy are coupled to the large scale outflow and enrichment of the intergalactic medium by heavy elements. Since the amount of SN energy available to drive the wind depends on the mean ISM density, models with different SN energy inputs have been discussed. If due to the development of superbubbles up to 60% of E_{SN} can be used to generate a galactic wind, almost all thermal gas is removed from a star burst region thereby self-regulating the history of star formation. The lower gravitational potential in smaller galaxies enables such outflows more easily as suggested for the case of dwarf galaxies.

Repeated SN explosions from an underlying OB-cluster produce shock waves propagating through the galactic wind. Such large scale structures can reaccelerate the galactic cosmic rays already so near to the galactic plane that they can diffuse back into the galaxy. Therefore is seems plausible that both populations of cosmic rays can be joined to a common spectrum at the so-called *cosmic ray knee* (e.g. Sokolsky, 1989) and even particles beyond 10^{15}eV can be produced by time-dependent galactic winds in the close vicinity of the galactic plane.

Acknowledgements

I would like to thank D. Breitschwerdt (MPE, Garching) for many enlightening discussions on galactic winds.

References

Axford, W.I., Leer, E. and Skadron, G.: 1977, *Proc. 15th Int. Cosmic Ray Conf.* (Plodiv) **11**, 132.
Axford, W.I.: 1981, *Proc. Int. School and Workshop on Plasma Astrophysics*, Varenna, ESA SP-161, 425.
Bell, A.R.: 1978a, *MNRAS* **182**, 147.
Bell, A.R.: 1978b, *MNRAS* **182**, 443.
Blandford, R.D. and Ostriker, J.P.: 1978, *ApJ* **221**, L29.
Blandford, R.D. and Ostriker, J.P.: 1980, *ApJ* **237**, 793.
Breitschwerdt, D., McKenzie, J.F. and Völk, H.J.: *A&A* **245**, 79.
Dawson, S., Spinrad, H., Stern, D., Dey, A., van Breugel, W., de Vries, W. and Reuland, M.: 2002, *ApJ* **570**, 92.
Dorfi, E.A.: 2000, *Ap&SS* **272**, 227.
Dorfi, E.A. and Breitschwerdt, D.: 2003, *A&A*, submitted.
Drury, L.O'C.: 1983, *Rep. Prog. Phys.* **46**, 973.
Drury, L.O'C., Markiewicz, W. and Völk, H.J.: 1989, *A&A* **225**, 179.
Ginzburg, V.L. and Ptuskin, V.S.: 1985, *Sov. Sci. Rev. E. Ap&SS* **4**, 171.

Koyama, K., Petre, R., Gotthelf, E.V., Hwang, U., Matsuura, M., Ozaki, M. and Holt, S.S.: 1995, *Nature* **378**, 255.
Krymsky, G.F.: 1977, *Dokl. Nauk. SSR* **234**, 1306. (Engl. trans. Sov. Phys. Dokl. 23, 327).
Leitherer, C., Schaerer, D., Goldader, J.D., Delgado, R.M.G., Robert, C., Kune, D.F., de Mello, D.F., Devost, D. and Heckman, T.M.: 1999, *ApJS* **123**, 3.
Lehnert, M.D., Heckman, T.M. and Weaver, K.A.: 1999, *ApJ* **523**, 575.
McKee, C.F. and Ostriker, J.P.: 1977, *ApJ* **218**, 148.
Muxlow, T.W.B., Pedlar, A., Wilkinson, P.N., Axon, D.J., Sanders, E.M. and de Bruyn, A.G.: *MNRAS* **266**, 455.
Pietsch, W., Roberts, T.P., Sako, M., Freyberg, M.J., Read, A.M., Borozdin, K.N., Branduardi-Raymont, G., Cappi, M., Ehle, M., Ferrando, Kahn, P.S.M., Ponman,T. J., Ptak, A., Shirey, R.E. and Ward, M.: 2001, *A&A* **365**, L174.
Simpson, J.A. and Garcia-Munoz, M.: *Space Sci. Rev.* **46**, 205.
Sokolsky, P.: 1989, *Introduction to Ultrahigh Energy Cosmic Ray Physics*, Addision-Wesley Publ. Comp., Redwood City.
Veilleux, S. and Rupke, D.S.: 2002, *ApJ* **565**, L63.
Wang, Q.D., Immler, S., Walterbos, R., Lauroesch, J.T. and Breitschwerdt, D.: 2001, *ApJ* **555**, L99.

EXTRAPLANAR ISM AND IGM

DIFFUSE IONIZED GAS IN HALOS OF SPIRAL GALAXIES

RALF-JÜRGEN DETTMAR
Astronomical Institute, Ruhr-University Bochum, Germany

Abstract. The presence of gaseous halos in star forming disk galaxies is reviewed in the context of a proposed disk-halo connection of the interstellar medium (ISM). Results from a new survey for H^+ halos of edge-on galaxies are presented. The data confirm that the presence of diffuse ionized gas (DIG) in the disk-halo interface of spiral galaxies is related to star formation processes in the underlying disk. A discussion allows us to establish a minimum energy release per unit area that is required to start the disk-halo mass exchange. By comparing some recent observational results for diagnostic emission lines with model predictions from photoionization we demonstrate that the origin and excitation of the ionized halo gas is still not completely understood and that a discussion gives important constraints for models of the ISM. In comparison with similar findings for the Milky Way the need for an additional heating source is established. Special emphasis is given to some recent developments. In particular, new kinematical information for the DIG layer in NGC 5775 from ESO/VLT long-slit spectra is discussed in connection with the magnetic field structure in the halo of this object as deduced from VLA radio-continuum polarization data (Tüllmann et al., 2001). Finally, the rôle of dust for the physical processes in the disk-halo interface is briefly addressed.

Keywords: interstellar medium, ionized gas, galactic halos, disk-halo interaction

1. Introduction

The various contributions for these proceedings discuss the complex interrelation of the different phases in the interstellar medium (ISM). The resulting picture of the ISM would be incomplete without addressing the importance of the widespread diffuse ionized medium and, more specifically, without discussing its extra-planar or halo component since it is this component that may let us directly witness the influence of the energy released by young and massive stars on the ISM and its feedback process controlling the long term evolution of the medium and thus the evolution of galaxies. This H^+ gas outside of classical HII regions is distributed in shells, loops, bubbles, filaments, and knots, as well as in a smoothly distributed, diffuse component. It is now frequently called *Diffuse Ionized Gas* (DIG) or *Warm Ionized Medium* (WIM) and can be identified with the *Reynolds-layer* of the Milky Way. Its existence and energy balance is closely related to the UV radiation field and the radiative transfer in galactic disks, e.g., the question how leaky HII regions are and how porous the ISM is. This specific question was addressed in several papers (Lehnert and Heckman, 1994; Ferguson et al., 1996; Hoopes et al., 1996) which all conclude that DIG contributes 25%–50% of the Hα flux in spiral galaxies. The general finding that DIG exists in the halos (or at least in the disk-halo

Astrophysics and Space Science **289:** 349–357, 2004.

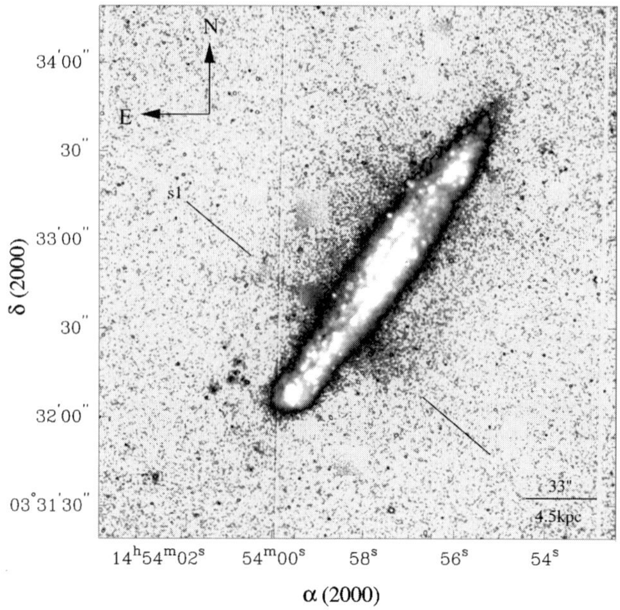

Figure 1. The galaxy NGC5775 exhibits a bright and thick layer of extra-planar H^+ with spurs that can be traced out to 10 kpc above the mid-plane (adopted from Tüllmann et al., 2001). The position of a VLT/FORS1 long-slit spectrum discussed in this paper is indicated.

interface)of many spiral galaxies corroborates scenarios of a large scale exchange of matter between the ISM in galactic disks and their halos driven by the energy input of star forming regions in the disk (e.g., Dahlem, 1997; Dettmar, 1998).

From this general idea the concept of the disk-halo interaction emerged in which the disk medium can connect to the halo. This is described in different model approaches in terms of galactic fountains, chimneys, or super-bubble-outbreak by theory. Models for the gas transport from the disk into the halo are proposed in several theoretical approaches such as the chimney scenario (Norman and Ikeuchi 1989), galactic fountains or superbubble blow-outs (e.g., McLow and Ferrara, 1999).

Among the different components of the ISM found in the halos of spiral galaxies – from cold HI to X-ray coronae and even molecular gas – the H^+ with a scale height of typically 1 kpc has some observational relevance since it is relatively easily observed in the optical (Dettmar 1992, 1998). Therefore, imaging in and spectroscopy of optical emission lines allow us to study the distribution and excitation of this Diffuse Ionized Gas with a spatial resolution not achievable for other phases of the ISM in external galaxies. To demonstrate the distribution of DIG in halos of galaxies we reproduce in Figure 1 a Hα-image of one of our target galaxies (Rossa and Dettmar, 2000).

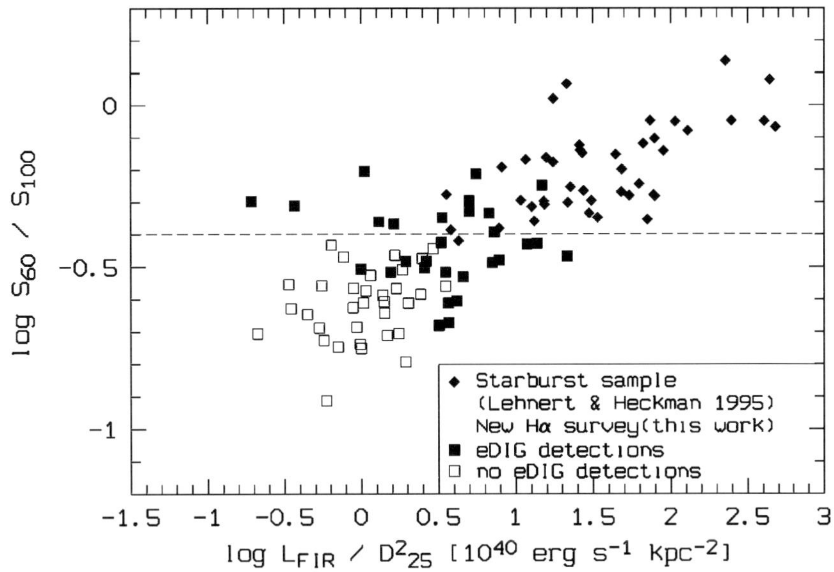

Figure 2. To demonstrate the dependence of DIG in halos of galaxies on the normalized star formation rate (per unit area) the ratio of the FIR fluxes at 60μm and 100μm (S_{60}/S_{100}) is plotted versus the ratio of the FIR luminosity to the optical galaxy diameter squared of the 25th mag in units of 10^{40} erg s^{-1} kpc^{-2}. The filled lozenges denote the starburst galaxy sample studied by Lehnert and Heckman (1995). Squares indicate our new observations with filled symbols for detections of H^+ gas in the halo and open squares for non-detections (from Rossa and Dettmar, 2003).

In the following we present results of an ongoing investigation of extra-planar DIG (eDIG) in edge-on galaxies. Based on a quantitative Hα survey and emission line spectroscopy we are trying to answer questions such as: How common are galactic gaseous halos? What is the minimal SFR per unit area which triggers gaseous galactic outflows? Is there a correlation between gaseous halos and other halo components such as radio continuum 'thick disks' that are seen in several edge–on galaxies? Is the magnetic field structure in galaxies related to outflows? What are the ionization and excitation processes for DIG at high galactic latitudes?

2. DIG in Halos and Its Relation to Star Formation in the Disk

A halo component of DIG for external galaxies was first discovered in NGC 891 a bit more than a decade ago (Dettmar, 1990; Rand et al., 1990). A few small samples have been studied since then (e.g., Pildis et al., 1994; Rand, 1996; Rossa and Dettmar, 2000), however, more complete and larger samples would help to clearly demonstrate the suggested correlations of halo gas properties with the star formation rate (SFR) in the disk. In particular, a broader range in SFR has to be covered since up to now emphasis was given to galaxies with high SFR or even starburst galaxies. An attempt to relate the presence of DIG in halos of spiral galaxies to their far infrared (FIR) properties as a measure of the SFR is shown in Figure 2. Here a sample of galaxies was chosen such that it supplements the previously studied 'starburst' objects at the low activity end. If the starformation rate per unit area is low as determined by the FIR luminosity normalized to the disk surface the presence of halo DIG indeed diminishes. The FIR luminosity per unit area therefore seems to be a promising indicator for the presence of halo H^+ gas and a minimum SFR per area seems to be required to drive the disk-halo interaction (Rossa and Dettmar, 2003).

3. The Ionization and Excitation of Halo DIG

It has been shown in the past (Domgörgen and Mathis, 1994; Golla et al., 1996; Tüllmann and Dettmar, 2000) that the emission line characteristics of DIG in galactic halos is not explained by photoionization in a straight forward manner. The detailed ionization structure rather remains a major problem for pure photoionization models and several recent studies for extra-planar DIG in the Milky Way and NGC 891 (Reynolds et al., 1999; Rand, 1997, 1998) therefore claimed a need for secondary heating and/or ionizing sources. Ideas for such supplemental mechanisms include shock ionization, turbulent dissipation, and magnetic reconnection. In order to examine possible ionization and/or excitation mechanisms of the eDIG we compared spectroscopically observed emission line ratios and diagnostic diagrams with model predictions (photoionization codes of Mathis 1986 and Domgörgen and Mathis, 1994) for a small sample of edge-on galaxies (Tüllmann and Dettmar, 2000). A short discussion of further work in this direction is given by Elwert and Dettmar (2003) in these proceedings. The relation between the [SII]/Hα and [NII]/Hα ratios which are both raising with increasing distance from the galactic mid-plane is also observed in the Milky Way. Here it is explained as an effect of an increase of temperature at higher latitudes (Reynolds et al., 1999). This behavior of emission lines for halo DIG is also observed in NGC 5775 as shown in Figure 3. The very deep spectrum of this galaxy, tracing diagnostic emission lines out to 10 kpc from the midplane, was observed with the FORS1 instrument at the ESO VLT-UT1. From the observed emission line ratios the temperature profile in the upper panel of the figure was deduced showing the same general behavior of an increasing temperature as it is discussed for the Milky Way (see above).

4. Kinematics of DIG in the Halo of NGC 5775

A particularly striking result is related to the kinematical information regarding the DIG layer which is extracted from the same deep VLT spectrum discussed in the

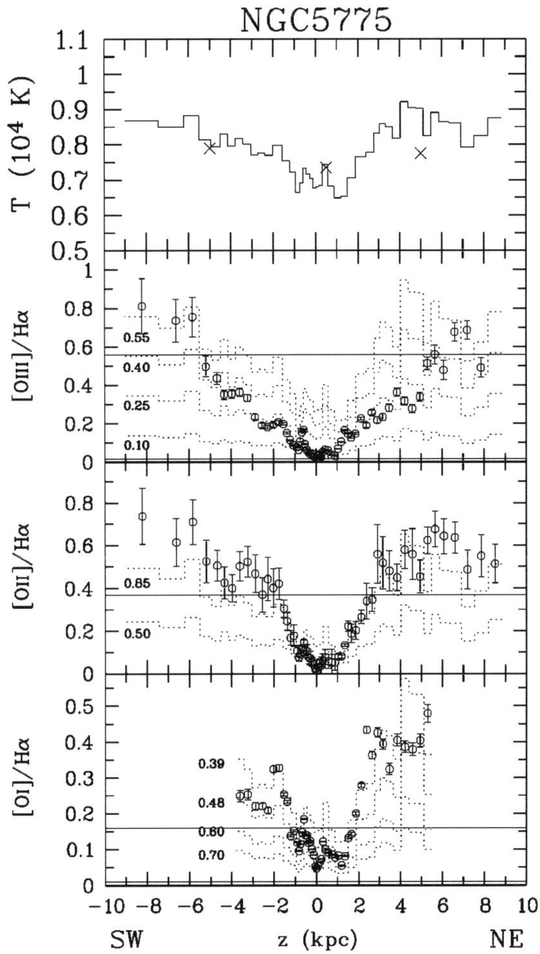

Figure 3. Emission line rations and temperature profile (uppermost panel) of the DIG in NGC 5775 along the slit marked in Figure 1 (adopted from Tüllmann et al., 2001). Horizontal lines in the lower panels indicate the range of emission line ratios allowed by photoionization models.

previous section. As shown in Figure 4 the velocity of the DIG at a galactocentric distance of $\approx$4 kpc is dropping from the rotational velocity in the disk plane to systemic at a distance of $\sim$10 kpc above the plane. This is most likely not just a local perturbance in the velocity field since similar spectra of NGC 5775 by Rand (2000) show the same phenomenon for two different slit positions. A model to explain such a velocity field reaching systemic velocity at large distances is currently not available. Observations of UV absorption line systems by Côtè et al. (2000), however, show a similar phenomenon and therefore ask for a more general solution. Theoretical attempts may require a reconsideration of all relevant processes such as the discussion of forces by magnetic tension as introduced by, e.g., Benjamin

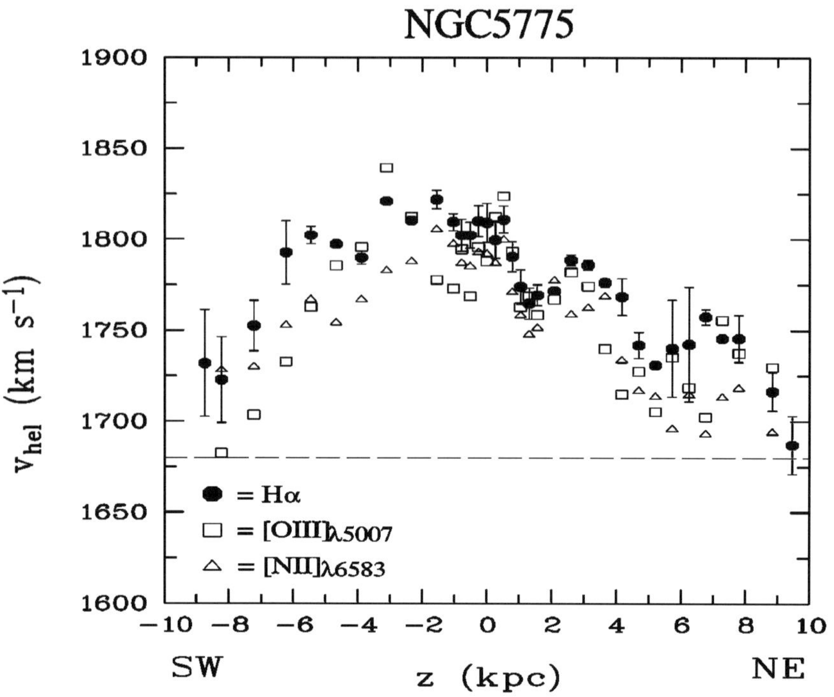

Figure 4. DIG kinematics along slit perpendicular to major axis. The velocity drops from rotational velocity at the midplane to systemic in the halo (adopted from Tüllmann et al. 2001).

(2000) in response to the observed slight drop in rotational velocity for halo gas in NGC 891 by Rand (1998) and Swaters et al. (1997).

5. Halo DIG and the Structure of Galactic Magnetic Fields

The presence of a disk-halo interaction in a galaxy traced by DIG in the halo is also supported by the frequently found simultaneous presence of other ISM constituents such as X-ray emitting plasmas or cosmic ray electrons observed in the radio continuum. The break-out condition for cosmic rays as a function of star formation rate per unit area is discussed by Dahlem et al. (1995) and a compilation of observations of edge-on galaxies in Hα, radio continuum, and X-rays is given in Dettmar (1998).

The previously discussed case of NGC 5775 also shows a prominent radio continuum halo (as well as a X-ray halo and extra-planar CO emission; Lee et al., 2001) and a study of its polarization characteristics with the VLA reveals even an highly structured large scale magnetic field with field lines opening up into the halo (Figure 5). It is well possible that the velocity structure of the halo gas described above in Sect. 4 is related to the origin of the magnetic field since the theory of

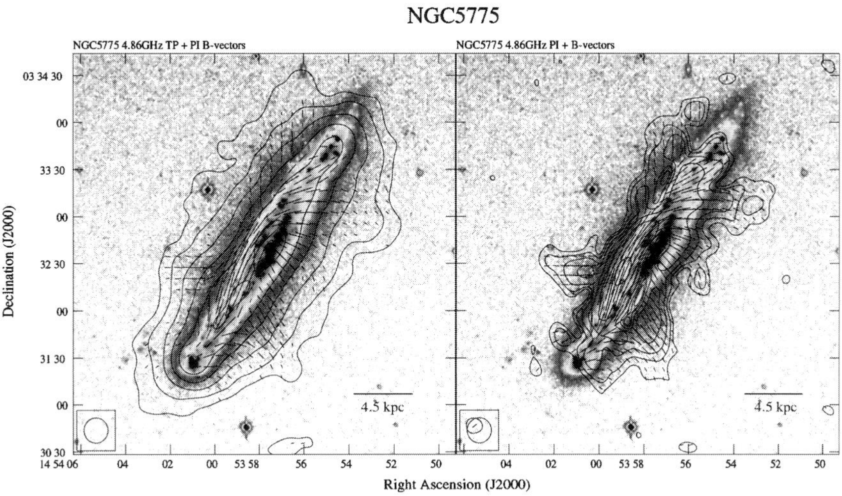

Figure 5. VLA radiocontinuum map of NGC 5775 at 6cm; the left panel gives total power while the right panel represents polarized intensity. The bars represent the magnetic field direction. (Adopted from Tüllmann et al. (2001).)

galactic dynamos relates the z-component of the magnetic field and a shear caused by the gradient dv/dz to the generation of an azimuthal magnetic component.

6. Radiatively Driven Dust in the Halo

Since the star formation process in the disks will also locally enhance the radiation field it was suggested more than a decade ago that dust could be elevated by radiation pressure from the disk into the halo (Franco et al., 1991). Ferrara (1997) modeled the evolution in time of the dust distribution under the influence of the radiation field caused by an OB association and showed that the dust would be dispersed into a large volume. This, however, is in contrast to the observed dust distribution in several galaxies. Sofue et al. (1994) demonstrate that the dust in the disk of NGC 253 is highly structured in large scale arches much resembling solar prominences in shape. Also the dust distribution in the edge-on galaxy NGC 891 studied, e.g., by Howk and Savage (1997) is organized in long filaments reaching from the disk into the halo. These filaments end in a level of dust sheets high above the disk plane. The case of NGC 891 is also studied by Rossa et al. (2002) with the high resolution imaging capability of HST and in Figure 6 we present a first image of the dust distribution from these WFPC observations. The possible influence of magnetic fields on the charged dust particles as an additional process to structure the dust distribution in galactic halos was put forward by Shchekinov et al. (1999). In the presence of large scale organized magnetic fields in galactic halos as demonstrated above for the case of NGC 5775 such processes have to be discussed in more detail in the future. It should be mentioned here that in such an environment also the dipole radiation of spinning dust (Ferrara and Dettmar, 1994; Draine and Lazarian, 1999) could contribute to the radioemission from galaxies at GHz frequencies.

7. Summary

Gaseous DIG halos are found in galaxies with sufficient SFR per unit area. Typically these layers of extraplanar DIG can be traced out to distances of z $\leq$1–2 kpc – sometimes even up to 5 kpc or more – from the mid-plane of the disk. The emission line ratios indicate an increase in the electron temperature of the DIG with increasing hight above the midplane. This behavior is similar to the Reynolds layer in the Milky Way and a heating mechanism in addition to photoionization is currently being discussed. These H^+ halos seem to be associated with halos of cosmic rays and X-ray plasma as expected if caused by the disk-halo interaction of the ISM. The polarized radiocontinuum emissions indicates an ordered magnetic field in the halo. The peculiar velocity structure of the halo gas in the case of NGC 5775 is possibly related to this ordered magnetic field and also the highly structure dust distribution in the halos of some galaxies could be influenced by magnetic fields due the coupling of charged dust to both the radiation and magnetic fields.

Acknowledgements

The author wishes to thank his students T. Elwert, J. Rossa, and R. Tüllmann as well as numerous colleagues involved in the presented work for a fruitful collaboration. This work is supported by Deutsche Forschungsgemeinschaft (DFG) and Deutsches Zentrum für Luft- und Raumfahrt (DLR). The discussed observational material was obtained at ESO (Chile), DSAZ/Calar Alto (Spain), and the VLA/NRAO (USA).

References

Benjamin, R.: 2000, *Rev. Mex. A and A* **9**, 256.
Côtè, S. et al.: 2000, *ASP Conf. Ser.* Vol. **215**, p. 287.
Dahlem, M.: 1997, *PASP* **109**, 1298.
Dahlem, M., Liesenfeld, U. and Golla, G.: 1995 *ApJ* **444**, 119.
Dettmar, R.-J.: 1990, *A&A* **232**, L15.
Dettmar, R.-J.: 1992: *Fund. Cosmic Phys.* **15**, 143.
Dettmar, R.-J.: 1998: in: D. Breidtschwerdt et al. (eds.), *The Local Bubble and Beyond*, Springer LNP 506, p. 527.
Domgörgen, H. and Mathis, J.S.: 1994, *ApJ* **428**, 647.

Draine, B.T. and Lazarian, A.: 1999, *ApJ* **512**, 740.
Elwert, T. and Dettmar, R.-J.: 2003, these proceedings.
Ferguson, A.M.N., Wyse, R.F.G. and Gallagher, J.S.: 1996, *AJ* **112**, 256.
Ferrara, A.: 1997, in: H. Lesch et al. (eds.), *The Physics of Galactic Halos*, Akademie Verlag, p. 189.
Ferrara, A. and Dettmar, R.-J.: 1994, *ApJ* **427**, 155.
Franco, J. et al.: 1991, *ApJ* **366**, 443.
Golla, G., Dettmar, R.-J., Domgörgen, H.: 1996, *A&A* **313**, 439.
Hoopes, C.G., Walterbos, R.A.M. and Greenawalt, B.E.: 1996, *AJ* **112**, 1429.
Howk, J.C. and Savage, B.D.: 1997, *AJ* **114**, 2463.
Lee et al.: 2001, *A&A* **377**, 759.
Lehnert, M.D. and Heckman, T.M.: 1994 *ApJ* **426**, L27.
Lehnert, M.D. and Heckman, T.M.: 1995, *ApJS* **97**, 89.
MacLow, M.M. and Ferrara, A.: 1999, *ApJ* **513**, 142.
Mathis, J.S.: 1986, *ApJ* **301**, 423.
Norman, C.A. and Ikeuchi, S.: 1989, *ApJ* **345**, 372.
Pildis, R.A., Bregman, J.N. and Schombert, J.M.: 1994, *ApJ* **427**, 160.
Rand, R.J.: 1996, *ApJ* **462**, 712.
Rand, R.J.: 1997, *ApJ* **474**, 129.
Rand, R.J.: 1998, *ApJ* **501**, 137.
Rand, R.J.: 2000, *ApJ* **537**, L13.
Rand, R.J., Kulkarni, S.R. and Hester, J.J.: 1990, *ApJ* **352**, L1.
Reynolds, R.J., Haffner, L.M. and Tufte, S.L.: 1999, *ApJ* **525**, L21.
Rossa, J. and Dettmar, R.-J.: 2000, *A&A* **359**, 433.
Rossa, J. and Dettmar, R.-J.: 2003, (submitted).
Rossa, J. et al.: 2003, in prep.
Shchekinov, Y.A. and Dettmar, R.-J.: 1999, *Proceedings 26th ICRC, session OG*, Vol. **4**, p. 298.
Sofue, Y., Wakamatsu, K.-I. and Malin, D.F.: 1994, *AJ* **108**, 2102.
Swaters, R.A., Sancisi, R. and van der Hulst, J.M.: 1997, *ApJ* **491**, 140.
Tüllmann, R. and Dettmar, R.-J.: 2000, *A&A* **362**, 119.
Tüllmann, R. et al.: 2001, *A&A* **364**, L36.

MAGNETIC RECONNECTION AND EXTRAPLANAR DIFFUSE IONIZED GAS

GUIDO T. BIRK and HARALD LESCH
Institute for Astronomy and Astrophysics, University of Munich, Germany
also: Centre of Interdisciplinary Plasma Science, Garching, Germany

THOMAS NEUKIRCH
School of Mathematics and Statistics, University of St. Andrews, Scotland, UK

Abstract. Photoionization by stars cannot be the only cause for the diffuse ionized gas component in the distant halos of galaxies. Quasistationary localized magnetic reconnection can efficiently contribute to the ionization of the extraplanar halo gas. The gas is heated and re-ionized in numerous current filaments fast enough to compensate recombination. The necessary magnetic energy is provided by the disk activity.

Keywords: diffuse ionized gas, galactic halo, magnetic reconnection

1. Re-Heating and Ionization by Magnetic Reconnection

Diffuse ionized gas (DIG) is an ubiquitous phenomenon in galactic halos (Dettmar, 1992). What are the energy sources that keep the DIG ionized? It can be ruled out that photoionization by OB stars is the only energy source (Dettmar, 1992; Dettmar and Schulz, 1992). Magnetic reconnection can efficiently re-heat and ionize the interstellar medium (Birk et al., 1998) The process we have in mind can be regarded as a quasi-stationary working cycle (Figure 1). In the first step extraplanar gas cools and recombines in the magnetized halo. This results in a locally lower thermal plasma pressure and in a higher rate of electron-neutral collisions. The plasma pressure that has to balance the stresses exerted by the magnetic field is reduced which leads to a contraction of the current carrying region. The increased number of electron-neutral collisions leads to a locally reduced electrical conductivity. In this situation localized dynamical magnetic dissipation (Parker, 1994) is likely to occur. During this process due to the local violation of ideal Ohm's law magnetic field lines are reconnected and magnetic flux is dissipated. The free magnetic energy is mainly converted into heat via Ohmic dissipation and the gas is re-heated and re-ionized.

Astrophysics and Space Science **289:** 359–362, 2004.

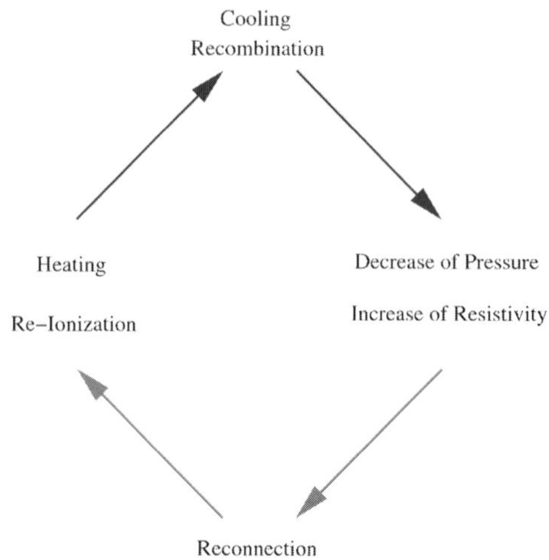

Figure 1. The quasistationary working cyle.

2. Quasistationary Heating: Plasma Density and Temperature Profiles

Let us consider a mainly hydrogen quasineutral partially ionized plasma in ionization equilibrium. The energetics are assumed to be governed by local Ohmic dissipation during the reconnection process and radiative losses caused by bremsstrahlung and recombination radiation. The stationary ionization and energy balance equations are

$$\alpha n_p^2 = \iota n_n \tag{1}$$

$$\mathbf{j}^2 = \sigma (L_{brems}^{rad} + L_{recom}^{rad}) \tag{2}$$

where n_p, n_n, $\mathbf{j}$, L_{brems}^{rad}, L_{recom}^{rad} denote the plasma and neutral gas particle density, the current density and the radiative loss functions (Huba, 1994) due to bremsstrahlung $L_{brems}^{rad} = 1.7 \cdot 10^{-25} n_p^2 T_e^{1/2}$erg s^{-1}cm^{-1} ($T_e$ is the electron temperature measured in eV) and recombination radiation $L_{recom}^{rad} = 1.7 \cdot 10^{-25} E_\infty n_p^2 T_e^{-1/2}$erg s^{-1}cm^{-1} ($E_\infty$ is the ionization energy, i.e. $E_\infty = 13.6$eV for hydrogen). The ionization frequency ι is given by $\iota = 10^{-5} n_p (T_e/E_\infty)^{1/2} \exp(-E_\infty/T_e)(6 + T_e/E_\infty)$ $E_\infty^{-3/2}$s^{-1} whereas the radiative recombination rate is $\alpha = 5.2 \cdot 10^{-14}(E_\infty/T_e)^{1/2} \times$ $(0.43{+}0.5\ln(E_\infty/T_e){+}0.5(E_\infty/T_e)^{-1/3})$cm^{3}s^{-1} (Huba, 1994). By σ we denote the collisional conductivity due to electron-neutral collisions $\sigma = 2 \cdot 10^{14} n_p e^2/m_e n_n$ $(kT_e/m_e)^{1/2}$s^{-1} where e, m_e and k are the elementary charge, the electron mass and the Boltzmann constant. The local filamentary structure of the magnetic field and the associated electrical current sheets can be expressed as (we assume the magnetic field to vary in the y-direction):

$$\mathbf{j} = 7.7 \cdot 10^{-16} \left(\frac{B}{\mu \mathrm{G}}\right) \left(\frac{\delta}{\mathrm{pc}}\right)^{-1} \cosh^{-2}\left(\frac{y}{\delta}\right) \mathbf{e}_z \text{ statamp cm}^{-2} \tag{3}$$

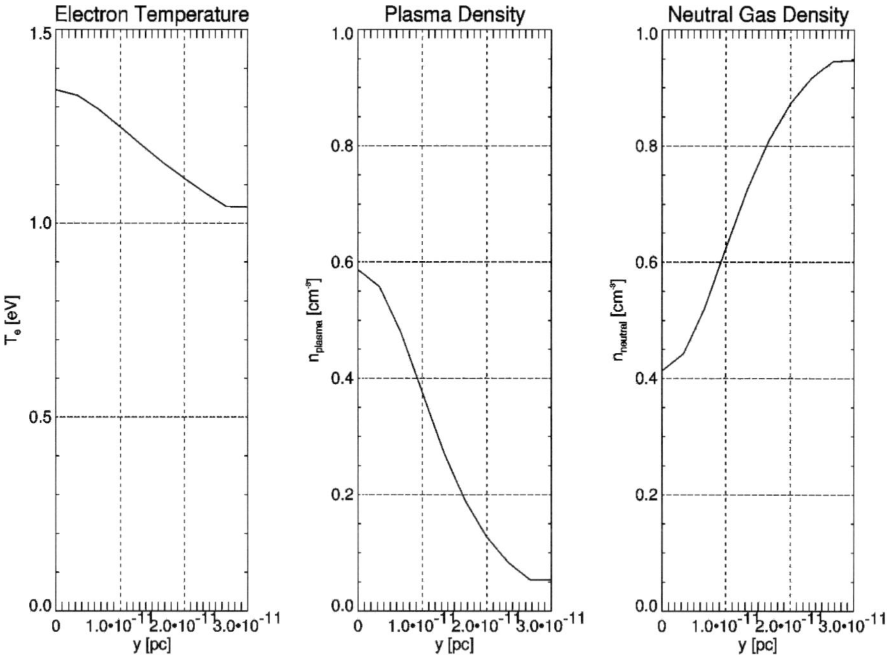

Figure 2. Spatial profiles of the electron temperature, the plasma density and the neutral gas density for $n_t = 1\text{cm}^{-3}$ and $\delta = 10^{-11}$pc.

where δ is the length scale of the inhomogeneity of the filamentary magnetic field, i.e. the half-thickness of the current layer. Thus, we have to solve for the set of non-linear algebraic equations:

$$n_p^3 + 2.7 \cdot 10^{-23} \frac{B^2(n_p - n_t)}{\delta^2 \cosh^4(y/\delta)(1 + E_\infty T_e^{-1/4})} = 0 \tag{4}$$

$$\left(0.43 + 0.5\ln\left(\frac{E_\infty}{T_e}\right) + 0.5\left(\frac{T_e}{E_\infty}\right)^{1/3}\right) n_p$$
$$-2 \cdot 10^8 \frac{T_e}{E_\infty^{5/2}} \frac{n_t - n_p}{6 + T_e/E_\infty^{1/4}} e^{-E_\infty/T_e} = 0 \tag{5}$$

The lower estimate for the magnetic field strength is about 5μG from radio data. The only free input parameters are the total particle density n_t and the half-thickness of the current sheet δ.

Figure 2 shows, for example, the spatial profiles of the electron temperature T_e, plasma density n_p and neutral gas density n_n for $n_t = 1\text{cm}^{-3}$ and $\delta = 10^{-11}$pc. The central ionization rate of $\sim$ 60% agrees well with observed values in some

Galactic DIG regions. Extensive parameter studies show (Birk et al., 1998) that magnetic reconnection may account for the existence of extraplanar ionized gas with observed densities $n_p \approx 0.01 - 0.5\text{cm}^{-3}$ provided that it takes place in filaments of the thickness of $\delta = 10^{-11} - 10^{-9}\text{pc}$. The dissipation of magnetic energy via reconnection is able to sustain a local recombination-reconnection equilibrium, if the dissipated magnetic flux is refreshed fast enough. This means that at most the magnetic dissipation rate Λ_{diss} has to equal the energy input rate Γ caused by disk activity

$$\Lambda_{\text{diss}} = \frac{B^2}{8\pi\,\tau_{\text{rec}}} f_{diss} = \Gamma = \frac{\dot{E}_{\text{disk}}}{V_{\text{diss}}} f_{in} \tag{6}$$

where f_{diss} and f_{in} denote the efficiency of magnetic field dissipation (which is about $f_{diss} = 0.1$ for the case of Petschek type reconnection) and the fraction of conversion of kinetic energy involved in the disk activity to magnetic energy. By $\dot{E}_{\text{disk}} \approx 10^{41}\text{ergs}^{-1}$ the kinetic disk luminosity is denoted and V_{diss} is the volume of energy conversion on the time scale τ_{rec} for which the recombination time $\approx 10^{13}$s gives an upper limit. For $\tau_{rec} \approx 10^{13}$s and $f_{in}/f_{diss} = 10$ one estimates $V_{diss} \approx$7kpc. In such a volume magnetic energy has to be dissipated in numerous individxsual reconnection events. The suggested scenario resembles very much the concept of coronal heating by nano-flares.

References

Birk, G.T., Lesch, H. and Neukirch, T.: 1998, Magnetic reconnection as a cause for the extraplanar diffuse ionized gas, *Mon. Not. R. Astr. Soc.* **296**, 165–172.

Dettmar, R.-J.: 1992, Extraplanar diffuse ionized gas and the disk-halo connection in spiral galaxies, *Fund. Cosm. Phys.* **15**, 143–208.

Dettmar, R.-J. and Schultz, H.: 1992, The ionization of the diffuse gas in NGC891 and the decaying dark matter theory, *Astron. Astrophys.* **254**, L25–L28.

Huba, J.D.: 1994, *NRL Plasma Formulary,* Naval Research Laboratory, Washington.

Parker, E.N.: 1994, *Spontaneous Current Sheets in Magnetic Fields*, Oxford University Press, chap. 10, 11, 12.

THE THIN LAYER OF WARM IONIZED GAS: TOWARDS A 3-D RECONSTRUCTION OF THE SPATIAL DISTRIBUTION OF HII REGIONS

ROBERTA PALADINI[1], ROD DAVIES[2] and GIANFRANCO DEZOTTI[3]
[1]*SISSA, International School for Advanced Studies, via Beirut 2-4, I-34014 Trieste, Italy*
[2]*University of Manchester, Jodrell Bank Observatory, Macclesfield - Cheshire SK11 9DL, UK*
[3]*INAF-Oss. Astro. Padova, Vicolo dell'Osservatorio 5, I-35122 Padova, Italy*

Abstract. HII regions are known to contribute to the so-called *thin layer* of the diffuse Warm Ionized Gas. In order to constrain this contribution, we reconstruct the 3-D distribution of the sources. A detailed spatial analysis of the largest up-to-date sample of HII regions is presented.

Keywords: WIM, HII regions

1. Introduction

The thin disk component of the Galactic free electron distribution is largely contributed by localized HII regions. In order to constrain this contribution, it is important to reconstruct the spatial distribution of known sources.

In a recent paper (Paladini et al., 2003, hereafter Paper I), we describe the construction of an extensive radio catalog (1442 sources) of Galactic HII regions by the combination of 24 published lists and catalogs of these objects. The final compilation consists of a Master Catalog (containing original data and corresponding errors) and a Synthetic Catalog at 2.7 GHz (which summarizes the basic information – flux density, angular diameter and V_{LSR} – for each source of the Master Catalog). The Synthetic Catalog has provided the source of information for the spatial analysis here presented.

2. 3-D Distribution of Galactic HII regions

In order to assess the level of completeness of our sample we study the derived log N – log S distribution (Figure 1).

This can be well fitted by a two-component power-law such as:

$$N(>S) \propto \begin{cases} S^{-\frac{3}{2}} & S_{max} > S > \sim 70\ Jy \\ S^{-1} & \sim 70\ Jy > S > \sim 1\ Jy \end{cases} \tag{1}$$

For $S_{max} > S >\sim 70$ Jy we are sampling well into the local spiral arm (spherical distribution approximation) while for $S_{max} > S >\sim 70$ Jy we sample into more

Astrophysics and Space Science **289:** 363–366, 2004.

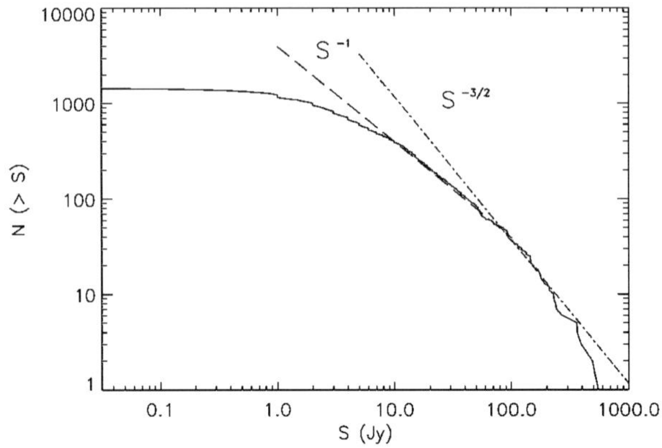

Figure 1. Cumulative counts N(> *S*) for the 2.7 GHz Synthetic Catalog.

distant regions into the Galactic disk (disk distribution approximation). Below ~ 7 Jy, the Synthetic Catalog starts missing sources which are in regions of high confusion level.

Velocities are given for ~ 800 of the 1442 HII regions in the Synthetic Catalog, corresponding to 60% of the total. Details about the kinematic data are given in Paper I. These data have been combined with the linear rotation model by Fich, Blitz and Stark 1989:

$$\Theta = (221.64 - 0.44R)\ km\ s^{-1} \tag{2}$$

to compute galactocentric (R) and solar distances (D). Taking into account the typical measurement error on radial velocities (a few km s^{-1}), we have removed from our sample all the sources with an observed velocity $< |10|$ km s^{-1}. These sources are characterized by a large distance-uncertainty. The recovered radial distribution is shown in Figure 2.

In computing solar distances, one has to consider the well known problem of distance degeneracy for sources lying inside the solar circle. Auxiliary data can be useful in order to solve this ambiguity. In particular, we have used absorption lines data (HI, H_2CO) and catalogued optical counterparts. HI data have been mainly taken from Kuchar and Bania (1994) and Caswell et al. (1975). Additional data are from Kerr and Knapp (1970) and Goss and Radhakrishnan (1969). H_2CO data are from Wilson 1980. The optical catalogs which have been used are the Marsálková (1974) Master Catalog, the Blitz et al. (1982) Catalog and the Brand and Blitz (1993) Catalog. Complementary data on individual sources are from Miller (1968) and Shaver et al. (1981). Through this method, we are able to assign a solar distance to 177 sources. These sources have to be added to 143 for which a unique distance can be computed from kinematic data. For the remaining 288 sources, we have worked out a method based on a distance indicator independent from kinematic data. This distance indicator (whose robustness is currently under analysis) has

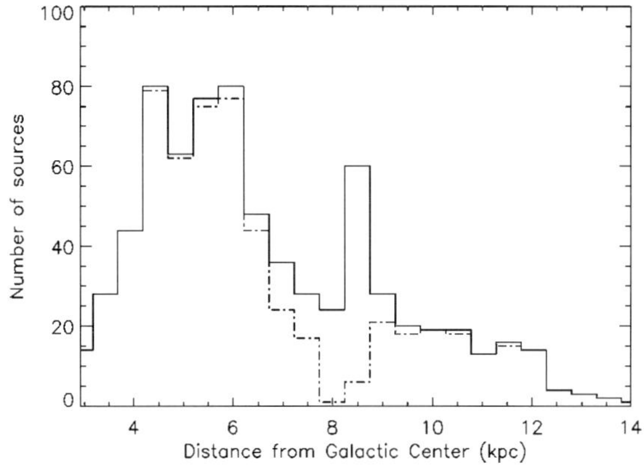

Figure 2. Radial distribution of the Synthetic Catalog HII regions (solid line). Overlaid (dashed-dotted line) is the distribution obtained by considering only sources with velocities $> |10|$ Km s^{-1}.

been found from a luminosity vs. physical diameter correlation obtained by exploiting the 2.7 and 5 GHz flux density and angular diameter data of the Master Catalog. Therefore, according to this correlation, we can compute the solar distance as:

$$D = 10^{\frac{a}{2-b}} \left(\frac{\theta}{1''}\right)^{\frac{b}{2-b}} \left(\frac{\nu}{1GHz}\right)^{\frac{-1}{2-b}} \left(\frac{S}{1Jy}\right)^{\frac{-1}{2-b}} \text{ kpc} \quad (3)$$

where the recovered values of the parameters a and b are, respectively, $\sim$ 30.6 and 1.1. The correlation-method turns out to be appliable to 256 HII regions ($\sim$ 89% of the total number of sources with a distance ambiguity) for which at least one observed value of flux density and angular diameter at 2.7 or 5 GHz is available. With these computed solar distances, we are also able to determine the thickness of the Galactic HII regions layer. A preliminary analysis has retrieved a value in the range $\sim$ 32-51 pc, depending on the inclusion, in the calculation, of only sources with a safely defined solar distance (former lower value) or also of sources with a correlation-assigned distance (latter one).

Acknowledgements

R. Paladini acknowledges financial support from ESA for the participation in the workshop.

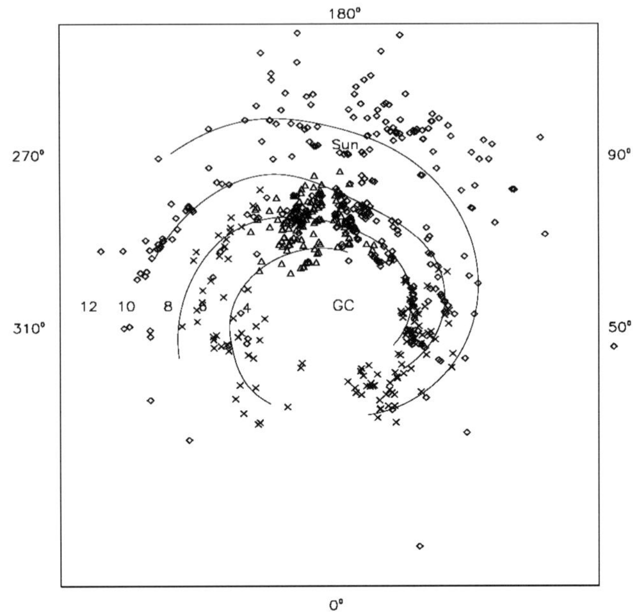

Figure 3. 2-D distribution of HII regions from the Synthetic Catalog at 2.7 GHz. The plot combines: 143 unambiguous sources (diamond) ; 177 sources having either optical or absorption auxiliary information (diamond) ; 256 *degenerate* sources for which near (triangle) or far (cross) solar distances have been assigned through the correlation method. Overlaid on the 2-D distribution is the spiral arms model by Taylor and Cordes 1993.

References

Blitz, L., Fich, M. and Stark, A.A.: 1982, *ApJS* **49**, 183.
Brand, J. and Blitz, L.: 1993, *A&A* **275**, 67.
Caswell, J.L., Murray, J.D., Roger, R.S., Cole, D.J. and Cooke, D.J.: 1975, *A&A* **45**, 239.
Goss, W.M. and Radhakrishnan, V.: 1969, *Astrophy. Lett.* **4**, 199.
Fich, M., Blitz, L. and Stark, A.A.: 1989, *ApJ* **342**, 272.
Kerr, F.J. and Knapp, Gillian R.: 1970, *AuJPA* **18**, 9.
Kuchar, T.A. and Bania, T.M.: 1994, *ApJ* **436**, 117.
Marsalkova, P.: 1974, *A&SS* **27**, 3.
Miller, J.S.: 1968, *ApJ* **151**, 473.
Paladini, R., Burigana, C., Davies, R. et al.: 2002, *A&A*, in press.
Shaver, P.A., Retallack, D.S., Wamsteker, W. and Danks, A.C.: 1981, *A&A* **102**, 225.
Taylor, J.H. and Cordes, J.M.: 1993, *ApJ* **411**, 674.
Wilson, T.L.: 1980, in: *Radio recombination lines* **205**.

X-RAY HALOS IN GALAXIES

GINEVRA TRINCHIERI
INAF-Osservatorio Astronomico di Brera

Abstract. A hot phase of the interstellar medium has now been detected and studied in several objects through its X-ray emission. A proper assessment of its characteristics is relevant for our understanding of several aspects of galaxy properties, from the large scale distribution of matter to the stellar and galaxian evolution, to the dynamics of systems and to the feeding of a central black hole. I will briefly summarize our current understanding of some of the main issues related to the hot gaseous component in galaxies that are fast evolving given the ever more striking and interesting details provided by the X-ray satellites currently operating. I hope to convince you that the X-ray characteristics of the hot gas are quite complex, both in morphology and spectra, in a wide range of objects, which should promote greater efforts in understanding the role played by this component in all galaxies.

Keywords: ISM, galaxies

1. Why is Hot Gas Important?

Because of the implications and relations that a hot phase of the interstellar/intergalactic medium has with several other galaxy-related issues, understanding its properties has become more relevant and intriguing. As just a few examples, gas is closely related to:

- galaxy structure. Gas can be traced at large galactic radii and has been used to study and measure the total gravitational mass in early type galaxies (e.g. Fabricant & Gorenstein 1983)
- evolution of stars, of galaxies and of larger systems. Metal enriched material that is formed in stars gives a measure of stellar evolution; metal gradients indicate how far (how?) metals have traveled from their formation sites; metal concentrations in specific locations could indicate local stripping phenomena
- the dynamics of systems. At high energies different phenomena such as stripping and shocks could be discovered and studied
- active nuclei, who are nourished by gas accretion. When hot, the gas could indeed starve the nucleus rather than feed it, and explain the lack of activity or periodic cycles

The main question is then to find the best candidates to efficiently study the properties of this phase of the InterStellarMedium (ISM).

Astrophysics and Space Science **289:** 367–376, 2004.

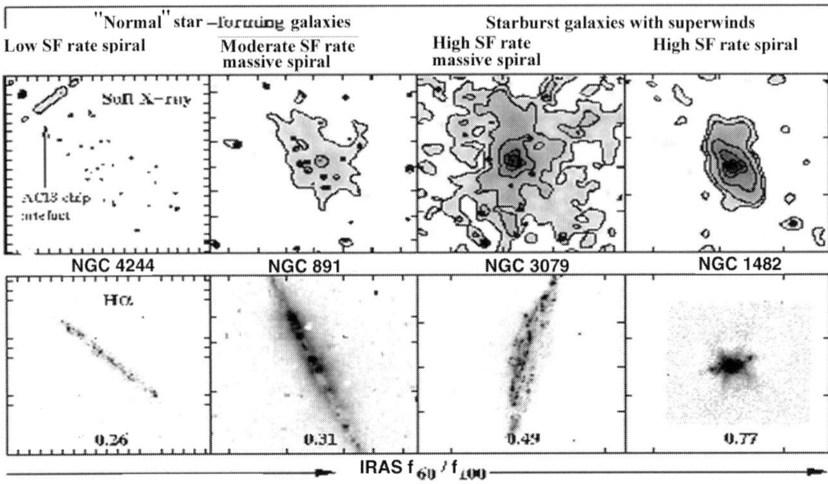

Figure 1. From Strickland et al. (2002b). Top panels: X-ray images in the 0.3-2.0 keV energy band from Chandra data for 4 galaxies from normal to actively forming stars. Lower panels: The same galaxies in the continuum-subtracted Hα+[NII] emission. The arrow at the bottom indicates the increasing IRAS f_{60}/f_{100} ratio, which is used as a measure of the star formation (SF) intensity. The increasingly strong soft X-ray emission associated with increasing SF activity is evident.

2. Where is the Hot Gas?

There is very little [if any] diffuse emission in normal spiral galaxies. Gas appears to be associated mostly with:

1. **star formation**. A figure from a talk by Strickland et al. (2002b) provides an excellent example and confirmation of this association. The X-ray images reveal more and more extended soft diffuse emission from hot gas as the line emission and star formation activity become more prominent (see Figure 1).
2. **gravity**. Large quantities of hot gas have been discovered in early type galaxies since *Einstein* observations, and are associated mostly with galaxies at the center of small groups. A long standing issue of whether the large scale gas should be associated with the galaxy or with the group is alive and unresolved as yet, but it is beyond the scope of this talk and it will be mostly [unjustly!] ignored here.

Smaller quantities of gas have recently been found associated with spiral bulges (e.g. M31, NGC 1291) and low luminosity early type galaxies (e.g. NGC 4697), but these are still hard to quantify and characterize. If this is a property of the whole class, a new and interesting field will open up and bring new issues to study and resolve.

It is evident that the characteristics of the gas might not be the same in both cases above, considering the clearly different origins of the two phenomena. It is however

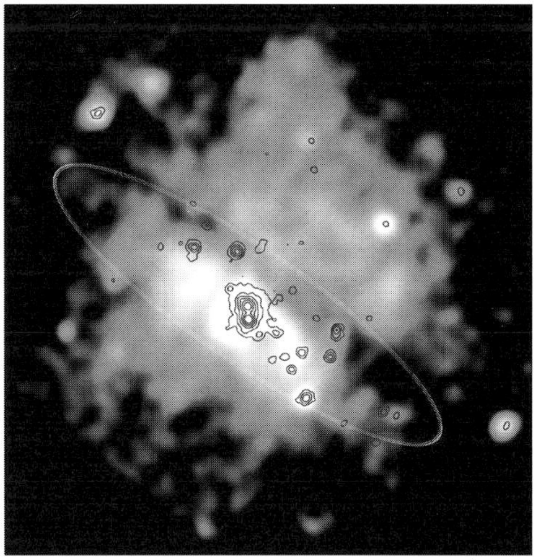

Figure 2. In this figure, the X-ray halo of NGC 253 in the soft (0.2–0.5 keV) energy band observed with XMM-Newton (courtesy of W. Pietsch) is shown, with the hard emission overlayed as contours. The oval shape indicates the approximate size and orientation of the optical disk. A few point sources are evident in the halo region, otherwise the area above and below the plane appears uniformly filled at these energies.

likely that different systems will be dominated by only one of the two components, thereby allowing us a relatively clean study of both phenomena separately.

3. Characteristics of the ISM in Starburst Galaxies/Regions

Starburst galaxies are the subject of separate talks in this conference (cf. talks by Dahlem and by Dettmar), so I will briefly review the X-ray evidence, mostly at large scale. The data are fast accumulating but they have not been properly digested yet, and most of the results are still rather qualitative and based on a few examples only.

A large fraction of the total emission of starburst galaxies in the soft X-ray band is due to the hot ISM. The fraction varies in different galaxies, with $> 20\%$ in NGC 3256 (Moran et al., 1999; Lira et al., 2002), to $\sim 50\%$ in the Antennae (Fabbiano et al., 2001), to $\sim 80\%$ in M82 (Zezas et al., 2001). Hot gas is found mainly in three regions: (A) the large scale distribution is associated with extra-planar emission, in the galaxy's **halo**, usually associated with evidence of halo emission at other wavelengths (HI, radio continuum, Hα outflows). The emission can be traced to quite large distances: the rather spectacular halo of NGC 253 observed with XMM-Newton in the very soft energy band (Figure 2) fills a very large region above and below the plane. The halo is usually fueled/fed by a nuclear/galactic outflow/wind. Often 'horns' or X-shaped morphologies are observed in galaxies viewed edge-on. A much finer filamentary structure is also observed

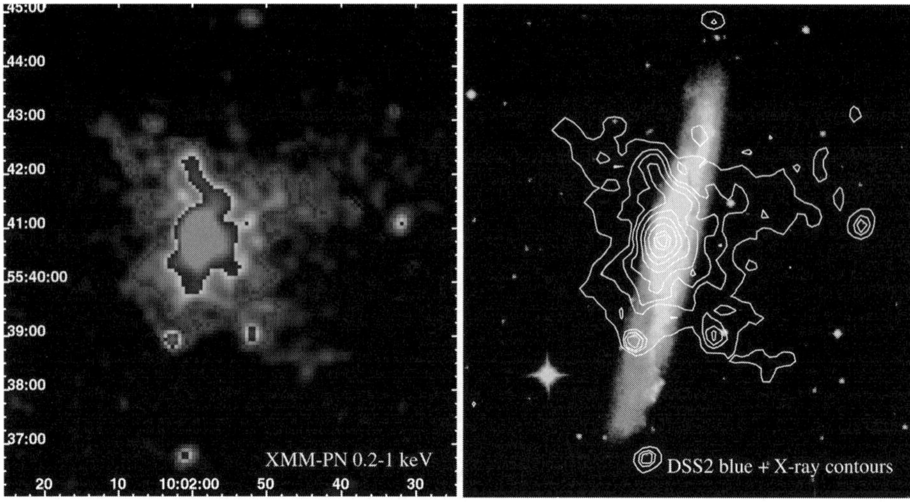

Figure 3. XMM-Newton data of NGC 3079 in the soft band indicate a large scale emission above/below the plane of the galaxy, as defined by the optical image. The filamentary structure of the emission is evident. Filaments are traced out to 13-17 kpc. Three main spurs plus a smaller one are responsible for the X-shaped morphology of the halo. From Breitschwerdt et al. (2002).

in halos, e.g. in NGC 3079's halo (Breitschwerdt et al., 2002) with XMM-Newton data and in NGC 253 itself with Chandra data, with spatial fluctuations of $\sim$130 pc in size (Strickland et al., 2002a). In both cases, these point to a close relation with similar structures in the line emission (cf. above and Figure 1). (B) The **disk** appears to contain a multi-phase hot ISM. To use the case of NGC 253 again, the data require two temperatures (0.15; 0.5 kT) with no additional absorption and show unquestionable signs of plasma lines, e.g. OVII, FeXVII, that point to emission from a hot ionized medium (Pietsch et al., 2001). It is possible that the multi-temperature of the gas in the disk is to be attributed in part to contamination from emission at higher galactic latitudes projected onto the disk. (C) Possibly the stronger emission is associated with the **nuclear/circumnuclear** regions, from which plumes and outflows are likely to be fueled. These regions appear to be hotter than the rest of the disk (e.g. kT$\sim$6 keV is found in the inner regions of NGC 253 [again!], Pietsch et al.2001), and might contain a higher metal fraction than the outer regions.

A much larger and better studied sample of objects needs to be examined for more quantitative analysis and a broader range of parameters need to be explored before a classification of the X-ray properties of the hot ISM in these galaxies can be finalized.

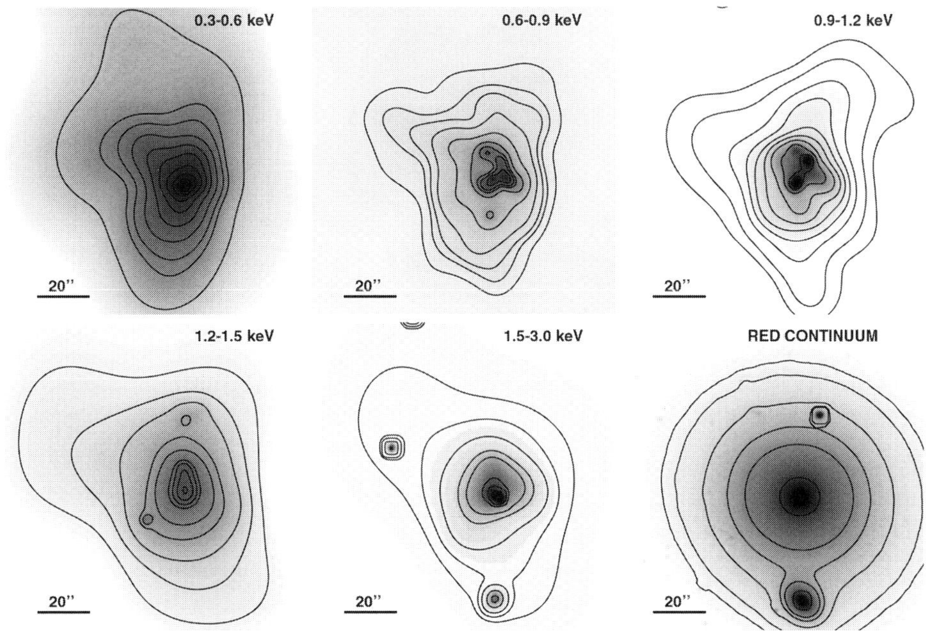

Figure 4. Adaptively smoothed X-ray data of the central region of NGC 5846 at different energies, as indicated in each panel. The red continuum image of the same region (data from the ESO 3.6m + EFOSC) is shown in the bottom right cannel. The comparison between the different images is a clear example of the complexity of the X-ray morphology, and shows how little it reproduces the stellar light distribution.

4. Early Type Galaxies and Groups

The large quantities of gas observed in this class of sources have now been studied in some detail and several general characteristics of the large scale emission are now relatively well established. However, little is known of the emission at intermediate to small scales. New observations have now shown that the gas distribution in the inner regions of early type galaxies can be quite spectacular. At small radii the emission is often disturbed, and arcs-tails-peculiar/unexpected shapes are coupled with inhomogeneous spectral characteristics of the gas. I will concentrate on a few examples from the literature that will illustrate how gas distribution in early type galaxies can be just as interesting as in later types, and rather more unexpected!

4.1. NGC 5846

This galaxy is at the center of a small group, and optically is a rather unperturbed object. The X-ray emission is very extended, roughly azimuthally symmetric and regular, with a large total X-ray luminosity of $L_X(0.2 - 2\text{keV})\ 7 \times 10^{41}$ erg s^{-1} within r=10$'$ (Finoguenov et al., 1999). The emission on the sub-arcminute scale instead is much more spectacular and complex: as shown by Figure 4, an arc struc-

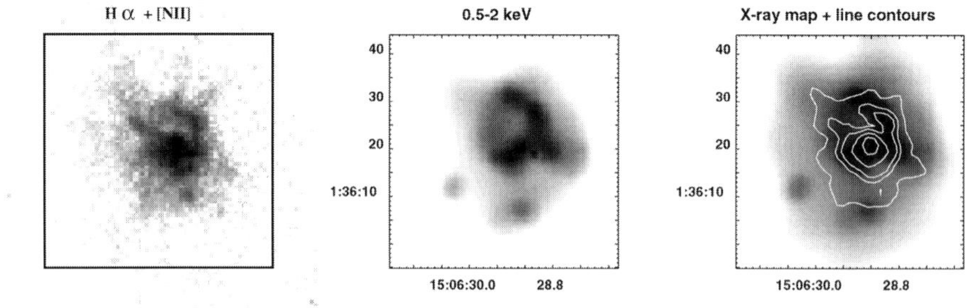

Figure 5. Hα+[NII] image of the inner regions of NGC 5846 (LEFT) on the same scale as the 0.2-2.0 keV emission observed with Chandra (MIDDLE). The Hα+[NII] contours on the X-ray smoothed image (RIGHT) indicate that both morphology and scales are very closely related in this region (from Trinchieri and Goudfrooij [2002]).

ture ('hook'), a tail pointing S and several knots are evident in the soft energy band, and their relative strengths are clearly energy-dependent.

There is also a curious correspondence with the optical line emission and with the dust distribution, both on scales of $1' - 2'$, as already discovered with the ROSAT HRI (Trinchieri et al., 1997), and on much smaller scales of $2'' - 4''$ (300-600 kpc) discovered with Chandra (Trinchieri and Goudfrooij, 2002). The correspondence is extremely good, but the physical link between the two emissions is still rather puzzling:

- there is no nuclear activity nor a strong Radio Source in the center that would justify/trigger gas excitation
- the origin of the two gas phases are not the same:
  - Hα is most likely of 'external' origin, acquired from merging or interaction with smaller, gas rich bodies together with dust. *Here* the internal kinematics supports external origin
  - X rays are 'internal', from the ISM of the galaxy presumably accumulated from stellar mass loss.

Why are their morphologies so similar then (Figure 5)? A possible link between Hα [+dust] and X rays is through thermal conduction (Cowie and McKee, 1977; Sparks, Macchetto and Golombek, 1989; de Jong et al., 1990). Cold gas+dust are acquired from the outside at the same time; dust/cold gas act simultaneously on the hot coronal gas, inducing local cooling + excitation into emission; this produces an expected $L_{H\alpha} \sim 1.5 \times 10^{39}\ T_7^{3/2}\ n_{0.01}$ erg s^{-1} (Cowie and McKee, 1977). For the temperature and densities derived from the X-ray observation e.g. in the 'hook', of T$\sim 6 \times 10^6$ K and $n \sim 0.35$ cm^{-3} respectively, the Hα luminosity should be $\sim 2.4 \times 10^{40}$ erg s^{-1}, to be compared to a measured $L_{H\alpha} \sim 1.3 \times 10^{40}$ erg s^{-1}. **Too good to be true!**

Can we then assume that this mechanism is responsible for the complex morphologies at small radii seen in other galaxies? The number of cases with both good X-ray and Hα data is limited, but there are a few examples – unfortunately

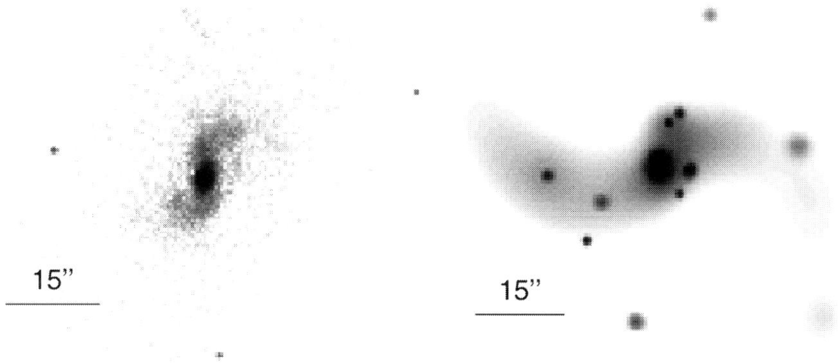

Figure 6. Inner regions on NGC 1553 in Hα+[NII] (left; ESO NTT) and 0.3-1.0 keV band (Chandra archive). In the very soft X-ray band a spiral structure roughly symmetric about a central point-like source is evident, with a possible NS oriented, 10$''$ long, inner bar, reminiscent of the Hα+[NII] structure. No such feature is evident in the optical continuum or the X rays above 1 keV.

so far NGC 5846 remains the best and perhaps only example for which such close link is observed. For other objects there seem to be a need of other explanations for the small scale morphological perturbation, and unfortunately again each object appears so far to be 'unique' (but the cases studies are very few!)

4.2. NGC 1553

The overall emission from NGC 1553 indicates an extended component, slightly elongated NW-SE, consistent with the optical axis (Trinchieri et al., 1997). Even with the lower resolution ROSAT data, several features could be identified that indicated structures at smaller, arcmin size, angular scales. New Chandra data now show that the inner regions are quite complex (Blanton et al., 2001): at energies below 1 keV there is a clear evidence of a spiral structure, a possible inner bar, and a twisted NW-SE elongation at larger radii (cf. Blanton et al. and Figure 6). At intermediate radii, ROSAT HRI data had already indicated an overall similarity with line emission, although not as close as in the NGC 5846 case (Trinchieri et al., 1997). This similarity appears to break down at small radii (Blanton et al., 2001): there could be a common NS 'bar' in the inner 10$''$ (Trinchieri et al., 1997; Rampazzo et al., 2002), but it is hard to envision thermal conduction to work along the spiral structure where the Hα does not extend. New high resolution Fabry-Perot interferometric data have further shown quite complex gas dynamics in NGC 1553 (Rampazzo et al., 2002), making the possible similarity with the hot phase more difficult to interpret. The origin of the spiral feature is better interpreted with adiabatic or shock compression of the ambient gas possibly due to interaction with the radio source (Blanton et al., 2001).

Figure 7. (From Jones et al. [2002]): Full resolution ACIS-S image of the central regions of NGC 4636 in the 0.5–2.0 keV energy range. A plus sign marks the galaxy center.

4.3. NGC 4636

A wealth of X-ray data has been accumulated for this X-ray bright galaxy since *Einstein* Observatory times, that indicate a very extended distribution of the X-ray emission, due almost entirely to hot gas at an average temperature of $\sim$ 1 keV and total luminosity of $L_X \sim 2 \times 10^{41}$ erg s^{-1} (Trinchieri et al., 1994; Matsushita et al., 1998; Buote, 2000). From an *Einstein* HRI observation, (Stanger and Warwick, 1986) had already proposed an asymmetric gas distribution at small radii. Chandra data now clearly show symmetric, 8 kpc long, clearly defined spiral-like twisted structures in the inner few arcmin region of this galaxy (Figure 4.3, Jones et al., 2002), coupled with a complex temperature structure of the gas, a weak but measurable temperature gradient (but no cooling flow!) in the inner regions and a number of discrete spectral lines measured with XMM-Newton data (Xu et al., 2002). To date there are no reports of a close morphological relation between the hot and the warm phases, although Hα is detected in this object (Demoulin-Ulrich et al., 1984) with chaotic gas kinematics and with evidence for a kinematically distinct inner region like NGC 5846 (Caon et al., 2000). Jones et al. propose that the X-ray morphology is formed as a result of shocks driven by a nuclear outburst in the recent past. These outbursts would also have implications both in the accumulation of material in the galaxy's center, and in the fueling of a central AGN.

5. Unexpected Phenomena

Another set of surprises come from the presence of 'unsuspected' large scale shocks, not seen/foreseen from the optical data. How many are we missing because we are not looking at the right wavelength?

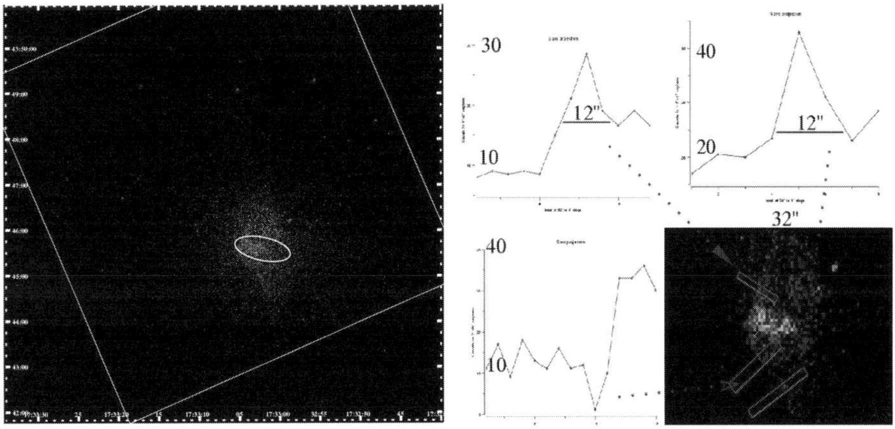

Figure 8. The full Chandra field of IC1262 (LEFT). The oval indicates the optical position of the galxy. A turmoil at the center of the more diffuse emission and a sharp discontinuity in surface brightness are evident even in the raw data, in cuts across the narrow features visible.

5.1. IC1262

This galaxy, brightest in a small poor group, became interesting because of its high L_x/L_b ratio as seen in the ROSAT All Sky Survey. X-ray emission was later measured to be very extended (>400 kpc), suggesting a large contribution from the group's potential (Trinchieri and Pietsch, 2000). However, more interesting, the HRI saw an 'arc' on $\leq$arcmin scales, that was interpreted as the signature of a shock.

New Chandra data have now shown a quite dramatic view of the system (Figure 8): a sharp discontinuity east of the central galaxy, with steep drops and a relatively narrow feature along a possible shock front, plus an arc to the NNW, are all indicative of a turmoil in the high energy component, possibly a trace of shocked material caused either by peculiar motions in the system or by a recent merger process. Although a better sampling of the velocity field is needed, neither the optical classification of the galaxy as a cD nor the small velocity dispersion for the group (300 km s^{-1}, (Wegner et al., 1999)) could have suggested anomalous motions of this nature. A proper assessment of the X-ray properties in this and similar systems might be crucial for a more complete understanding of the dynamical and evolutionary properties of small galaxy systems.

6. What Can We Conclude From All of This?

It is probably too early to draw conclusions from the evidence accumulated so far. However, it is becoming clearer that the small scale gas morphology indicates that the 'early-type' galaxy population is far from 'quiescent'. Same can be said for 'cD' galaxies in small/sparse groups. This would indicate that there is ground for more detailed studies of the high energy emission in galaxies; these might reveal unsuspected phenomena like shocks that are crucial for our understanding of the evolution of these systems. The multiwavelength approach is the only one that can provide a complete picture of all the forces at play, and the study of the gaseous components, though complex, has a very high potential for easy discovery of extreme, but perhaps not uncommon, phenomena.

References

Blanton, E., Sarazin, C. and Irwin, J.: 2001, *ApJ* **552**, 106.
Breitschwerdt, D., Pietsch, W., Read, A., Trinchieri, G. and Vogler, A.: 2002, in prep.
Buote, David A.: 2000, *MNRAS* **311**, 176.
Caon, N., Macchetto, D. and Pastoriza, M.: 2000, *ApJS* **127**, 39.
Cowie, L.L. and McKee, C.F.: 1977, *ApJ* **211** 135.
de Jong, T., Norgaard-Nielsen, H., Jorgensen, H. and Hansen, L.: 1990, *A&A* **232** 317.
Demoulin-Ulrich, M.-H., Butcher, H.R. and Boksenberg, A.: 1984, *ApJ* **285**, 527.
Fabbiano, G., Zezas, A. and Murray, S.S.: 2001, *ApJ* **554** 1035.
Fabricant, D. and Gorenstein, P.: 1983, *ApJ* **267**, 535.
Finoguenov, A., Jones, C., Forman, W. and David, L.: 1999, *ApJ* **514**, 844.
Jones, C., Forman, W. Vikhlinin et al.: 2002, *ApJ* **567**, L115.
Lira, P., Ward, M., Zezas, A., Alonso-Herrero, A. and Ueno, S.: 2002, *MNRAS* **330**, 259.
Matsushita, K., Makishima, K., Ikebe, Y. et al.: 1998, *ApJ* **499L**, 13.
Moran, E.C., Lehnert, M.D. and Helfand, D.J.: 1999, *ApJ* **526**, 649.
Pietsch, W., Roberts, T.P., Sako, M. et al.: 2001, *A&A* **365**, 174.
Rampazzo, R., Plana, H, Longhetti, M. et al.: 2002, *MNRAS*, submitted.
Sparks, W.B., Macchetto, F. and Golombek, D.: 1989 *ApJ* **345**, 153.
Stanger, V.J. and Warwick, R.S.: 1986, *MNRAS* **220**, 363.
Strickland, D.K., Heckman, T.M., Weaver, K.A. et al.: 2002, *ApJ* **568**, 689.
Strickland, D.K., Heckman, T.M., Colbert, E.J.M., Hoopes, C.G. and Weaver, K.A.: 2002, in: K.A. van der Hucht, A. Herrero and C. Esteban (eds.), *IAU symposium* **212**.
Trinchieri, G., Kim, D.-W., Fabbiano, G. and Canizares, C.R.C.: 1994, *ApJ* **428**, 555.
Trinchieri, G., Noris, L. and di Serego Alighieri, S.: 1997, *A&A* **326**, 565.
Trinchieri, G. and Goudfrooij, P.: 2002, *A&A* **386**.
Trinchieri, G. and Pietsch, W.: 2000, *A&A* **353**, 487.
Wegner, G., Colless, M., Saglia, R.P. et al.: 1999, *MNRAS* **305**, 259.
Xu, H., Kahn, S.M., Peterson, J.R. et al.: 2002, *ApJ* **579**, 600.
Zezas, A., Fabbiano, G., Ward, M., Prestwich, A. and Murray, S.S.: 2001, *ASP Conference Proceedings*, Vol. **249**, p. 425.

THE H I HALO OF SPIRAL GALAXIES

FILIPPO FRATERNALI
Kapteyn Astronomical Institute, RUG, Groningen, NL

TOM OOSTERLOO
ASTRON, Dwingeloo, NL

RENZO SANCISI
Astronomical Observatory of Bologna, INAF, Italy
Kapteyn Astronomical Institute, RUG, Groningen, NL

Abstract. A deep H I survey with the VLA of the spiral galaxy NGC 2403 has revealed the existence of a thick, low density layer of neutral gas surrounding the thin 'cold' disk. This layer has a mean rotation velocity 25–50 km s^{-1} lower than that of the disk and a 10–20 km s^{-1} inflow towards the centre of the galaxy. In the central parts there are velocity differences from rotation of up to 150 km s^{-1}. *Chandra* observations of NGC 2403 show a diffuse, hot X-ray emitting gas component with a temperature of a few 10^6 K. These results point at galactic fountain type of flows between disk and halo. 'Halo' gas with similar characteristics has also been observed in other spiral galaxies (e.g. NGC 6946, NGC 891). Such gas is probably similar to the IVCs and to some of the HVCs of the Milky Way.

Keywords: galaxies: individuals (NGC 2403, NGC 891), galaxies: ISM, galaxies: structure and dynamics, galaxies: halo

1. Introduction

In recent years, high sensitivity H I observations of spiral galaxies at different inclination angles have been made with the aim of studying the gas 3D distribution and kinematics. The study of face-on galaxies has led to the detection of vertical motions and of 'holes' in the H I distribution (e.g. see NGC 6946, Boomsma et al., 2002). The study of edge-on galaxies has revealed the presence of H I emission up to several kpc from the plane and, in the case of NGC 891, also a lower rotation velocity (about 25 to 100 km s^{-1}) for the halo gas with respect to the gas in the disk (Swaters et al., 1997). Optical spectroscopy has shown the presence of very thick layers of diffuse ionized gas and a vertical gradient in the rotation velocity (Rand, 2000). Finally, halo X-ray diffuse emission has been observed, with typical temperatures of a few 10^6 K, in some nearby edge-on spiral galaxies (Wang et al., 2001). All these results suggest a complex circulation of gas, at different phases, between disk and halo of spiral galaxies.

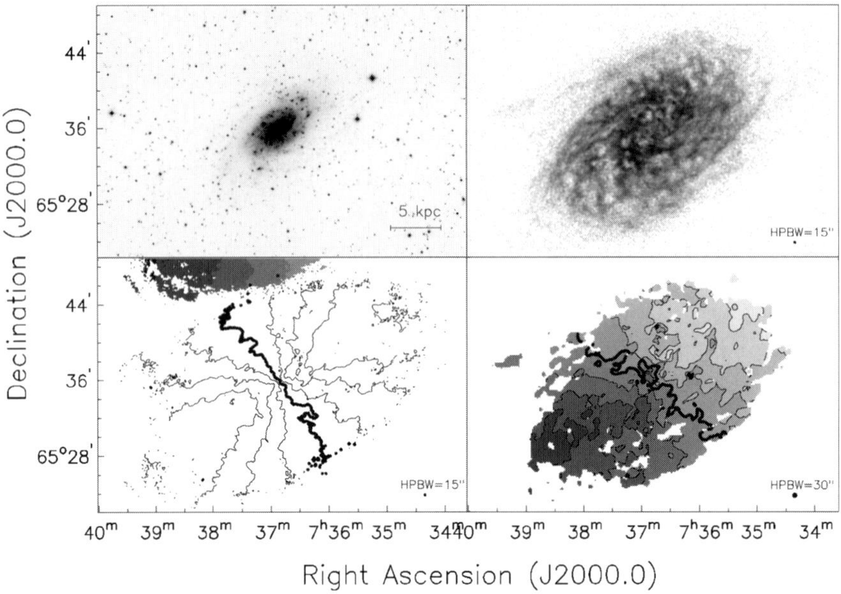

Figure 1. NGC 2403. Upper panels: optical image (DSS) and total H I density map (VLA). Bottom panels: velocity field of the cold thin disk (left) and of the anomalous 'halo' gas (right). All plots are on the same scale.

2. The Spiral Galaxy NGC 2403

NGC 2403 (see Figure 1) is a non-interacting (Sc) spiral, very similar to M 33 and located at a distance of about 3 Mpc. Deep H I observations with the VLA (Fraternali et al., 2002a) have revealed the presence of wings of emission in the line profiles along the major axis of the galaxy (Figure 2). The gas responsible for such emission (here referred to as the anomalous gas) has systematically lower rotation velocities as compared to the galaxy rotation (white squares). Such a pattern is different from what expected for a thin disk (see the model in Figure 2). Part of the anomalous gas, located in the inner parts of the galaxy, shows large non-circular motions ('forbidden' gas) with projected differences of up to 150 km s^{-1} from the rotation velocity.

The anomalous H I forms a thick (1–3 kpc) layer with a mass of 3×10^8 $M_\odot$ (about 1/10 of the total H I mass) and a mean rotation velocity 20–50 km s^{-1} lower than that of the gas in the disk. Its velocity field (Figure 1) shows that the projected kinematical minor axis is rotated with respect to that of the thin disk. This has been interpreted as due to a radial flow (of 10–20 km s^{-1}) of this gas towards the centre of the galaxy (Fraternali et al., 2001).

Chandra observations (36 ks) of NGC 2403 show diffuse X-ray emission in the soft (0.4–1 keV) band (Figure 3). Its spectrum, after subtracting the point sources, indicates that it is produced by hot ($2–8 \times 10^6$ K) gas, probably heated by star

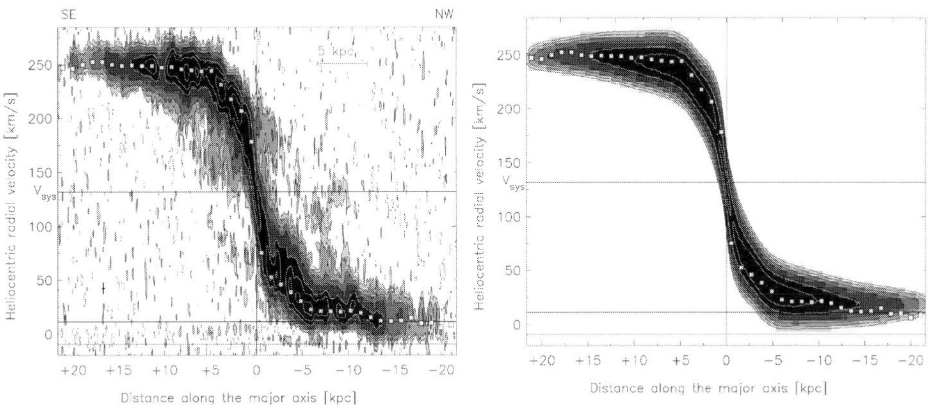

Figure 2. H I position-velocity diagram along the major axis (p.a.=124°) of NGC 2403. The beam size is $\sim 15''$, the velocity resolution is 10.3 km s^{-1}. The white squares show the (projected) rotation curve. The right panel shows a model for a thin H I disk with a Gaussian velocity dispersion.

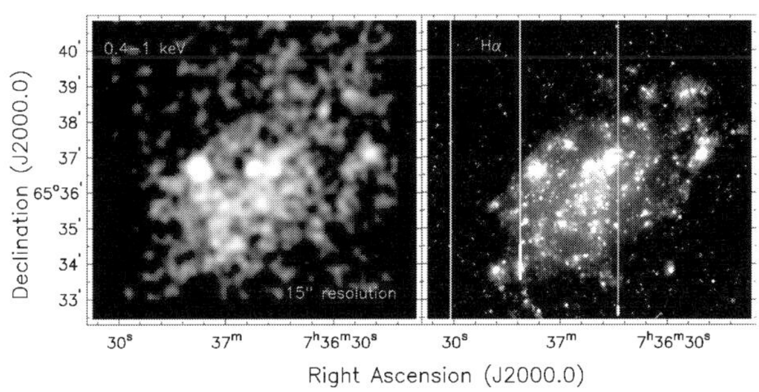

Figure 3. Chandra X-ray image (0.4–1 keV) of NGC 2403 after the subtraction of point sources (left panel) and Hα image (right panel). Note the great similarity between the distributions of hot and ionized gas.

formation activity in the disk (Fraternali et al., 2002b). Its distribution shows a great similarity with that of the Hα emission.

3. Discussion and Future Work

The anomalous gas discovered in NGC 2403 may be explained by a galactic fountain mechanism (Bregman, 1980) and would then be the final 'infalling' stage of the fountain. The discovery of hot diffuse X-ray emitting gas seems to support this interpretation. A second possibility is that of the accretion of extragalactic, maybe *primordial* gas (Oort, 1970).

A number of spiral galaxies at different inclination angles have been recently observed with the Westerbork Synthesis Radio Telescope with high sensitivity. They show similar patterns of motion as observed in NGC 2403 (e.g. the spiral galaxy NGC 6946, Boomsma et al., 2002). Also new very deep observations of the

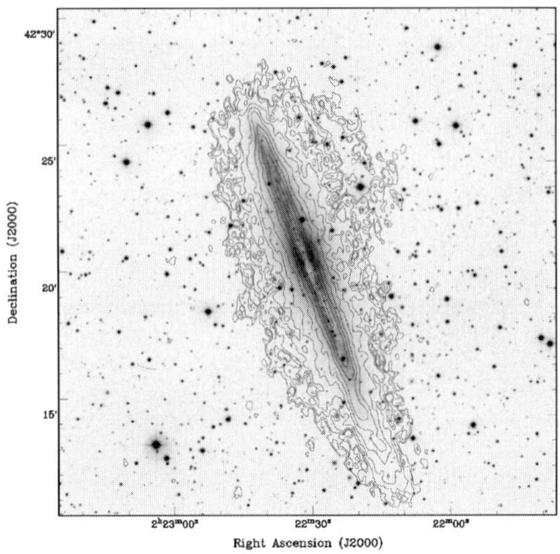

Figure 4. WSRT total H I map of NGC 891 (Swaters, Fraternali, Oosterloo, & Sancisi 2002, in preparation) overlaid on an optical DSS image. Note the presence of H I emission far outside the plane of the disk. The lower contour is 1.7×10^{19} cm^{-2}.

edge-on spiral galaxy NGC 891 have been obtained (Swaters, Fraternali, Oosterloo and Sancisi, 2002, in preparation). Figure 4 shows a preliminary total H I map (in contours) overlaid on an optical DSS image. H I emission is found far above the plane up to distances of 5−10 kpc.

The anomalous gas found in the halo region of these nearby spiral galaxies is probably similar to some of the High Velocity Clouds observed *in* the Milky Way (Wakker and van Woerden, 1997).

References

Bregman, J.N.: 1980, *ApJ* **236**, 577.
Boomsma, R., Oosterloo, T., Sancisi, R. and Van der Hulst, J.M.: 2002, this conference.
Fraternali, F., Oosterloo T., Sancisi, R. and Van Moorsel, G.: 2001, *ApJ* **562**, L47.
Fraternali, F., Van Moorsel, G., Sancisi, R. and Oosterloo, T.: 2002a, *AJ* **123**, 3124.
Fraternali, F., Cappi, M., Sancisi, R. and Oosterloo, T.: 2002b, *ApJ* **578**, 109.
Oort, J.H.: 1970, *A&A* **7**, 381.
Rand, R.J.: 2000, *ApJ* **537**, L13.
Swaters, R.A., Sancisi, R. and van der Hulst, J.M.: 1997, *ApJ* **491**, 140.
Wakker, B.P. and van Woerden, H.: 1997, *ARA&A* **35**, 217.
Wang, Q.D., Immler, S., Walterbos, R., Lauroesch, J.T. and Breitschwerdt, D.: 2001, *ApJ* **555**, L99.

RECENT DEVELOPMENTS CONCERNING HIGH-VELOCITY CLOUDS

BART P. WAKKER
University of Wisconsin-Madison

Abstract. I summarize the new developments in HVC research that have occurred in the past few years. These include a) the determination of approximate distances to some IVCs and HVC complex A; b) the measurement of the metallicity of the Magellanic Stream and the conclusion that tidal forces are dominant in creating the Stream; c) the derivation that the metallicity of complex C is 0.1 solar, with some evidence for metallicity variations; d) the result that N/O in complex C is 0.2 times the solar ratio, a value otherwise only seen in dwarf galaxies and Damped Lyα systems; e) the discovery of high-velocity O VI absorption, both associated with previously known H I HVCs and in directions without a H I HVC; f) the proposal that HVCs form a population of objects permeating the Local Group.

1. Introduction

High-velocity clouds (HVCs) were discovered 40 years ago (Muller et al., 1963). Historically, H I clouds with $|v_{\rm LSR}|>100\,{\rm km\,s^{-1}}$ are called 'high-velocity clouds' or HVCs, while gas clouds with $|v_{\rm LSR}|$ between 40 and $100\,{\rm km\,s^{-1}}$ are called 'intermediate-velocity clouds' or IVCs. Wakker (1991) introduced the 'deviation velocity', $v_{\rm dev}$, which gives the difference between $v_{\rm LSR}$ and the maximum velocity that can be understood with a simple model of differential galactic rotation.

Since by definition it is not possible to use velocities to estimate a 'kinematic distance', many different hypotheses have been put forward to explain the origin of the HVCs. These include local supernova remnants ($d\sim$100 pc), gas in the lower Galactic Halo associated with a Galactic Fountain ($d\sim$1–10 kpc), distant spiral arms ($d\sim$10–20 kpc), infalling intergalactic gas ($d\sim$10–50 kpc), tidal features ($d\sim$50–100 kpc), Local Group clouds ($d\sim$100–1000 kpc), and protogalaxies ($d>$1 Mpc).

In recent years much progress has been made, based on single-dish radio telescopes (Dwingeloo/Effelsberg/Villa Elisa/Parkes), optical absorption-line studies with the Herschel telescope (*WHT*), H I maps made with the Westerbork and *AT* interferometers, detections of Hα at three different observatories, and UV absorption line measurements with the Hubble Space Telescope (*HST*) and the Far Ultraviolet Spectroscopic Explorer (*FUSE*). The resulting new H I maps of IVCs and measurements of the distance, metallicity and ionization conditions in several HVCs show that examples can be found for several of the proposed origins. Below I will summarize these developments.

Astrophysics and Space Science **289:** 381–390, 2004.

2. IVCs and Galactic Fountain-like models

The 'Leiden-Dwingeloo Survey' (LDS) of Hartmann and Burton (1997) provides the first completely sampled map of H I at declinations $>-35°$ with sufficient velocity resolution to easily separate the IVCs from the low-velocity gas. Using the LDS data, Wakker (2001) presents maps of several IVCs and introduces complexes K and gp. Earlier, Kuntz and Danly (1996) had already classified the main northern IVCs and termed these the 'Intermediate-Velocity Arch', 'Low-Latitude Intermediate-Velocity Arch' and 'Intermediate-Velocity Spur'. Figure 1 shows a map of the IVCs in the LDS. Data similar to the LDS exist for the southern sky (Arnal et al., 2000), but are not easily available.

Wakker (2001) summarizes the many observations of UV and optical absorption line observations of IVCs, and derives distance brackets for some. An upper limit to the distance results from detecting absorption at the velocity of the IVC in the spectrum of a background star with known distance. A lower limit requires a non-detection, combined with evidence that the absorption line would have been detected if the cloud were in front of the star. This method is summarized in more detail by Wakker and van Woerden (1997). Wakker (2001) finds that the IV-Arch lies between 0.7 and 1.7 kpc above the plane, the LLIV-Arch lies at about 0.6 kpc, complex gp lies between 0.3 and 2 kpc, while for complex K only an upper limit of $\sim$4 kpc can be set.

Measurements of the metallicity of IVCs are still rare. Richter et al. (2001a, b) find that the IV-Arch and LLIV-Arch have near-solar metallicity. No reliable data exist for other IVCs.

The origin of the IVCs may lie in the Galactic Fountain (Shapiro and Field 1976). In this picture hot (10^6 K) gas from superbubbles vents into the halo, cools radiatively, and rains back down to the plane, setting up a 'hydrological cycle'. The details will depend on the gravitational field and the distribution and size of the energy input. Although this is a compelling picture, there are alternative ways of creating cool gas in the Galactic Halo, such as cosmic ray winds (Breitschwerdt et al., 1993), or spiral-arm shocks (Goméz and Cox, 2002). Such models produce a fountain-like flow, but the gas is never in a hot phase.

The Fountain picture explains the presence of highly-ionized atoms (C^{+3}, N^{+4}, O^{+5}) with a scaleheight of 2–4 kpc (Sembach and Savage, 1992, Savage et al., 1997, 2003). Savage et al. (2003) find that toward extra-galactic targets observed by *FUSE* in the direction of IVCs, O VI absorption can be seen in 2/3rds of the cases. There is no apparent correlation between N(O VI) and the presence or absence of an IVC. This is consistent with the idea that the IVCs represent returning halo gas, while the O VI represents cooling halo gas at a different location.

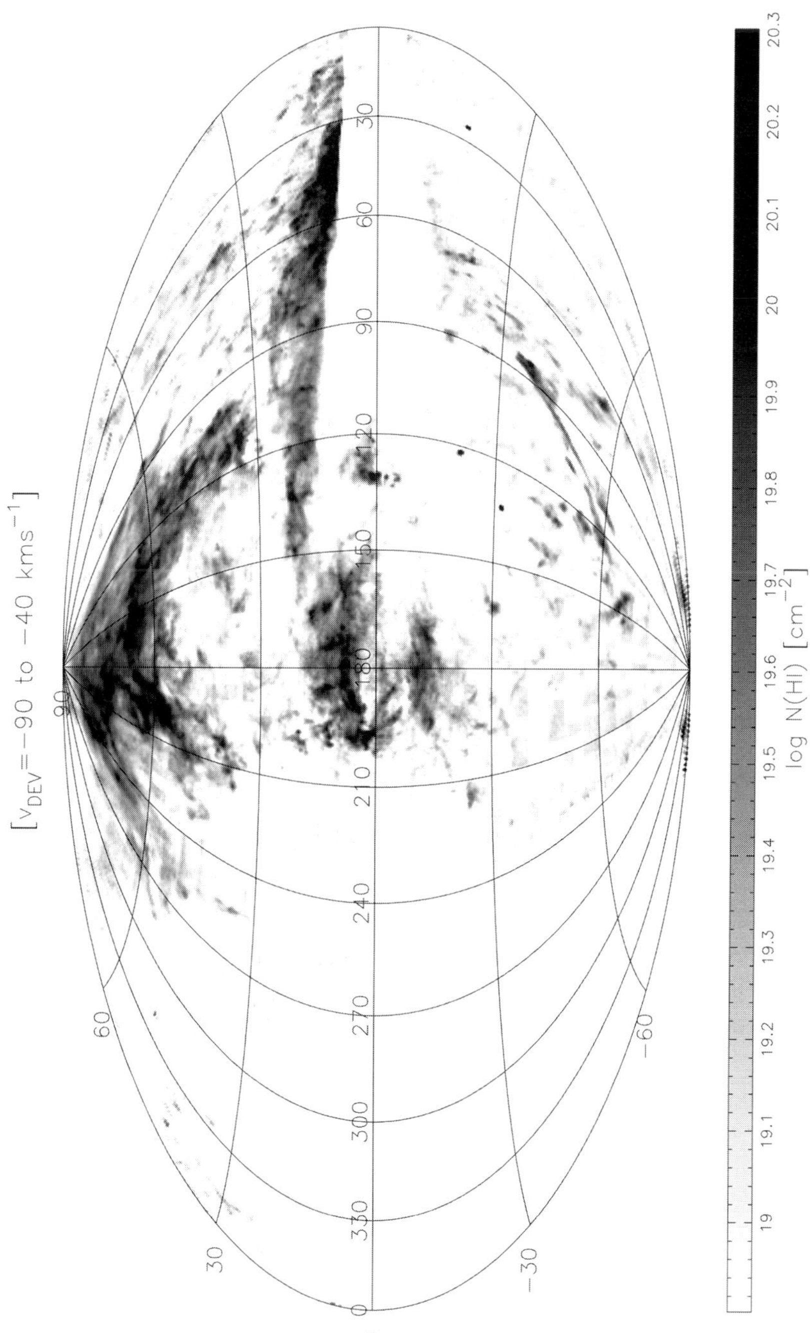

Figure 1. Map of the H I column density of the Intermediate-Velocity Clouds with v_{dev}=−90 to −40 km s^{-1}, based on the LDS. Column densities are displayed on a logarithmic scale. The blank area in the lower left corresponds to declinations <−35°. The feature running from l=20° to 150° at b∼15° is the upper part of the Outer Arm (or the Galactic warp); a slightly more realistic model of galactic rotation can explain its velocities.

3. The Magellanic Stream

The Magellanic Stream (Figure 2) was discovered in the early 1970s (Mathewson et al., 1974). Its velocity field and location can be explained as the result of a strong tidal interaction between the Milky Way and the two Magellanic Clouds that took place 2.2 Gyr ago. In the best model (Gardiner and Noguchi 1996) this interaction ripped off some of the gas in the outer parts of the SMC, forming a trailing and a leading arm. The velocity field and angular distribution of the trailing arm are relatively well reproduced. Gibson et al. (2000) measured the S II abundance of the trailing arm in the sightline to Fairall 9. Using H I data from Parkes (17′ beam), they found an abundance of 0.3 times solar, confirming that the Stream originates from the Magellanic Clouds.

Since early efforts at modeling the Stream could only roughly explain the observed velocities, several alternative origins were proposed. For instance, Moore and Davis (1994) suggested that gas might be stripped from the LMC and SMC by the ram pressure provided by an as-yet undetected hot Galactic corona with a density of 10^{-4} cm^{-3} or less. However, two new pieces of evidence strongly support the tidal model.

Lu et al. (1998) combined S II absorption toward NGC 3783 with a high-resolution H I map of HVC WW 187 (Wakker et al., 2002) to derive an abundance of 0.25 times solar. This cloud has a position and velocity near that expected for the leading arm, although the model of Gardiner and Noguchi (1996) does not predict the correct parameters in detail. Further supporting evidence for the presence of the leading arm was provided by Putman et al. (2000), who used Parkes data to map high-velocity H I in the southern sky and found clear evidence for neutral gas streaming ahead of the Magellanic Clouds, which can only be explained if the Stream has a tidal origin.

The discovery by Sembach et al. (2003) of O VI associated with the Magellanic Stream suggests that it is embedded in a hot ($>10^6$ K) Galactic corona after all. However, the success of the tidal model shows that the density of this corona is not high enough to fundamentally change the orbits of the stripped gas, though it may influence them. The O VI may result from turbulent mixing between the hot coronal gas and the cool Stream gas, or from a conductive interface. However, these models predict O VI column densities to be $\sim 10^{13}$ cm^{-2} for each interface, whereas the observed values are more like 10^{14} cm^{-2}. Maybe there are many small scale clumps, each of which contributes an interface. Alternatively, the O VI may originate by the cooling of the hot coronal gas, induced by the passage of the Stream. These two origins can be separated when data from C IV and N V absorption lines become available for the sightlines where O VI has been detected.

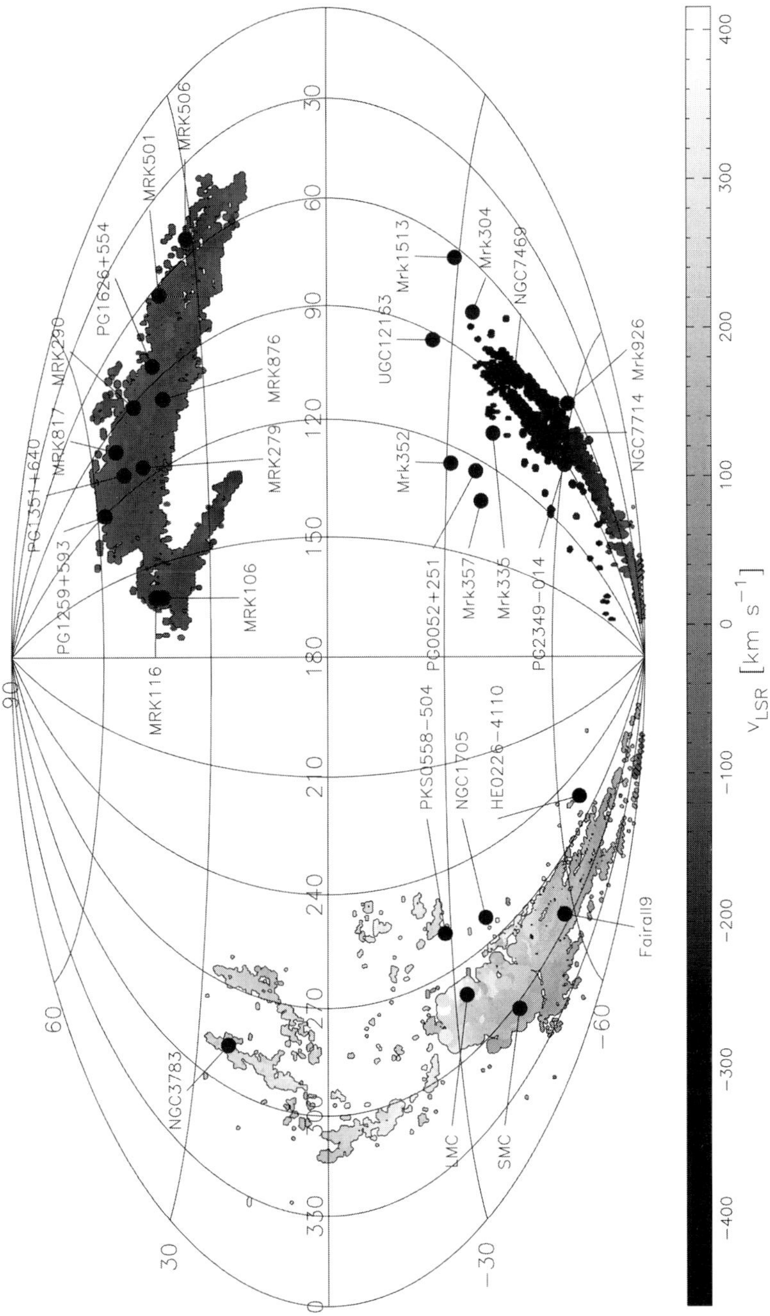

Figure 2. Map of the velocity field of the Magellanic Stream, and complexes A and C. Toward Fairall 9 and NGC 3783 the S II abundance has been determined to be 0.25 times solar; toward the other labeled AGNs Sembach et al. (2003) report O VI at velocities similar to those seen in the Magellanic Stream. Abundances and O VI data have been measured for all labeled targets projected on complex C.

4. Infalling Intergalactic Gas

The first HVCs discovered were complexes A and C (Muller et al., 1963, Hulsbosch and Raimond, 1966) ('B' is now considered part of A). These large (40° and 90° long, respectively), elongated (aspect ratio 1:10 and 2:10), northern-sky (b>25°) high-velocity clouds (v_{LSR}=$-$200 to $-$100 km s^{-1}) are considered the 'canonical' HVCs, and most of the proposed hypotheses for the origin of the HVCs take them as good examples.

Using *HST* data, Wakker et al. (1996) find a lower distance limit of 4 kpc for complex A (z>2.6 kpc above the plane). Wakker et al. (2003) improve this to 8 kpc, using a non-detection of C II toward PG 0832+675. Van Woerden et al. (1999a) reported Ca II absorption in the spectrum of the RR Lyrae star AD UMa, implying an upper limit of 10 kpc (z<7 kpc). This 8–10 kpc distance bracket applies to the higher-latitude part of complex A. The lower-latitude part may be more distant, as it is narrower. The filamentary nature of complex A suggests that tidal forces may play a role in shaping it.

The distance of complex C remains unknown, and only a lower limit of z>4.3 kpc has been set (van Woerden et al., 1999b). Indirect arguments suggest that the distance probably is less than 25 kpc. On the other hand, measurements of its metallicity are now plentiful. First, Wakker et al. (1999) used a combination of S II, H I and Hα data to derive a sulphur abundance of 0.09$\pm$0.02 solar in the direction toward Mrk 290. Combining an estimate of the mass of complex C with its infall velocity, Wakker et al. (1999) derived that this HVC represents an infall of 0.1–0.2 $M_\odot$ yr^{-1}.

An present-day infall rate of 1 $M_\odot$ yr^{-1} is theoretically required to explain the 'G-dwarf problem', but such infall was not previously observed. The 'G-dwarf problem' (van den Bergh, 1962) refers to the fact that if one takes a box filled with interstellar gas and lets it evolve, the theory predicts a continuous increase in the heavy element abundance of stars over time. This is not observed and the favored explanation has been that the interstellar gas is being diluted by more pristine material, in such a way that after an initial quick rise, the heavy element abundances in the gas stay more or less constant over time.

The low metallicity of complex C was corroborated by Richter et al. (2001b), who used FUSE to derive O I/ H I=0.1 solar and S II/ H I=0.14 solar toward PG 1259 +593. Intriguingly, Richter et al. (2000) also find that N I/ H I=0.01 solar. Unpublished FUSE data show that N II/ H I is also ~0.01 solar. Similar ratios of N/O are seen in some Damped Lyα systems and some dwarf galaxies. The low value of N/O suggests that the metals in complex C were mostly produced by high mass stars, with little contribution from evolved stars.

A complication with this interpretation is that Gibson et al. (2001) derived higher metallicity values (S II/ H I=0.2 solar toward Mrk 817). Some, but not all, of the difference can be explained by ionization effects. Tripp et al. (2003) analyze the complex C absorptions toward 3C 351, and after a proper ionization correction they also find that O I/ H I=0.2 solar in this direction. Considering all the evidence, Tripp et al. (2003) suggest that these apparent metallicity variations may result from the early phases of the mixing of the original complex C gas with higher metallicity Milky Way gas in the Galactic Halo.

5. Highly-Ionized HVCs

Wakker et al. (2003) and Sembach et al. (2003) report on the *FUSE* sample of 100 extragalactic targets. They find 85 high-velocity O VI components in 58 of these directions.

For all nine sightlines projected onto complex C, O VI is seen with centroid velocities within 20 km s^{-1} of those of the H I. Figure 3 shows the four cases with the highest S/N ratio. N(H I) and N(O VI) do not correlate, but N(O VI) is larger at higher longitudes and latitudes. Fox et al. (2003, in preparation) add observations of N V and C IV. They conclude that the O VI is most likely produced in a conductive interface between the neutral gas in complex C and a hot Galactic halo.

O VI associated with the Magellanic Stream is discussed in Section 3.

In the region l=175° to 360°, b>20° it is common to see O VI absorption at velocities between +100 and +250 km s^{-1}. Such a component is seen in 17 of the 27 sightlines in this part of the sky, often as an extended wing. Figure 3 shows two good examples. Sembach et al. (2003) offer a number of possible explanations for these components. One possibility is that they trace 10^5 K gas flowing out from the Galaxy, maybe even as part of the Galactic Fountain. Although most Fountain models assume that the outflowing gas is hotter (>10^6 K), some warm gas could be entrained, or the Fountain may have a lower temperature, as was proposed by Houck and Bregman (1990). Such a low-temperature fountain can also explain the kinematics of the IVCs. Alternatively, the broad O VI absorption wings might represent tidal debris associated with the Magellanic Stream or gas in the Local Group. Additional observations of other ions (C III, C IV, N V) should help discriminate between these models.

Finally, in the region l=30° to 180°, b<−20° there are many O VI absorptions with high negative velocities. The bottom row of Figure 3 shows two representative cases. Some of these sightlines show two components, one at $v\sim-150$ km s^{-1} and one at $v\sim-300$ km s^{-1}. The latter could be associated with the Magellanic Stream, but the former have velocities that are too low to fit in a tidal model. However, these velocities are similar to those of Local Group galaxies in this region of the sky, and therefore Sembach et al. (2003) suggest that they could represent hot gas in the Local Group.

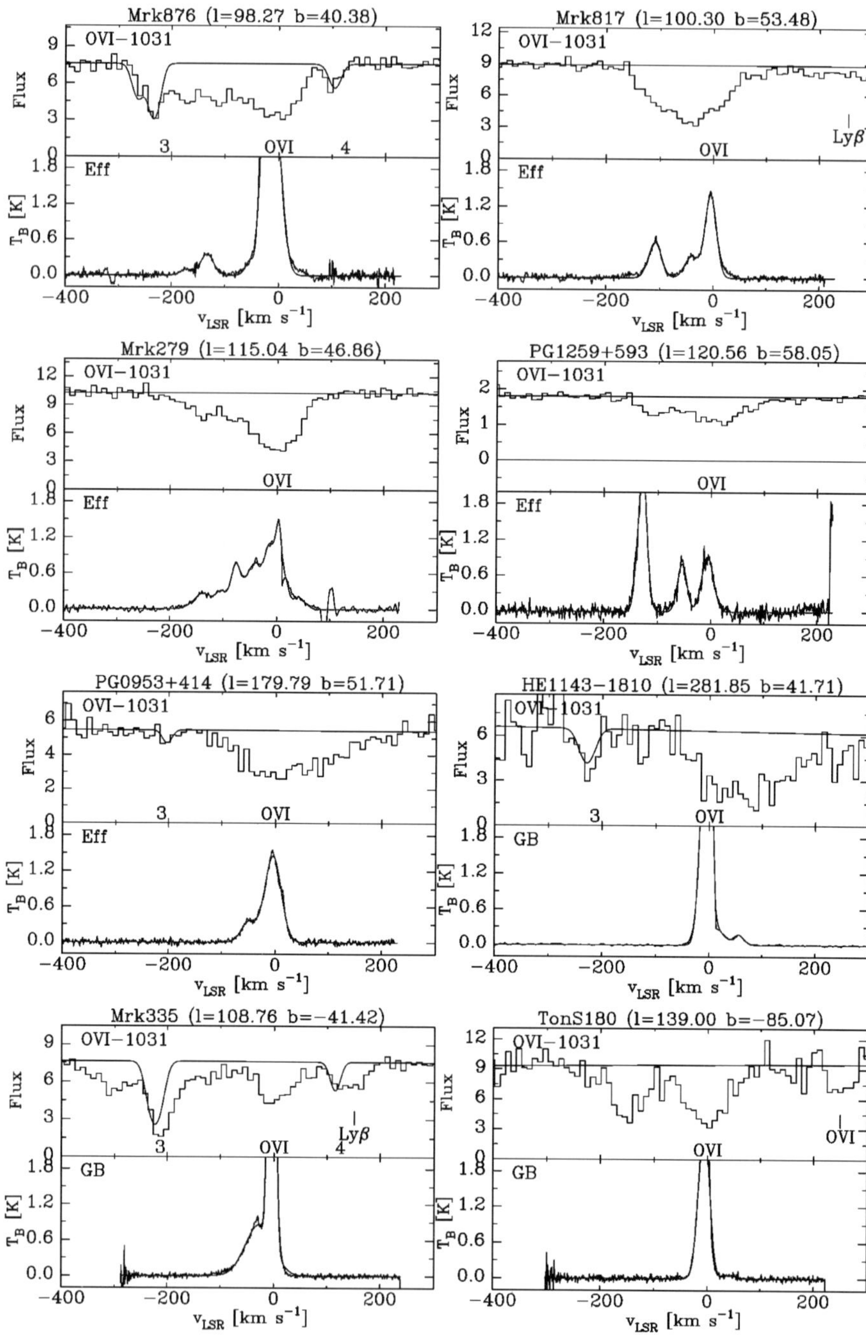

Figure 3. Examples of high-velocity O VI absorption, compared to the H I spectrum. The top four sightlines intersect complex C, and clearly show the association of O VI with H I. The bottom row is representative of sightlines with high-negative velocity O VI absorption.

6. HVCs in the Local Group

Blitz et al. (1999) resuscitated the idea that HVCs may be objects floating around in the Local Group, as remnants left over of its formation. The idea that HVCs are Local Group objects had already been put forward thirty year earlier (Kerr and Sullivan, 1969; Verschuur, 1969), In these early versions the observed 21-cm fluxes were combined with line widths, radii and the virial theorem to estimate a mass and distance. Since the clouds would only be stable at distances of several Mpc (and then have masses of $>10^9$ $M_{\odot}$), this suggestion was no longer considered valid. Blitz et al. (1999) proposed that 90% of the HVC's masses consists of dark matter, and that their 'stability distances' would then be 100–1500 Mpc, and their masses $\sim 10^{7-8}$ $M_{\odot}$. However, their calculations are not quite correct: I find that an average stability distance of $\sim$1 Mpc requires a dark matter fraction of >95%.

Blitz et al. (1999) made a number of predictions that would test their hypothesis. First, they suggested that Hα observations could discriminate between dense HVCs close to the Milky Way, bathed in ionizing radiation, and tenuous HVCs in the Local Group. Only the former should be detectable in Hα emission. So far, most observed HVCs have been found to show Hα emission (Weiner et al., 2001; Tufte et al., 2002). However, most of these are large clouds. and it is still possible that some of the HVCs with small angular sizes are more distant. The second proposed test was that of HVC metallicity, which was predicted to be similar to that of intergalactic clouds (0.1 solar and below). So far, a metallicity has been measured for the Magellanic Stream (0.25 solar) and for complex C (0.1 solar). Thus, complex C fits the model, although it is not very distant. For one small cloud (WW84), observations of the sightline to Mrk 205 seem to show a metallicity <0.05 solar, and this HVC may indeed be intergalactic.

Even if some HVCs are found to truly be far from the Milky Way, the Blitz et al. (1999) model is not necessarily correct. It is more likely that both the Milky Way and M 31 are surrounded by a 200 kpc diameter halo containing up to 100 clouds with masses $<10^7$ $M_{\odot}$, as proposed by de Heij et al. (2002). If the numbers or masses of HVCs were greater at larger distances, they should have been picked up in surveys for HVC-like objects around external galaxies, which was the third test proposed by Blitz et al. (1999). The deepest such surveys are the ones by Zwaan et al. (1997) and Zwaan (2001), but no starless H I clouds were found, even though Zwaan (2001) was sensitive to objects with M(H I)$>7\times10^6$ $M_{\odot}$. Further, Charlton et al. (2000) show that if the HVCs were as large and common as Blitz et al. (1999) proposed, there would be many more intergalactic Mg II absorbers.

References

Arnal, E.M., Bajaja, E., Larrarte, J.J., Morras, R. and Pöppel W.: 2000, *A&ApS* **142**, 35.
Blitz, L., Spergel, D., Teuben, P., Hartmann, D. and Burton, W.: 1999, *ApJ* **514**, 818.
Breitschwerdt, D., McKenzie, J.F. and Völk, H.J.: 1993, *A&Ap* **269**, 54.

Charlton, J.C., Churchill, C.W. and Rigby, J.R.: 2000, *ApJ* **544**, 702.
de Heij, V., Braun, R. and Burton, W.B.: 2002, *A&A* **392**, 417.
Gardiner, L.T. and Noguchi, M.: 1996, *MNRAS* **278**, 191.
Gibson, B., Giroux, M., Penton, S., Putman, M., Stocke, J. and Shull, J.: 2000, *AJ* **120**, 1830.
Gibson, B., Giroux, M., Penton, S., Stocke, J., Shull, J. and Tumlinson, J.: 2001, *AJ* **122**, 3280.
Goméz, G.C. and Cox, D.P.: 2002, *ApJ* **580**, 235.
Hartmann, D. and Burton, W.B.: 1997, *Atlas of Galactic Neutral Hydrogen*, Cambridge University Press (LDS).
Houck, J.C. and Bregman, J.N.: 1990, *ApJ* **352**, 506.
Hulsbosch, A.N.M. and Raimond, E.: 1966, *BAN* **18**, 413.
Kerr, F.J. and Sullivan, III W.T.: 1969, *ApJ* **158**, 115.
Kuntz, K.D. and Danly, L.: 1996, *ApJ* **457**, 703.
Lu, L., Savage, B., Sembach, K., Wakker, B., Sargent, W. and Oosterloo, T.: 1998, *AJ* **115**, 162.
Mathewson, D.S., Cleary, M.N. and Murray, J.D.: 1974, *ApJ* **190**, 291.
Moore, B. and Davis, M.: 1994, *MNRAS* **270**, 209.
Muller, C.A., Oort, J.H. and Raimond, E.: 1963, *C.R. Acad. Sci. Paris* **257**, 1661.
Putman, M.E. et al.: 1998, *Nature* **394**, 752.
Richter, P., Savage, B., Wakker, B., Sembach, K. and Kalberla, P.: 2001a, *ApJ* **549**, 281.
Richter, P., Sembach, K.R., Wakker, B.P., Savage, B.D. et al.: 2001b, *ApJ* **559**, 318.
Savage, B.D., Sembach, K.R. and Lu, L.: 1997, *AJ* **113**, 2158.
Savage, B.D., Sembach, K.R., Wakker, B.P. et al.: 2003, *ApJS*, astro-ph/0208140.
Sembach, K.R. and Savage, B.D.: 1992, *ApJS* **83**, 147.
Sembach, K.R., Wakker, B.P., Savage, B.D. et al.: 2003, *ApJS*, astro-ph/0207562.
Shapiro, P.R. and Field, G.B.: 1976, *ApJ* **205**, 762.
Tripp, T.M., Wakker, B.P. et al.: 2003, *ApJ*, submitted.
Tufte, S., Wilson, J., Madsen, G., Haffner, L. and Reynolds, R.: 2002, *ApJ* **572**, L153.
van den Bergh, S.: 1962, *AJ* **67**, 486.
van Woerden, H., Schwarz, U.J, Peletier, R.F., Wakker, B.P. and Kalberla, P.M.W.: 1999a, *Nature* **400**, 138.
van Woerden, H., Peletier, R.F., Schwarz, U.J., Wakker, B.P. and Kalberla, P.M.W.: 1999b, in: B.K. Gibson, T.S. Axelrod and M.E. Putman (eds.), *ASP Conf. Ser.* **165**, p. 469.
Verschuur, G.L.: 1969, *ApJ* **156**, 771.
Wakker, B.P.: 1991, *A&Ap* **250**, 499.
Wakker, B.P.: 2001, *ApJS* **136**, 463.
Wakker, B.P. and van Woerden, H.: 1997, *ARA&A* **35**, 217.
Wakker, B., Howk, C., Schwarz, U., van Woerden, H., Beers, T., Wilhelm, R., Kalberla, P. and Danly, L.: 1996, *ApJ* **473**, 834.
Wakker, B.P., Howk, J.C., Savage, B.D., van Woerden, H., Tufte, S.R., Schwarz, U.J., Benjamin, R., Reynolds, R., Peletier, R. and Kalberla, P.: 1999, *Nature* **400**, 388.
Wakker, B.P., Oosterloo, T.A. and Putman, M.E.: 2002, *AJ* **123**, 1953.
Wakker, B.P., Savage, B.D., Sembach, K.R. et al.: 2003, *ApJS*, astro-ph/0208009.
Weiner, B.J., Vogel, S.N. and Williams, T.B.: 2002, in: J.S. Mulchaey and J.T. Stocke (eds.), *ASP Conf. Proc.* **254**, p. 256.
Zwaan, M.A., Briggs, F.H. and Sprayberry, D.: 1997, *ApJ* **490**, 173.
Zwaan, M.A.: 2001, *MNRAS* **325**, 1142,

DYNAMICAL EVOLUTION OF HIGH VELOCITY CLOUDS

CHRISTIAN KONZ, GUIDO T. BIRK and HARALD LESCH
Institute for Astronomy and Astrophysics, University of Munich, Germany
also: Centre of Interdisciplinary Plasma Science, Garching, Germany

Abstract. HI observations of high-velocity clouds (HVCs) indicate that they are interacting with their ambient medium. The question on the dynamical and thermal stabilization of a cold dense neutral cloud in a hot, thin, and magnetized ambient halo plasma is investigated by plasma-neutral gas simulations. The simulations show the formation of a comet-like head-tail structure combined with a magnetic barrier which exerts a stabilizing pressure on the cloud and hinders hot plasma from diffusing into the cloud.

Keywords: high velocity clouds, galactic halos, Magellanic clouds, galactic magnetic fields

1. The Stability of HVCs

Absorption line studies indicate that HVCs mostly appear as almost pure neutral atomic hydrogen clouds. The bright and very extended HVC complexes consist of processed material that shows $\leq 1/3$ of the solar abundances (Wakker, 2001). Thus, the question of the dynamical and thermal stabilization of the cold dense neutral clouds that run through the Galactic hot plasma halo arises. The lifetime of HVCs, e.g, in the Magellanic Stream would be very limited, if pressure confinement played an important role. In this case, energy transfer would be very efficient and would not allow the clouds to survive long enough (Murali, 2000; Cowie and McKee, 1977). Magnetic dynamical confinement can be regarded as a promising alternative stabilization mechanism of HVCs (Konz et al., 2002). The interaction of the clouds with the magnetized halo results in the formation of magnetic barriers comparable to the phenomena that result from the interaction of the solar wind with unmagnetized planets and comets. This view is supported by the observational evidence for head-tail structures exhibited by numerous HVCs (Braun and Burton, 1999; Brüns et al., 2001).

2. Numerical Results

We study the dynamical evolution of a fast moving, cool, dense neutral gas cloud in a magnetized, hot ambient medium by two-dimensional ($\partial/\partial z = 0$) plasma-neutral gas fluid simulations. The balance equations for the mass and momentum densities and the pressures for both the neutral and the ionized fluid components as

Astrophysics and Space Science **289:** 391–394, 2004.

well as the magnetic field are numerically integrated by an explicit algorithm (for technical details see Birk and Otto, 1996). In our simulations (for details see Konz et al., 2002) a dense neutral gas cloud is moving in the y-direction impinging on a homogeneous magnetic field lying in the x-direction. Typical physical parameters for HVCs in the Magellanic Stream are chosen. The normalizing quantities for the simulations are given by the mass density $\hat{\rho} \approx 1.67 \times 10^{-27}$ g cm^{-3}, the Alfvén velocity $v_{\rm A} \approx 210$ km s^{-1}, the Alfvén transit time $\tau_{\rm A} \approx 416\,000$ yr, the temperature $\hat{T} \approx 1.3 \times 10^6$ K for a cloud with an extension of 100pc and a halo magnetic field of $B = 3\mu G$. The calculations are done in a domain with x going from -20 to 20 and y going from -40 to 20 and a uniform grid with 103 grid points in the x- and 153 grid points in the y-direction.

The initial configuration is characterized by a dense neutral gas cloud at rest centered at the origin $(x, y) = (0, 0)$ of the system having the form

$$\rho_{\rm n}(r) = \rho_{\rm min} + \frac{\rho_{\rm n_0}}{\cosh ar} . \tag{1}$$

Here, r gives the distance from the origin of the system while $a = 1/2$ defines the scale length for the neutral gas density gradient. The constants $\rho_{\rm min}$ and $\rho_{\rm n_0}$ define the minimum and the maximum neutral gas density and are chosen to be $\rho_{\rm min} = 1$ and $\rho_{\rm n_0} = 25$, respectively. The neutral gas is assumed to be initially in pressure equilibrium at a homogeneous pressure $p_{\rm n_0} = T_{\rm n_0}\left(\rho_{\rm min} + \rho_{\rm n_0}\right)$ where $T_{\rm n_0} = 1$ is the minimum temperature of the neutral gas cloud. The plasma initially has a homogeneous density $\rho_0 = 1$ and a temperature profile that is comparable to that of the neutral gas with a maximum of $T_0 = 10$. The friction between the plasma and the neutral gas is measured by the effective collision frequency $\nu = \hat{\nu}\rho_{\rm n}T^{1/2}$ with $\hat{\nu} = 8 \cdot 10^{-3}{\rm s}^{-1}$. Diffusivity is negligible. However, a small constant background diffusivity $\eta_0 = 10^{-5}$ is applied for numerical reasons. Initially, the plasma streams in the y-direction with the velocity profile

$$v_y(r) = -v_{y_0}\left(1 - \frac{1}{\cosh ar}\right) \tag{2}$$

with the amplitude $v_{y0} = 0.1$ in normalized units. The plasma flow is maintained at the upper y-boundary. The plasma is streaming past the neutral cloud in the negative y-direction and across an initially homogeneous magnetic field in the x-direction: $B_x = B_{x0} = 0.01$, $B_y = B_z = 0$. A complete representation of the dynamically evolved system after about 330 Alfvén times is shown in Figure 1. The arrows represent the flow of the plasma around the neutral gas cloud. The plasma flow is partly redirected sideways similar to a hydrodynamic stream around a solid obstacle. Since the magnetic field lines are frozen in the plasma, magnetic flux is transported toward the leading edge of the cloud while the field lines on both sides of the cloud are draped around the cloud. Thus, a magnetic barrier forms at the front and a magnetotail at the wake of the HVC. The magnetic barrier dynamically stabilizes the HVC, i.e., the stability of the neutral cloud results from the magnetic

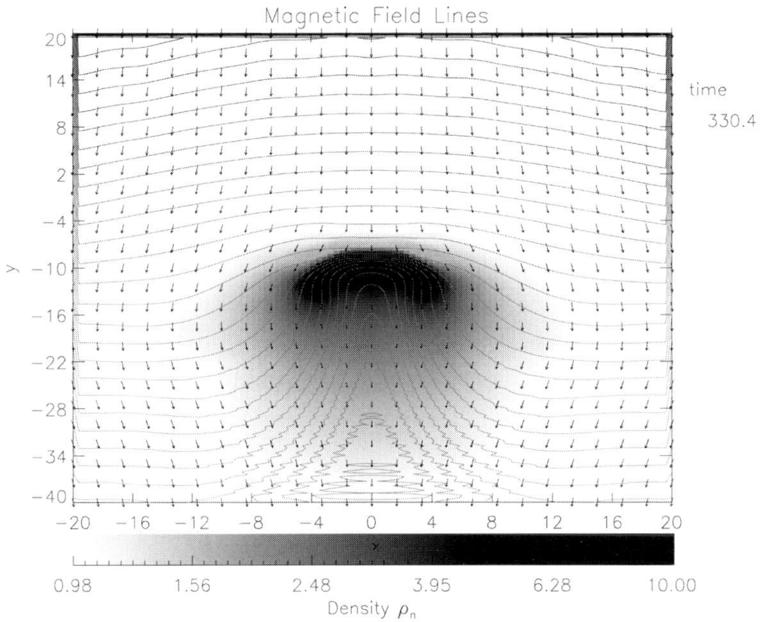

Figure 1. The HVC in the halo after $t \approx 330\tau_A$. The grey scale, the arrows and the solid lines indicate the neutral gas density, the plasma flow and the magnetic field, respectively.

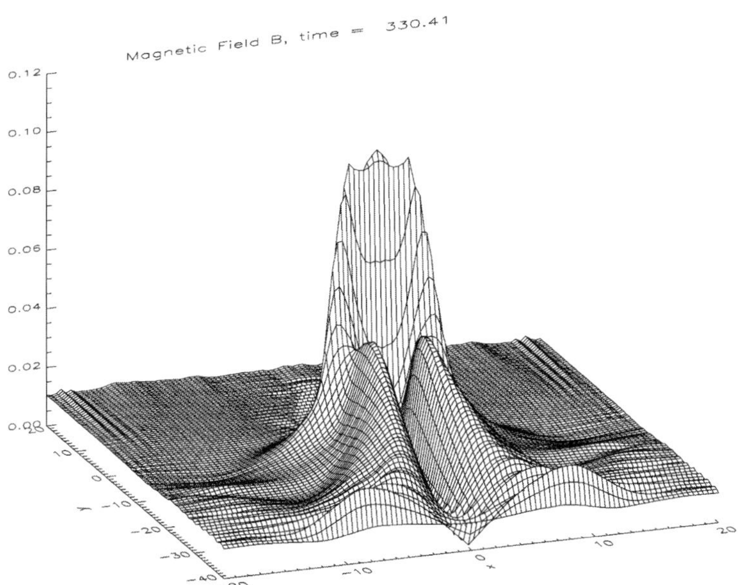

Figure 2. The magnetic barrier after $t \approx 330\tau_A$.

confinement by the accumulated magnetic flux. Even for the chosen relatively weak initial magnetic field unstable Kelvin-Helmholtz modes are suppressed at the boundaries of the cloud by the magnetic field components parallel to the flow. At the front of the HVC, the magnetic field strength is increased by a factor 10, in the tail by a factor 4 (Figure 2). The effect of ionization by relative cloud–halo motions (Konz et al., 2001), which is not included in the present simulations, is expected to further enhance the magnetic barrier and thereby the cloud stabilization. The tail formation is consistent with the observations of head-tail structures.

It should be noted that, in principle, the thermal conduction perpendicular to the magnetic field should be drastically reduced compared to the one parallel to the field. The inclusion of thermal conduction in the energy equation as well as ionization/recombination effects are principal tasks for future work.

References

Birk, G.T. and Otto, A.: 1996, A three-dimensional plasma-neutral gas-fluid code, *J. Comp. Phys.* **125**, 513–525.

Braun, R. and Burton, W.B.: 1999, The kinematic and spatial deployment of compact, isolated high-velocity clouds, *Astron. Astrophys.* **341**, 437–450.

Brüns, C., Kerp, J. and Pagels, A.: 2001, Deep HI observations of the compact high-velocity cloud HVC 125+41-207, *Astron. Astrophys.* **370**, L26–L30.

Cowie, L.L. and McKee, C.F.: 1977, The evaporation of spherical clouds in a hot gas. I. Classical and saturated mass loss rates, *Astrophys. J.* **211**, 135–146.

Konz, C., Lesch, H., Birk, G.T. and Wiechen, H.: 2001, The critical velocity effect as a cause for the Hα emission from the Magellanic stream, *Astrophys. J.* **548**, 249–252.

Konz, C., Brüns, C. and Birk, G.T.: 2002, Dynamical evolution of high velocity clouds in the intergalactic medium, *Astron. Astrophys.* **391**, 713–723.

Murali, C.: 2000, The Magellanic stream and the density of coronal gas in the galactic halo, *Astrophys. J.* **529**, L81–L84.

Wakker, B.P.: 2001, Distances and metalicities of high- and intermediate-velocity clouds, *Astrophys. J. Supp.* **136**, 463–535.

INFALL OF GAS IN GALAXIES AND TRIGGERED STAR FORMATION

FERNANDO COMERÓN
European Southern Observatory, Karl-Schwarzschild-Str. 2, D-85748 Garching bei München, Germany

Abstract. While the importance of merging, accretion, and infall processes in determining galactic evolution is well established both theoretically and observationally, details on how such processes are taking place nowadays even in our own Galaxy are still relatively poorly known, especially due to large remaining uncertainties on the location and origin of high velocity clouds. In this paper we focus on the possible role that galactic outflows and gas infall may have on directly triggering star formation in the halo and in galactic disks. While compelling evidence has been accumulating in recent years suggesting that some level of star formation directly triggered by outflows is very likely to exist in the halo of some galaxies, the evidence for star formation dynamically triggered by infall is far more elusive due to confusion with other, more efficient large-scale star forming mechanisms operating in the galactic disk. Despite of increasingly realistic simulations of the gas circulation betwen the gas and the halo and of high velocity cloud impacts on galactic disks, the efficiency of star formation directly triggered by such impacts remains an open question.

Keywords: infall, large-scale star formation, galactic outflows, high velocity clouds

1. Infall: Need and Evidence

Ample evidence exists nowadays on the importance of aggregation processes in the buildup and evolution of structures in the Universe (e.g. Bertschinger, 1998). The observation of protogalactic clumps at high redshifts, and the remarkable differences in galaxy morphologies between the early universe and the local one (Ferguson et al., 2000; Venemans et al., 2002), testify to the importance of mergers in progressively giving shape to the present-day population of galaxies (Freeman and Bland-Hawthorn, 2002; Keel, 2002). Hierarchical merging, by which dwarf galaxies are accreted by larger ones, has been recognized to play an important role in the formation and evolution of galactic haloes up to the present day, most notably in the case of our own Milky Way (e.g. Unavane et al., 1996; Johnston, 1999; Helmi and de Zeeuw 2000; Bullock et al., 2001; Palma et al., 2002).

In the context of these aggregation processes, we can regard the infall of gas on galactic disks as the lower end in the accretion hierarchy. Extended infall of external gas on galactic disks is needed by current galactic evolution models as a way to maintain star formation rates similar to the presently observed ones in normal spirals (Takeuchi and Hirashita, 2000; Semelin and Combes, 2002), as well as to explain the observed patterns in the chemical evolution of the Milky Way (e.g. Dwek, 1998; Chiappini et al., 1997; Wakker et al., 1999; Chiappini, 2002; Geiss et

Astrophysics and Space Science **289:** 395–403, 2004.

al., 2002) that require a continued supply of low metallicity or even primordial gas. At least to some extent, the observational counterpart of this gas accretion stream is likely to be linked to the existence of High Velocity Clouds (HVCs) observed in ours and other galaxies (see reviews by Wakker and van Woerden, 1997; Wakker et al., 1999). However, even four decades after their discovery, the characteristics, origin, and role of high velocity clouds in the context of galaxy evolution are far from being well established. In particular, their ability to provide with the needed amount of gas at sufficiently low metallicity, and their possible role in shaping the gaseous layer of galactic disks, possibly even directly triggering star formation, are still unclear.

2. Primordial Infall vs. Galactic Fountain

The external origin of HVCs as leftovers of the building blocks of the local group was proposed shortly after their discovery in the early 1960s (Oort, 1970). A related scenario is the tidal stripping and further accretion of low-metallicity gas from dwarf irregular companions, which in the case of the Magellanic Stream is directly observed (Matthewson et al., 1974; Putman et al., 1999; Wakker et al., 2002). On the other hand, the dynamical picture of the interstellar medium developed in the 1970s, with the recognition of the important role played by localized events such as supernova explosions and winds from massive stars (McKee and Ostriker, 1977), led to the realization that the circulation of gas between the disk and the halo should be a natural consequence of the energetic processes taking place in galactic disks, thus favoring the interpretation of infall as gas returning to the disk. In the *galactic fountain* scenario (Shapiro and Field, 1976; Bregman, 1980; Shapiro and Benjamin, 1991; Kahn, 1994; de Avillez, 2000), gas infall, and particularly HVCs, result from thermal instabilities in a hot gaseous halo that is continuously replenished with hot gas produced by supernovae and stellar winds in the disk. With a variety of refinements that take into account factors such as the clustering of the main sources of hot gas in galactic disks (Ikeuchi, 1988; Norman and Ikeuchi, 1989; Li and Ikeuchi, 1990; de Avillez and Berry, 2001), the galactic fountain is an integral part of current global models of the interstellar medium (de Avillez, 1999). From the observational point of view, the outgoing component of the galactic fountain has been studied in a number of nearby edge-on galaxies, in particular those having central starbursts, which display outflows of hot gas directly observable at different wavelengths (Lehnert and Heckman, 1995, 1996, and the review by Dahlem, 1997); M 82 is probably the best documented example (e.g. Lehnert et al., 1999).

While it is well established that the properties of gaseous halos are correlated with the star forming activity in the underlying disk (Dahlem, 1997), thus supporting the link between both, analogous correlations seem to exist between the high velocity HI masses of external galaxies (interpreted as HVC) and their star forming activity as traced by the integrated far-infrared flux (Schulman et al., 1994, 1996), providing further evidence for the link between HVCs and the galactic fountain.

Theoretical developments supporting the viability of the galactic fountain model, the determination of upper limits to the distances of some high and, especially, intermediate velocity clouds placing them within a few kpc from the Sun, and the determination of non-primordial metallicities by means of their detection in absorption against background probes (Wakker, 2001) favored the interpretation of HVCs as components of the galactic halo during the 1980s and 1990s. However, interest in HVCs as possible members of the local group unrelated to the Milky Way has been recently stirred by the work of Blitz and collaborators (Blitz et al., 1999; Blitz, 2002), leading to a reassessment of the arguments for and against an origin of HVCs in the galactic fountain. Arguments in favor of an extragalactic origin are the realization of the importance of accretion of dwarf galaxies in understanding galaxy evolution, the prediction by CDM-based structure formation models of a low-mass tail in galaxy groups represented by dwarf irregulars and possibly HVCs, the subsolar metallicity of most HVCs for which a reliable metallicity determination is available (Wakker et al., 1999; Wakker, 2001; Gibson et al., 2001a; Sembach et al., 2002) that could account for the low-metallicity gas infall required by galaxy evolution models as noted in the previous Section, and the identification of a population of compact HVCs with kinematical properties apparently relating them to the Local Group rather than to the Milky Way. The evidence is far from unambiguous however, as discussed by Gibson et al. (2001b, 2002), and a definitive conclusion is severely hampered by the very limited data on metallicities, and especially distances, existing so far. Intense observational efforts currently underway trying to detect a stellar population associated to the compact, isolated HVCs that lend the strongest support to Blitz et al.'s hypothesis for a Local Group origin have so far produced only upper limits that leave this question unsolved for the time being (Davies et al., 2002; Simon and Blitz, 2002).

3. Outflow/Inflow and Star Formation

The evidence for the exchange of large amounts of gas between the disk and the halo, be it by means of accretion of primordial material or through a galactic fountain-like mechanism, has led to speculations on the possible occurrence of star formation triggered by the dynamical interaction of gas supersonically entering the galactic disk.

3.1. Outflow-Triggered Star Formation in Galactic Haloes

More recently however, the possibility that gas outflows may also trigger star formation in exotic environments far removed from usual star forming sites such as the giant molecular cloud complexes in the disks of spiral galaxies has been the subject of considerable interest. In our own Milky Way, the existence of high-latitude B-type stars sharing all the characteristics of normal early-type main sequence stars, but lying at distances of the disk too large to have been achieved by formation in the disk and subsequent ejection as runaway (e.g. Brown et al., 1989; Conlon et al., 1992; Hambly et al., 1993; Little et al., 1995) has led to the suggestion of *in situ* formation, perhaps in instabilities within HVCs or via collisions among them (Dyson and Hartquist, 1983; Keenan and Lennon, 1984). Examples of out-of plane (and perhaps even out-of-galaxy!) star formation are now observed or suspected in other galaxies as well, and in such cases the relationship to outflow phenomena seems more clearly established. The best studied case is that of NGC 5128 (Centaurus A), in which a group of blue stars and HII regions has been identified (Graham, 1998; Graham et al., 2002) far from the galaxy in the region where the relativistic jet produced at the center of that galaxy intersects a cloud of HI probably left over from a tidally disrupted companion (Charmandaris et al., 2000). Other examples have been identified or suggested of AGN-triggered star formation in nearby clouds (Pinkney et al., 1996; van Breugel and Dey, 1993), or in galaxies intersecting the path of the AGN jet (van Breugel et al., 1995; Best et al., 1997; Scannapieco et al., 2000). Evidence for star formation in galactic haloes possibly triggered by the superwinds produced at the central starburst of NGC 253 has been recently presented by Comerón et al. (2001, 2003). Gerhard et al. (2002) have presented observations of a HII region in the outer halo of NGC 4388, a Seyfert 2 galaxy in the Virgo cluster, suggesting that triggering of star formation in the galactic halo does not need an external source as extreme as the Centaurus A jet. The possible detection of halo B-type stars in M 31 by Hambly et al. (1995) may confirm that star formation in such conditions is possible even in the absence of any energetic central powering engine in the galaxy, with chimneys distributed across the galactic disk pouring streams of hot gas into the halo providing an alternative mechanism.

3.2. Inflow-Triggered Star Formation in Galactic Disks

The injection of momentum in the halo by jets, starburst superwinds, or chimneys powered by massive stellar association in the disk is likely to have a counterpart in the form of HVCs colliding with the galactic disk. The magnitude of such a momentum injection, and therefore its gas-dynamical consequences on the galactic disk, can be poorly assessed with currently available data given the very limited present knowledge on HVC masses and the sensitivity of the consequences of HVC impacts to the volume filling factors of the different gaseous phases present in the galactic disk. On the other hand, even if star formation were a common consequence of HVC impacts on galactic disks and a mode of star formation much more common than outflow-induced star formation as discussed earlier, presenting unambiguous evidence for it would require convincingly discarding all other much more efficient star formation mechanisms at work in galactic disks. Observational evidence of the violent interaction of high velocity clouds with the galactic disk has been proposed in a number of instances, both galactic and extragalactic (Heiles, 1984; Kulkarni and Mathieu, 1986; Mirabel and Morras, 1990; Meyerdierks, 1991; Saito et al., 1992; Stark et al., 1994; Kerp et al., 1996; Rand and Stone, 1996; Brosch et al., 1999; Tamanaha, 1995, 1997). In our own Galaxy, a number of studies have suggested that past HVC collisions with the galactic disk may have played an important role in shaping the recent star formation history of the solar neighbourhood (Franco et al., 1988; Comerón and Torra, 1992, 1994; Lépine and Duvert, 1994; Edvardsson et al., 1995; Cabrera-Caño et al., 1995), in particular the Gould Belt (Pöppel, 1997). The support for HVC collisions as the cause of star formation in these instances is usually circumstantial, and typically based on the peculiar morphological arrangement of the resulting star forming complexes (an exception to this is Edvardsson et al.'s (1995) support for a HVC impact of the Orion complex, based on the slightly subsolar metallicity of its stellar population which may have resulted from the mixing between solar-metallicity galactic gas and low metallicity HVC gas). In distant regions of our Galaxy and in extragalactic complexes the interpretation is made more difficult by the lack of a three-dimensional picture of the object under study, which can be however obtained in the solar neighborhood, and support for the HVC impact hypothesis is usually based on energetic considerations and the comparison with the required input from supernovae. In recent years interpretations based on localized events more energetic than supernovae, such as Gamma-Ray Bursts, have been seriously considered as possible alternatives (Efremov et al., 1999).

3.3. Modelling Infall-Induced Star Formation

The collision of HVCs with a galactic disk and the properties of the structures formed as a consequence of it have been considered by a number of authors (Tenorio-Tagle, 1980, 1981; Tenorio-Tagle et al., 1986, 1987; Comerón and Torra, 1992; Santillán et al., 1999) usually by means of increasingly detailed and realistic numerical simulations, both in terms of spatial resolution and of input physics. Numerical simulations typically follow the evolution of a HVC with an initial size, density, and velocity starting from the time when it encounters either the boundary between the warm HI galactic layer and the hot halo gas, or a density in the former that is high enough to begin producing significant dynamical effects in the HVC. At that stage a shock starts running through the HVC and a shocked layer of dense gas, contributed by both the HVC and the galactic disk, develops. Numerical simulations usually follow the evolution of the gas until some time after all the shock has run across the whole HVC, leaving as a result a dense layer that may eventually become unstable against its own gravity and condense in a large star forming complex (Franco et al., 1988; Comerón and Torra, 1994). Early simulations concentrated on the case of a head-on collision, in which the HVC falls perpendicular to the galactic plane. Comerón and Torra (1992) introduced a velocity component parallel to the galactic plane, finding some important differences with respect to the head-on collision case concerning the evolution of the shocked layer (Comerón and Torra, 1994). Oblique impacts have also been considered by Santillán et al. (1999), and are included naturally in self-consistent 3-D models of the evolution of the galactic HI layer as ejected gas in the halo becomes thermally unstable and returns to the disk (de Avillez, 1999; de Avillez and Berry, 2001).

The emphasis of numerical simulations normally lies in the description of the dynamical consequences that the collision has on the galactic layer of warm gas and the way in which different components of the interstellar medium, such as the presence or absence of a large-scale magnetic field, affect the shape and extent of the resulting disturbances. By comparison, relatively little attention has been paid to the initial characteristics of the HVC, for which models have usually explored a rather restricted range of parameters. It is interesting however to note that existing observations, although not determining the initial HVC parameters due to the large distance uncertainty, do impose correlations among them that constrain their possible effects. For instance, although obviously a large and dense HVC should be able to produce a major disturbance in the gas layer near the galactic midplane (Comerón and Torra, 1992), the observed column densities and projected extent on the sky imply that, if HVCs are distant, they are large but not dense, while the opposite is true if they are at short distances. Both cases may however produce a similar outcome in preventing that any significant disturbances reach the proximities of the galactic plane. Recent numerical simulations (Comerón, 2004, in prep.) take as input parameters those derived from actual observations of HVC Complex A, a chain of roundish condensations for which a distance fork, and thus limits on size and density, are actually available (Wakker, 2001). The main preliminary result of that work is that the exponential tail in the vertical distribution of HI is in principle a very efficient protection preventing that the effects of a ‘Complex A-like’ impact may be noticeable in the proximities of the galactic midplane. Additional resistance is built in the galactic HI layer by the pressure and tension of the field lines of the large-scale magnetic field, as shown by Santillán et al. (1999), which may even be able to reverse the momentum of the infalling gas and turn the impact into a low-velocity outflow. Nevertheless, several factors may increase the ability of the effects of the collision to penetrate deeper in the galactic HI layer. On one side, the complex structure of HVCs, with hot cores embedded in diffuse haloes (e.g. Wakker and van Woerden, 1997; Brüns et al., 2000) and the large density inhomogeneities that form by instabilities and phase transitions in the shocked layer make more complex the dynamics of the collision, make it possible that individual dense clumps formed at large heights over the galactic plane reach its proximities. On the other, assuming an initially plane-parallel layer of warm HI in hydrostatic equilibrium, as models usually do, is known to be an oversimplification with potentially important consequences on the assessment of the effect of the HVC collision; self-consistent models of the duty cycle of the warm gas and its circulation between the disk and the halo, and actual observations of the large-scale structure of the interstellar medium, suggest a fairly different picture (de Avillez, 2000), with aspects such as the volume filling factor of the different phases of the interstellar gas, their clumpiness, and the vertical structure of the density, pressure, and magnetic field probably having a major importance as well.

4. Concluding Remarks

While the importance of infall in galactic evolution is well acknowledged, the details as to how infall takes place and the properties of the HVCs which most likely are the observational signature of infall are still far from being well established. The distances to HVCs continue to be the single most important factor of uncertainty regarding their origin, their location, their mass, and their effects.

Here we have focused on a particular aspect of the consequences of gas infall on galactic disks, namely the possibility that they directly trigger star formation. While few but compelling examples discovered in the last decade suggest that galactic *outflows* can directly trigger star formation in the exotic environment of the galactic halo, the evidence for infall-induced star formation by HVC impacts on galactic disks is more unclear. Suggestive examples do exist, even in the solar neighborhood, but confirmation is elusive and requires both multiwavelength observations and robust modeling of the observed structures in order to rule out other competing, unrelated possibilities. With the powerful observing facilities at all wavelengths that have entered operations in recent years, and the not less powerful computational resources nowadays available, this is certainly an exciting challenge for observers and modellers alike.

References

de Avillez, M.A.: 1999, *ASP Conf. Ser.* **166**, 103.
de Avillez, M.A.: 2000, *Ap&SS* **272**, 23.
de Avillez, M.A. and Berry, D.L.: 2001, *MNRAS* **328**, 708.
Bertschinger, E.: 1998, *ARA&A* **36**, 599.
Best, P.N., Longair, M.S. and Röttgering, H.J.A.: 1997, *MNRAS* **292**, 758.
Blitz, L., Spergel, D.N., Teuben, P.J., Hartmann, D. and Burton, W.B.: 1999, *ApJ* **514**, 818.
Blitz, L.: 2002, *ASP Conf. Ser.* **254**, 215.
Bregman, J.N.: 1980, *ApJ* **236**, 577.
van Breugel, W.J.M. and Dey, A.: 1993, *ApJ* **414**, 563.
van Breugel, W.J.M., Filippenko, A.V., Heckman, T. and Miley, G.: 1985, *ApJ* **293**, 83.
Brosch, N., Almoznino, E., Wszolek, B. and Rudnicki, K.: 1999, *MNRAS* **308**, 651.
Brown, P.J.F., Dufton, P.L., Keenan, F.P., Boksenberg, A., King, D.L. and Pettini, M.: 1989, *ApJ* **339**, 397.
Brüns, C., Kerp, J., Kalberla, P.M.W. and Mebold, U.: 2000, *A&A* **357**, 120.
Bullock, J.S., Kravtsov, A.V. and Weinberg, D.H.: 2001, *ApJ* **548**, 33.
Cabrera-Caño, J., Moreno, E., Franco, J. and Alfaro, E.J.: 1995, *ApJ* **448**, 149.

Charmandaris, V., Combes, F. and van der Hulst, J.M.: 2000, *A&A* **356**, L1.
Chiappini, C.: 2002, *Ap&SS* **281**, 253.
Chiappini, C., Matteuci, F. and Gratton, R.: 1997, *ApJ* **477**, 765.
Comerón, F. and Torra, J.: 1992, *A&A* **261**, 94.
Comerón, F. and Torra, J.: 1994, *A&A* **281**, 35.
Comerón, F., Torra, J., Méndez, R.A. and Gómez, A.E.: 2001, *A&A* **366**, 796.
Comerón, F., Gómez, A.E. and Torra, J.: 2003, *A&A* **400**, 137.
Conlon, E.S., Dufton, P.L., Keenan, F.P., McCausland, R.J.H. and Holmgren, D.: 1992, *ApJ* **400**, 273.
Dahlem, M.: 1997, *PASP* **109**, 1298.
Davies, J., Sabatini, S., Davies, L., Linder, S., Roberts, S., Smith, R. and Evans, R.: 2002, *MNRAS* **336**, 155.
Dwek, E.: 1998, *ApJ* **501**, 643.
Dyson, J.E. and Hartquist, T.W.: 1983, *MNRAS* **203**, 1233.
Edvardsson, B., Pettersson, B., Kharrazi, M. and Westerlund, B.: 1995, *A&A* **293**, 75.
Efremov, Y., Ehlerová, S. and Palous, J.: 1999, *A&A* **350**, 457.
Ferguson, H.C., Dickinson, M. and Williams, R.: 2000, *ARA&A* **38**, 667.
Franco, J., Tenorio-Tagle, G., Bodenheimer, P., Różyczka, M. and Mirabel, I.F.: 1988, *ApJ* **333**, 826.
Freeman, K. and Bland-Hawthorn, J.: 2002, *ARA&A* **40**, 487.
Geiss, J., Gloecker, G. and Charbonnel, C.: 2002, *ApJ* **578**, 862.
Gerhard, O., Arnaboldi, M., Freeman, K.C. and Okamura, S.: 2002, *ApJ* **580**, L121.
Gibson, B.K., Giroux, M.L., Penton, S.V., Stocke, J.T., Shull, M.J. and Tumlinson, J.: 2001a, *AJ* **122**, 3280.
Gibson, B.K., Giroux, M.L., Stocke, J.T. and Shull, M.J.: 2001b, *ASP Conf. Ser.* **240**, 493.
Gibson, B.K., Fenner, Y., Maddison, S.T. and Kawata, D.: 2001, *ASP Conf. Ser.* **254**, 225.
Graham, J.A.: 1998, *ApJ* **502**, 245.
Graham, J.A. and Fassett, C.I.: 2002, *ApJ* **575**, 712.
Hambly, N.C., Conlon, E.S., Dufton, P.L., Keenan, F.P., Little, J.E. and Miller, L.: 1993, *ApJ* **417**, 706.
Hambly, N.C., Fitzsimmons, A. Keenan, F.P., Dufton, P.L., Brown, P.J.F., Irwin, M.J. and Rolleston, W.R.J.: 1995, *ApJ* **448**, 628.
Heiles, C.: 1984, *ApJS* **55**, 585.
Helmi, A. and de Zeeuw, P.T.: 2000, *MNRAS* **319**, 657.
Ikeuchi, S.: 1988, *Fund. Cosm. Phys.* **12**, 255.
Johnston, K.V.: 1999, *ASP Conf. Ser.* **165**, 64.
Kahn, F.D.: 1994, *Ap&SS* **216**, 325.
Keel, W.C.: 2002, *The Road to Galaxy Formation*, Springer.
Keenan, F.P. and Lennon, D.J.: 1984, *A&A* **130**, 179.
Kerp, J., Mack, K.-H., Egger, R., Pietz, J., Zimmer, F., Mebold, U., Burton, W.B. and Hartmann, D.: 1996, *A&A* **312**, 67.
Kulkarni, S.R. and Mathieu, R.: 1986, *Ap&SS* **118**, 531.
Lehnert, M.D. and Heckman, T.M.: 1995, *ApJS* **97**, 89.
Lehnert, M.D. and Heckman, T.M.: 1996, *ApJ* **462**, 651.
Lehnert, M.D., Heckman, T.M. and Weaver, K.A.: 1999, *ApJ* **523**, 575.
Lépine, J.R.D. and Duvert, G.: 1994, *A&A* **286**, 60.
Little, J.E., Dufton, P.L., Keenan, F.P., Hambly, N.C., Conlon, E.S., Brown, P.J.F. and Miller, L.: 1995, *ApJ* **447**, 783.
Li, F. and Ikeuchi, S.: 1990, *ApJS* **73**, 401.
Matthewson, D.S., Cleary, N.M. and Murray, J.D.: 1974, *A&A* **190**, 291.
McKee, C.F. and Ostriker, J.P.: 1977, *ApJ* **218**, 148.
Meyerdierks, H.: 1991, *A&A* **251**, 269.
Mirabel, I.F. and Morras, R., *ApJ* **356**, 130.

Norman, C.A. and Ikeuchi, S.: 1989, *ApJ* **345**, 372.
Oort, J.H.: 1970, *A&A* **7**, 381.
Palma, C., Majewski, S.R. and Johnston, K.V.: 2002, *ApJ* **564**, 736.
Pinkney, J., Holtzmann, J., Garasi, C. et al.: 1996, *ApJ* **468**, L13.
Pöppel, W.G.L.: 1997, *Fund. Cosm. Phys.* **18**, 1.
Putman, M.E., Gibson, B.K. and Staveley-Smith, L.: 1999, *ASP Conf. Ser.* **165**, 113.
Rand, R.J. and Stone, J.M.: 1996, *AJ* **111**, 190.
Saito, M., Sasaki, M., Ohta, K. and Yamada, T.: 1992, *PASJ* **44**, 593.
Santillán, A., Franco, J., Martos, M. and Kim, J.: 1999, *ApJ* **515**, 657.
Scannapieco, E., Ferrara, A. and Broadhurst, T.: 2000, *ApJ* **536**, L11.
Schulman, E., Bregman, J.N. and Roberts, M.S.: 1994, *ApJ* **423**, 180.
Schulman, E., Bregman, J.N., Brinks, E. and Roberts, M.S.: 1996, *AJ* **112**, 960.
Sembach, K.R., Gibson, B.K., Fenner, Y. and Putman, M.E.: 2002, *ApJ* **572**, 178.
Semelin, B. and Combes, F.: 2002, *A&A* **388**, 826.
Shapiro, P.R. and Field, G.B.: 1976, *ApJ* **205**, 762.
Shapiro, P.R. and Benjamin, R.A.: 1991, *PASP* **103**, 923.
Simon, J.D. and Blitz, L.: 2002, *ApJ* **574**, 726.
Stark, R., Dickey, J.M., Burton, W.B. and Wennmacher, A.: 1994, *A&A* **281**, 199.
Takeuchi, T.T. and Hirashita, H.: 2000, *ApJ* **540**, 217.
Tamanaha, C.M.: 1995, *ApJ* **450**, 638.
Tamanaha, C.M.: 1997, *ApJS* **109**, 139.
Tenorio-Tagle, G.: 1980, *A&A* **88**, 61.
Tenorio-Tagle, G.: 1981, *A&A* **94**, 338.
Tenorio-Tagle, G., Bodenheimer, P., Różyczka, M. and Franco, J.: 1986, *A&A* **170**, 107.
Tenorio-Tagle, G., Franco, J., Bodenheimer, P., Różyczka, M. and Franco, J.: 1987, *A&A* **179**, 219.
Unavane, M., Wyse, R.F.G. and Gilmore, G.: 1996, *MNRAS* **278**, 727.
Venemans, B.P., Kurk, J.D., Miley, G.K., Röttgering, H.J.A., van Breugel, W., Carilli, C., L., de Breuck, C.L., Ford, H., Heckman, T., McCarthy, P. and Pentericci, L.: 2002, *ApJ* **569**, L11.
Wakker, B.P.: 2001, *ApJS* **136**, 463.
Wakker, B.P. and van Woerden, H.: 1997, *ARA&A* **35**, 217.
Wakker, B.P., van Woerden, H. and Gibson, B.K.: 1999, *ASP Conf. Ser.* **166**, 311.
Wakker, B.P., Oosterloo, T.A. and Putman, M.E.: 2002, *AJ* **123**, 1953.

STARBURST GALAXIES – AN OBSERVER'S VIEW

MICHAEL DAHLEM
European Southern Observatory, Casilla 19001, Vitacura, Santiago 19, Chile★

Abstract. Starbursts are the most efficient producers of metals in the Universe at low redshifts. They produce enough energy to drive outflows of material from their disks. This makes them important objects to study in order to understand the chemical evolution not only of the interstellar medium (ISM) in the starburst galaxies themselves, but also of the intergalactic medium (IGM) in their vicinity. However, several key quantities of starbursts that are needed as input to models of their ISM are still ill-constrained. Some of these critical parameters are e.g. the metal abundances of hot ionized gas, the ionization state of warm ionized gas, the amount of energy deposited into the ambient by a starburst, the efficiency of its conversion into mechanical energy and thus the total kinetic energy of the star formation-driven outflows and their kinematics. The latter are important when considering under which circumstances matter energized by a starburst will reach the so-called 'blowout' condition, i.e. supersede the threshold energy starting at which local energy injection into the ISM can drive an outflow first into the halo (where metal re-distribution might be very efficient) and eventually out into intergalactic space. I will discuss here a few of these quantities, how we can measure them better than in the past, and in which way some of our observing techniques need to be improved in order to obtain better constraints from the data.

Keywords: Galaxies, starbursts, interstellar medium, disk-halo interactions, intergalactic medium

1. Overview

In this talk multi-wavelength observation techniques of starburst galaxies are presented, with an emphasis on problem areas where we need to improve the quality of our results in order to place tighter constraints on models. I will concentrate here on observations of the interstellar medium (ISM) in starburst galaxies. The topics addressed here are: Important ISM properties, measurable quantities, problem areas, a few new observing techniques and information (to be) derived from these.

Although I will try to give a broad overview, the topic is too general for this presentation to be complete. For previous reviews on the ISM in starburst galaxies, their kinematics and superwinds, see e.g. Combes (1987), Heckman et al. (1990), Dettmar (1992), Dahlem (1997), Heckman (2001) and references therein.

Models of the ISM in galaxies, including spiral galaxies, were developed e.g. by Field et al. (1969), Cox and Smith (1974), Shapiro and Field (1976), McKee and Ostriker (1977), Habe and Ikeuchi (1980), Ikeuchi (1988), Rosen and Bregman

★ Current address: Australia Telescope National Facility, Paul Wild Observatory, Locked Bag 194, Narrabri NSW 2390, Australia; email: mdahlem@atnf.csiro.au

Astrophysics and Space Science **289:** 405–414, 2004.

(1995), Norman and Ferrara (1996). For the most recent results, see Mac Low or Korpi (this volume).

Articles on stellar populations of starbursts have been published by e.g. Leitherer, Maeder, Salpeter, Scalo, Schaerer and others. For recent results, see e.g. Kobulnicky (1999) and González Delgado (2002).

2. Important Quantities

In order to fully understand the workings of starbursts in galaxies we need to get a handle on several fundamental processes, such as:

1. Mechanism(s) of gas accretion leading to the onset of a starburst
Possible accretion mechanisms are e.g. non-axisymmetric perturbations of the galaxy potential caused by either a bar or a far-field interaction, which might lead to a disturbance of the gas kinematics, but hardly any visible distortion of the stellar disk (e.g. Combes 1987).

2. Amount of energy produced by starbursts and deposited into the ambient ISM
At present we usually use empirical numbers, such as the first-order estimate of the total energy released by a type II supernova of 10^{51} erg (McKee and Ostriker, 1977), multiplied by the supernova rate inferred from either far-infrared (FIR) or radio continuum data. Much remains to be done in terms of direct measurements of the total energy released and its conversion into kinetic energy (see Chevalier, 1974; Heckman, 2001).

3. Ionization states and emission mechanisms of various phases of a violent ISM
Fully consistent radiation transport calculations are extremely demanding and the various transitions through which the energy produced by high-mass star formation (SF) processes goes are not fully explored yet.

4. Propagation mechanisms and energy losses of particles
Particles like cosmic-rays also play an important role in distributing the energy produced by a starburst throughout the ISM in galaxies. Therefore, understanding their propagation and loss mechanisms is crucial (e.g. Breitschwerdt et al., 1991, 1993).

5. Chemical evolution of such a violent medium
Starbursts deposit not only radiation, but also metal-enriched material into their ambient medium, thereby influencing the chemical evolution of the star-forming galaxies. In the current Universe, starburst galaxies are the most efficient producers of metals. For recent results, see e.g. Mouhcine & Contini (2002) and references

therein.

6. Contribution of starbursts to the metal enrichment of the IGM
Radiation, energy and metals produced by starbusts are not only injected into the ambient ISM, but – once an energy threshold is surpassed – can leave the galaxy itself and contribute to the enrichment of the intergalactic medium (IGM; e.g. Heckman, 2001).

7. Cooling mechanisms of gas energized by star formation processes
With the advent of new infrared satellites, such as the Infrared Space Observatory (ISO), we are beginning to make progress in our understanding of the most important FIR cooling lines, which reprocess stellar UV radiation, in addition to the FIR dust continuum.

The [C II] (158 μm) and [O I] (63 μm) lines are most important in PDRs, while cooling via the [O III] (88 & 52 μm) and [N II] (122 μm) lines dominates in H II-regions (e.g. Malhotra et al. 2002).

8. Negative feedback of SF into the surrounding ISM
Our understanding of starbursts is not complete without knowledge about the negative feedback that SF activity causes by heating and diluting the ambient medium, thus inhibiting subsequent SF (e.g. Meurer et al., 1997).

Measurable quantities with which these processes can be characterized must be determined with the lowest possible inaccuracy in order to constrain models describing them in a self-consistent manner.

2.1. Example starburst galaxy: NGC 253

One example of a galaxy driving a starburst superwind is NGC 253. Although at a distance of only 2.6 Mpc, and although other tracers of halo emission have been found before (Schulz and Wegner, 1992; Carilli et al., 199;, Pietsch, 1993), only now an extended diffuse *warm* ionized medium in its halo could be found (see Figure 1).

This extraplanar diffuse ionized gas (eDIG) exhibits the typical exponential decay away from the disk plane, as also observed in other galaxies and their various gaseous halo components (e.g. Dahlem et al., 2001), and shown in Figure 2. Scale heights of 1.5 kpc (on the SE side) and 0.9 kpc (on the NW side) are measured, respectively.

The fact that this emission could be detected only now in such an actively star-forming nearby galaxy is just one example proving how demanding the acquisition of high signal-to-noise observations of the ISM in starburst galaxies can be. In the case of NGC 253, also a sufficiently large field-of-view is required to correctly

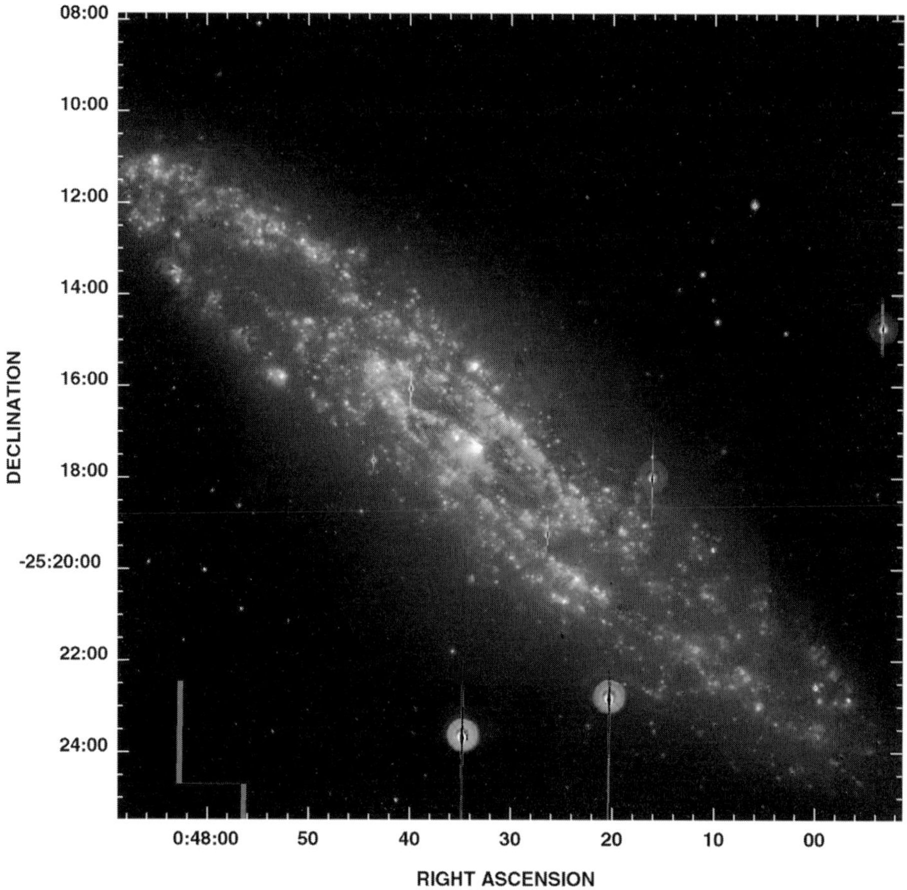

Figure 1. Previously unpublished mosaiced Hα image of NGC 253 obtained with the CTIO 4-m telescope, exhibiting the presence of a diffuse ionized medium in the halo of this starburst galaxy. In order to emphasize the faint halo emission, a logarithmic colour lookup table is used, ranging from 0 (black) to the maximum observed surface brightness (white).

determine the background sky level and thereby make the very faint halo emission visible.

3. Problem Areas

Several areas are observationally particularly difficult to tackle and therefore up to now poorly understood. Examples are:

1. Sources of ionization of warm ionized medium (WIM)
The relative importance of various processes (for different levels of energy input) is not clear. In some cases the existing data can be explained by pure photo-ionization

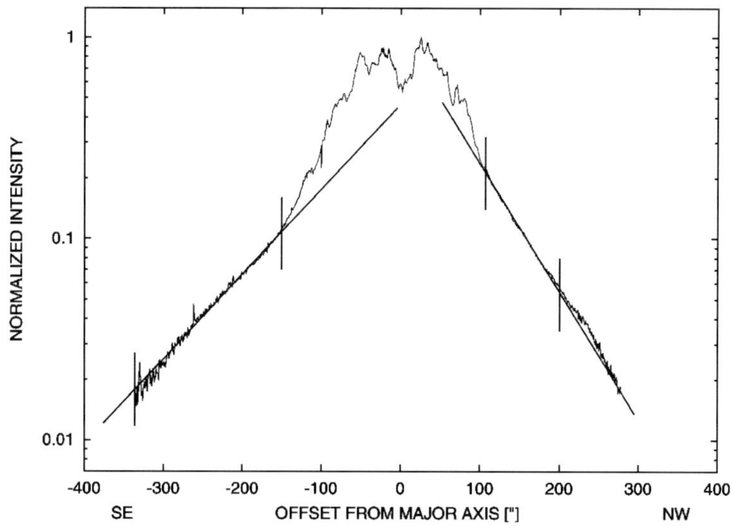

Figure 2. Cut through the Hα image in Figure 1, perpendicular to the disk plane of NGC 253, averaged over a radial range of ±1.7 kpc. Measurements of the exponential scale height of the extraplanar diffuse warm ionized gas were performed between the vertical lines (see text).

(e.g. NGC 2188; Domgörgen and Mathis, 1994; Domgörgen and Dettmar, 1997), while in others shocks and/or turbulent mixing layers must play a significant role too (Dahlem et al., 1997). It is also unknown yet exactly which role the presence of dust plays in photo-ionization dominated excitation of the WIM.

2. Temperature, ionization state and metallicity of hot ionized gas

Although most calculations and, in particular, X-ray spectral fits are performed based on the assumption that the gas is in local thermo-dynamic equlibrium (LTE), this is most likely not correct (Breitschwerdt and Schmutzler, 1994). The hot gas being in a non-LTE state has serious consequences on the interpretation of our observational results. The temperature of a non-LTE gas, for example, by comparing models with data, will turn out to be best fitted at a lower value than that of an LTE gas. Also the best-fitting metallicity will be influenced (see D. Breitschwerdt; this volume). Technically, metallicities are already quite uncertain because of a dichotomy between the fitted metallicity, Z, and the absorbing gas density, $N(\mathrm{H})$, as described by Dahlem et al. (2000).

3. Role of magnetic fields

The most important question about magnetic fields in starburst galaxies is whether their energy density is higher than that of other components of the ISM or not. In other words, do magnetic fields determine the direction of the flow of gas or does the flow of gas determine the B-field orientation? This connects to the question how important galactic dynamos are, see e.g. R. Beck or M. Hanasz (this volume).

4. Filling factors of various phases of the ISM
It is crucial for our understanding of the ISM in starburst galaxies to know how important the hot and warm ionized gaseous phases are and how much volume they occupy. This is relevant in the determination of the energy balance of the ISM (see e.g. E. Berkhuijsen; this volume). Also the small-scale structure of the ISM is important to know, because this bears consequences on the interdependence of different gaseous phases (e.g. R.-J. Dettmar; this volume).

5. Local energy input rates into the ISM
One of the determining parameters for the onset of disk-halo interactions and outflows of gas from a starburst into the IGM via a superwind is the amount of energy produced per unit time and unit surface area ($\dot{E}/A_{\rm SF}$) and for which energy input levels halos are observed. This threshold energy input was first quantitatively constrained by Dahlem et al. (1995). $\dot{E}/A_{\rm SF}$ and also the conversion efficiency of energy into kinetic energy are hard to measure for a seizable galaxy sample.

6. Measurement of superwind velocities
While halos can be detected most easily in edge-on galaxies, the measurement of outflow velocities in such systems is difficult, because motions occur perpendicular to our line of sight, thereby leading to small projected velocity components along the line of sight. In face-on galaxies, on the other hand, it is possible to measure the velocities of outflows, but the detection of a halo is very difficult.

7. Role of dust (cooling)
The fact that starburst galaxies are FIR-bright objects indicates that a significant fraction of the ionizing radiation produced by massive stars is absorbed and re-emitted before it leaves the galaxies. The amount of radiation re-processed this way, which will not be available for ionizing the ISM, is an important quantity to know; also because the influence of dust changes the spectrum of the ionizing radiation responsible for the heating of interstellar gas, which could explain some of our difficulties with the interpretation of the observed optical/UV line ratios.

8. Cause of gas condensation leading to a starburst
Where do starbursts form in the first place? Classical circumnuclear starbursts are located near the Inner Lindblad Resonance (ILR) of galaxies, which basically coincides with the location of the turnover of galactic rotation. Widespread high-level SF distributed over large areas of a galaxy disk is not called a starburst, but can still produce the same amount of energy, over a larger area. This type of SF occurs in spiral arms of the galaxies. An important, difficult to measure quantity in this context is the critical gas density at which the onset of a starburst will occur, i.e., how much gas must accumulate at a galactocentric radius of 500 pc, typically, for

a circumnuclear starburst ring to be created?

4. A Few Promising New Observing Techniques

In this section I present a few exemplary new observing techniques that can help to overcome some of the shortcomings mentioned above.

4.1. MULTI-SLIT OR INTEGRAL FIELD SPECTROSCOPY

While up to now we were often limited to spectroscopy with a long slit, new techniques have become available allowing us to produce three-dimensional datacubes, which can provide a more complete picture of the kinematics of starbursts and the excitation conditions of the gas therein. Interferometric data, for example, are normally taken in full imaging mode, with the velocity as the third dimension.

CO, H I and other lines, tracing high-density gas, provide crucial information on the kinematics of neutral gas in starburst galaxies, which can be used as inputs to models describing the large-scale gas kinematics and dynamics (e.g. in order to determine how gas accretes to form a circumnuclear starburst ring). Combined observational and theoretical studies provide information on the importance of non-axisymetric disturbances of the galaxies' gravitational potential in creating classical circumnuclear starburst rings (e.g. Combes, 1987; Lesch et al., 1990).

Integral field optical line studies provide line strengths and ratios along a large number of lines of sight, which can be used as inputs to models of the ionization of the ISM.

Radio or optical fully two-dimensional line studies can also be used to measure outflow velocities of gas associated with starbursts in face-on systems, providing inputs to models of the kinematics of starburst-driven superwind outflows. Such data are needed e.g. to assess whether escape velocity is reached, but also to measure the dynamical timescales for backfall of material onto the disk.

In general: Absorption lines provide high sensitivity, with which one can obtain information on gas in more distant galaxies than accessible via emission processes (Heckman et al., 2000).

4.2. OBSERVATIONS OF EXPANDING SHELLS IN GALAXY DISKS

If expanding structues like gaseous shells can be identified, line data cubes, e.g. of the CO line, offer an independent measure of the local kinetic energy input into the ISM. Comparing these values with measures of the total energy injected, they also provide information on the conversion efficiency into kinetic energy.

The examples of expanding molecular shells in external starburst galaxies identified so far (Irwin and Sofue, 1996; Weiss et al., 1999; Rand, 2000; Walter et al.,

2002) are located in rings, near the Inner Lindblad Resonances of the galaxies, from where the superwinds emerge.

4.3. High-fidelity broad-band X-ray observations

The new generation of X-ray satellites, *XMM-Newton* and *Chandra*, produce data that can pinpoint metallicities much better than earlier missions, due to their wide bandpass and superior angular and spectral resolution (and the combination of these technical improvements!).

Imaging provides information about the distribution of hot gas, which is required for the determination, amongst other important quantities, of the volume filling factor of the hot ionized medium. Spectroscopy yields metallicity measurements, which lead to the resolution of the Z vs. N(H) dichotomy and thus improved input to models of the chemical evolution of starburst galaxies. As an outstanding example I want to mention the work by Martin et al. (2002) on the loss of oxygen from NGC 1569 into the IGM. Such new input leads to a significant improvement of the input spectral models, which are needed for a correct interpretation of the hot gaseous phase in starburst galaxies (see Dahlem et al., 2000; Persic and Rephaeli, 2002).

4.4. Far-infrared/(sub-)mm emission

The FIR regime is one of the technically most demanding and, accordingly, the one in which we need to make a lot more progress. Good angular resolution is required to obtain FIR images and integral field spectroscopy of starburst galaxies. This will render the distribution of dust in galaxies, its temperature and other key parameters, information needed to assess the amount of ionizing radiation reprocessed by dust – an important parameter of the cooling function in ISM models.

4.5. High-sensitivity, multi-frequency radio imaging

The strength of magnetic fields in the ISM of starburst galaxies can reach levels of several ten μGauss. The technical means of observing B-fields is by multi-frequency *polarization* radio continuum imaging. This provides not only a measure of the field strengths, but also information on their structure and on the relative importance of ordered and turbulent field components (see e.g. R. Beck; this volume).

High-sensitivity radio continuum total power maps show the distribution of cosmic-rays (CRs) and provide a precise measure of the level of SF (and thus energy input) into galaxy disks.

Radio spectral index maps, which are obtained from multi-waveband continuum imaging, yield information on the propagation of CRs and the importance of inverse Compton and synchrotron losses and on the amount of thermal radio continuum emitted directly by SF regions.

There are, of course, many more – equally meritous – new techniques that cannot all be mentioned here.

5. Summary

The above points out individual areas where observations of starburst galaxies are still difficult and how some problems can be overcome. The most important message is the power of multi-wavelength observations, which do not just provide single bits and pieces of information, but, put together, improve our overall view of these objects. They serve as inputs to constrain theoretical models of the ISM in starburst galaxies and of the superwinds driven by them (e.g. Mac Low or Korpi; this volume).

Acknowledgements

It is a pleasure to thank my collaborators T. Heckman, M. Ehle, U. Lisenfeld, K. Weaver, and D. Strickland for their contributions. Special thanks to N. Stuhrmann for reducing the Hα data of NGC 253.

References

Breitschwerdt, D., McKenzie, J.F. and Völk, H.J.: 1991, *A&A* **245**, 79.
Breitschwerdt, D., McKenzie, J.F. and Völk, H.J.: 1993, *A&A* **269**, 54.
Breitschwerdt, D. and Schmutzler, T.: 1994, *Nature* **371**, 774.
Carilli, C.L., Holdaway, M.A., Ho, P.T.P. and De Pree, C.G.: 1992, *ApJ* **399**, L59.
Chevalier, R.A.: 1974, *ApJ* **188**, 501.
Combes, F.: 1987, in: R. Pudritz and M. Fich (eds.), *Proceedings NATO Conference on Galactic and Extragalactic Star Formation*, Reidel, Dordrecht Boston.
Cox, D.P. and Smith, B.W.: 1976, *ApJ* **203**, 361.
Dahlem, M.: 1997, *PASP* **109**, 1298.
Dahlem, M., Lazendic, J.S., Haynes, R.F., Ehle, M. and Lisenfeld, U.: 2001, *A&A* **374**, 42.
Dahlem, M., Lisenfeld, U. and Golla, G.: 1995, *ApJ* **444**, 119.
Dahlem, M., Parmar, A., Oosterbroek, T. et al.: 2000, *ApJ* **538**, 555.
Dahlem, M., Petr, M.G., Lehnert, M.D. et al.: 1997, *A&A* **320**, 731.
Dettmar, R.-J.: 1992, *Fund. Cosm. Physics* **15**, 143.
Domgörgen, H. and Dettmar, R.-J.: 1997, *A&A* **322**, 391.
Domgörgen, H. and Mathis, J.S.: 1994, *ApJ* **428**, 647.
Field, G.B., Goldsmith, D.W. and Habing, H.J.: 1969, *ApJ* **155**, L49.
González Delgado, R.M.: 2002, in: Pérez, González Delgado and Tenorio-Tagle (eds.), *Star Formation through Time*, Astronomical Society of the Pacific, ASP Conference Proceedings, San Francisco, (in press).
Habe, A. and Ikeuchi, S.: 1980, *Prog. Theor. Physics* **64**, 1995.
Heckman, T.M.: 2001, in: J.E. Hibbard, M. Rupen and J.H. van Gorkom (eds.), *Gas and Galaxy Evolution*, Astronomical Society of the Pacific, ASP Conference Proceedings 240, San Francisco, p. 345.

Heckman, T.M., Armus, L. and Miley, G.K.: 1990, *ApJS* **74**, 833.
Heckman, T.M., Lehnert, M.D., Strickland, D.K. and Armus, L.: 2000, *ApJS* **129**, 493.
Ikeuchi, S.: 1988, *Fund. Cosm. Physics* **12**, 255.
Irwin, J.A. and Sofue, Y.: 1996, *ApJ* **464**, 738.
Kobulnicky, H.A.: 1999, in: K.A. van der Hucht, G. Koenigsverger and P.R.J. Eenens (eds.), *Proceedings of the 193rd Symposium of the International Astronomical Union*, Astronomical Society of the Pacific, San Francisco, p. 670.
Lesch, H., Biermann, P.L., Crusius, A. et al.: 1990, *MNRAS* **242**, 149.
Malhotra, S., Kaufman, M.J., Hollenbach, D. et al.: *ApJ*, in press (astro-ph/0106485).
Martin, C.L., Kobulnicky, H.A. and Heckman, T.M.: *ApJ* **574**, 663.
McKee, C.F. and Ostriker, J.P.: 1977, *ApJ* **218**, 148.
Meurer, G.R., Heckman, T.M., Lehnert, M.D. et al.: 1997, *AJ* **114**, 54.
Mouhcine, M. and Contini, T.: 2002, *A&A* **389**, 106.
Norman, C.A. and Ferrara, A.: 1996, *ApJ* **467**, 291.
Persic, M. and Rephaeli, Y.: 2002, *A&A*, in press, (astro-ph/0112030).
Pietsch, W.: 1993, in: G. Hensler et al. (eds.), *Panchromatic View of Galaxies – Their Evolutionary Puzzle*, Éditions Frontières, Gif-sur-Yvette Cedex, p. 137.
Rand, R.J.: 2000, *ApJ* **535**, 663.
Rosen, A. and Bregman, J.N.: 1995, *ApJ* **440**, 634.
Schulz, H. and Wegner, G.: 1992, *A&A* **266**, 167.
Shapiro, P.R. and Field, G.B.: 1976, *ApJ* **205**, 762.
Tomisaka, K. and Ikeuchi, S.: 1986, *PASJ* **38**, 697.
Walter, F., Weiss, A. and Scoville, N.: 2002, *ApJ* **580**, L21.
Weiss, A., Walter, F., Neininger, N. and Klein, U.: 1999, *A&A* **345**, L23. 1999.

MID-INFRARED PROPERTIES OF EARLY-TYPE GALAXIES ⋆

E.M. XILOURIS, S.C. MADDEN and L. VIGROUX
DAPNIA/Service d'Astrophysique, CEA/Saclay, 91191 Gif-sur-Yvette Cedex, France;
E-mail: xilouris@cea.fr

Abstract. We have conducted a survey of early-type galaxies using the ISOCAM instrument on board the *Infrared Space Observatory* (ISO) in order to probe their dust content. We present Mid-Infrared photometric properties for a sample of 20 galaxies observed at 4.5, 7 and 15 μm. Using diagnostic color-color diagrams, we can delineate the different Mid-Infrared emission contribution from evolved stars, non-thermal emission from AGNs/radio sources, and dust. Detailed modelling of the Spectral Energy Distribution (SED) is performed showing that even though the emission from the evolved stellar population may account for a large part of the Mid-Infrared output, emission from the hot dust and the aromatic hydrocarbon bands (e.g. PAHs) is the main contributor in many cases. Modelling of the surface brightness distribution is also made, showing that most of the galaxies follow a de Vaucouleurs profile. The physical nature and spatial distribution of the dust that we find from our studies of early-type galaxies will help to constrain the origin of the dust in those galaxies.

Keywords: mid-infrared astronomy, elliptical galaxies, dust

1. Introduction

IRAS has shown that up to 50% of early-type galaxies contain warm dust ($\sim$ 30 K) with dust masses of the order of $10^5 - 10^6$ solar masses (Knapp et al., 1992). Still, with the advent of ISO observations, the origin of the dust remains controversial. Some authors attribut the Mid-Infrared (MIR) excess to photosphere emission of AGB stars (e.g., Athey et al., 2002), others to the presence of interstellar hot dust (e.g., Ferrari et al., 2002) and others to both (Malhotra et al., 2002). Our goal is to model the physical properties of the dust and to determine the distribution of the dust in early-type galaxies in order to constrain the origin of the dust in those galaxies.

2. Observations

Our sample consists of 20 early-type galaxies with Hubble types ranging from lenticulars to ellipticals. Observations were made with the ISOCAM instrument on board the ISO satellite. All galaxies were observed using 3 broad-band filters centered at 4.5 (stellar emission), 6.7 (stellar emission + PAHs) and 15 μm (stellar

⋆ Partially supported by EAS

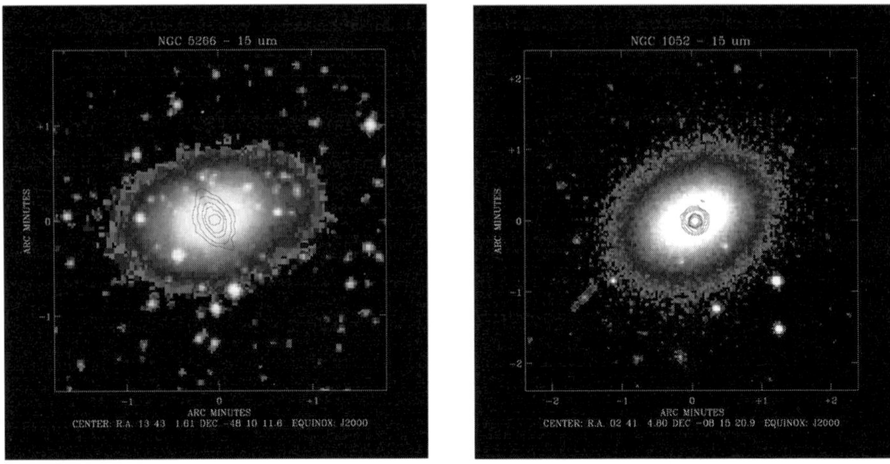

Figure 1. ISOCAM 15 μm emission maps (contours) superimposed on the optical images of NGC 5266 (left) and NGC 1052 (rigt).

emission + hot dust + radio continum) and a 6" pixel size. Additional observations were made with a 3" pixel size at 6.7 and 15 μm. At a distance of $\sim$ 20 Mpc (the mean distance of the galaxies in our sample) the 3" configuration corresponds to $\sim$ 300 pc while for the closest ones (0.7 Mpc) the 3" configuration correspond to $\sim$ 10 pc.

3. Discussion

Two examples of ISOCAM images are shown in Figure 1 for NGC 5266 (left panel) and for NGC 1052 (right panel). In both images the 3" maps of the 15μm emission are overplotted as contours on the optical images of the galaxies. A warped dusty disk is evident in NGC 5266 while a very compact source is seen in the radio galaxy NGC 1052.

The azimuthally averaged emission shows that the 4.5 μm stellar emission follows the de Vaucouleurs profile resonably well, while the 6.7 and 15 μm emission shows deviations from this distribution.

In Figure 2 we show the MIR SEDs for NGC 5266 (left) and for NGC 1052 (right). For each galaxy, the total SED is shown in upper panel, while the SED of the inner 20" aperture is shown in the lower panel. NGC 5266 shows enhanced 6.7 μm emission in the nuclear region, which is an unambiguous signature of the presence of PAHs. For NGC 1052 we see that the SED is dominated by the central part of the galaxy with an excess of 15 μm flux. The MIR SED of NGC 1052 is characteristic of a source dominated by radio continuum.

In Figure 3 we present the ISOCAM color-color diagram. The circles in this diagram show the colors of the total galaxy while the crosses show the colors of

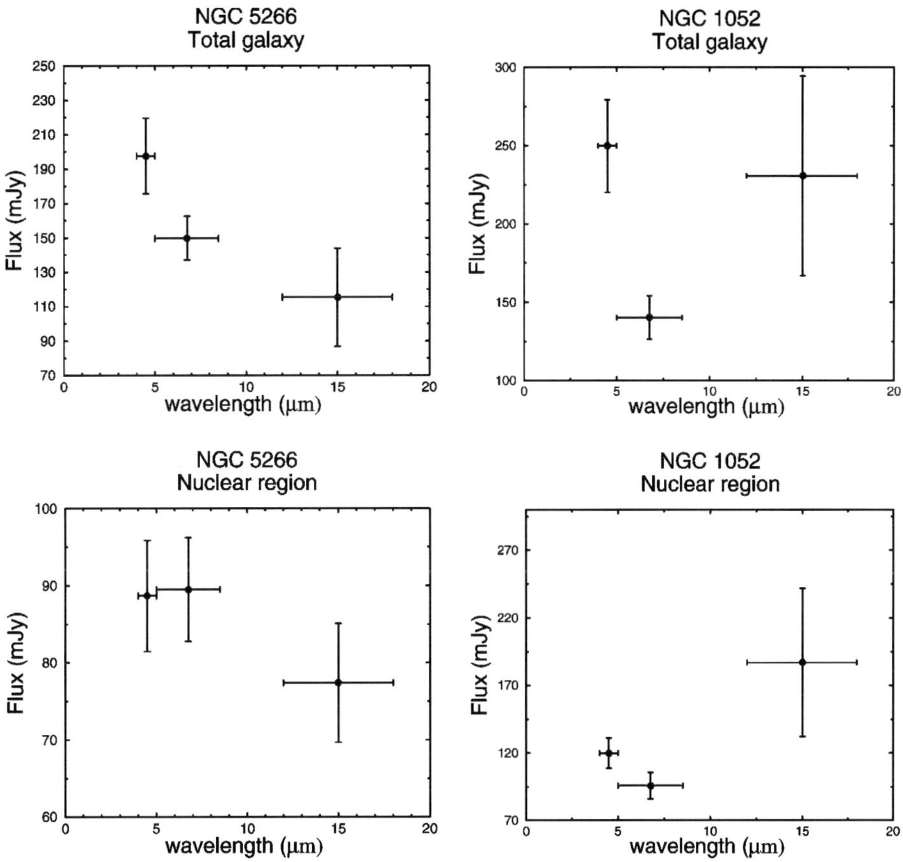

Figure 2. MIR SEDs for NGC 5266 (left) and for NGC 1052 (right). For each galaxy the plots on the top show the SED for the total galaxy while the plots on the bottom show the nuclear (within 20" aperture) SEDs.

the nuclear (inner 20") region of the galaxies. The straight line indicates points in this color-color diagram with black body temperatures ranging from 1000 K to 6000 K. Galaxies that lie close to this line are dominated by stellar emission. Galaxies occupying the left part of the diagram possess an excess of 15 μm flux and thus contain more hot dust, while galaxies occupying the bottom-right part of the diagram are dominated by PAHs. The nuclear colors of these galaxies tend to lie in the bottom-left part of this diagram indicating an excess of both hot dust and PAHs in the centers of the galaxies. A few active galaxies (such as NGC 1052) show strong excess of the 15 μm emission (typical of radio galaxies) and thus occupying a region outside this diagram.

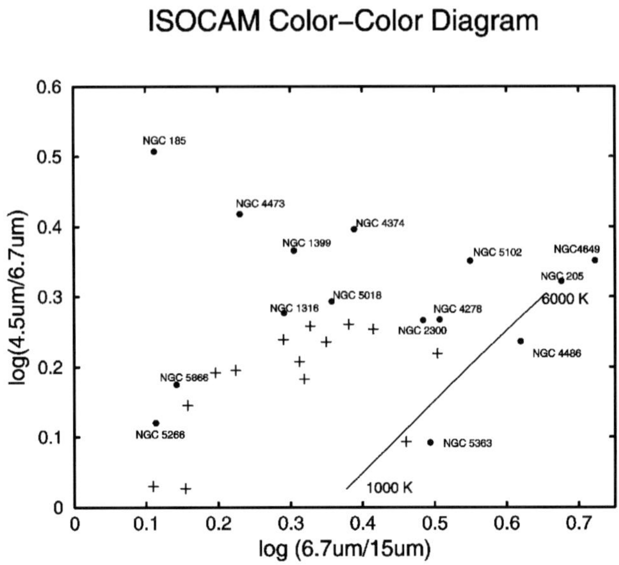

Figure 3. The ISOCAM color-color diagram of early-type galaxies for the total galaxy fluxes (circles) and for the nuclear fluxes (crosses).

4. conclusions

From the analysis of a sample of 20 early-type galaxies observed in the Mid-Infrared with ISO we derive the following results.

1. Most of the galaxies follow the de Vaucouleurs profile with some deviations which are especially evident in the 7 μm and 15 μm bands due to disks, dust lanes, compact sources, etc.
2. From color-color diagrams we can delineate different MIR emission contribution from evolved stars, warm dust, PAHs and non-thermal emission from strong AGN and radio sources.
3. The SEDs show that even though the emission from the evolved stars may account for a large part of the MIR output, emission from the warm dust and from PAHs is the main contributor in many cases.
4. Comparing the SEDs of the central region of the galaxy with the SEDs of the total galaxy we see a larger concentration of warm dust and of PAHs in the central region.

References

Athey, A., Bregman, J., Bregman, J. et al.: 2002, *ApJ* **571**, 272.
Ferrari, F., Pastoriza, M.G., Macchetto, F.D. et al.: 2002, *A&A* **389**, 335.
Knapp, G.R., Gunn, J.E. and Wynn-Williams, C.G.: 1992, *ApJ* **399**, 76.
Malhotra, S., Hollenbach, D., Helou, G. et al.: 2000, *ApJ* **543**, 634.

GAS IN GROUPS AND CLUSTERS OF GALAXIES

SABINE SCHINDLER
Institut für Astrophysik, Universität Innsbruck

Abstract. Groups and clusters contain a large fraction of hot gas which emits X-ray radiation. This gas yields information on the dynamical state and on the total mass of these systems. X-ray spectra show that heavy elements are present in the gas. As these metals must have been produced in the cluster/group galaxies and later transported into the gas, the metallicity is a good tracer for the transport processes. Several possible processes, that transport gas from the small potential wells of the galaxies into the clusters and groups, are discussed.

1. Gas Properties

Clusters and groups of galaxies do not only contain galaxies but all the space between the galaxies is filled with hot gas. This gas is also often referred to as intra-cluster medium (ICM) or intra-group medium (IGM). The high temperature of the gas of 1–10 keV is in good correspondence with the depth of the potential wells of groups and clusters (see Table I). According to the shallower potential of groups also the temperature is lower in groups ($\approx$ 1 keV) compared to clusters (3–10 keV). The gas is almost fully ionised and has very low densities of $10^{-2} - -10^{-4}$ cm^{-3} decreasing outwards. It is optically thin. In the spectra of both, groups and clusters, lines of heavy elements have been detected corresponding to metallicities of about 0.2 to 0.4 in solar units. These metals indicate that the gas cannot be of purely primordial origin, but part of it must have been previously the inter-stellar medium (ISM) in galaxies and then it must have been transported from the galaxies into the ICM and IGM, respectively, by certain processes. The processes will be discussed in the next section.

TABLE I
Properties of the gas in cluster and groups of galaxies

	Clusters	Groups
Temperature	3–10 keV	$\approx$ 1 keV
Extent	few Mpc	0.1–1 Mpc
Metallicity	0.2–0.4 solar	0.2 solar
X-ray luminosity	10^{43-45} erg/s	10^{42-43} erg/s

Astrophysics and Space Science **289:** 419–428, 2004.

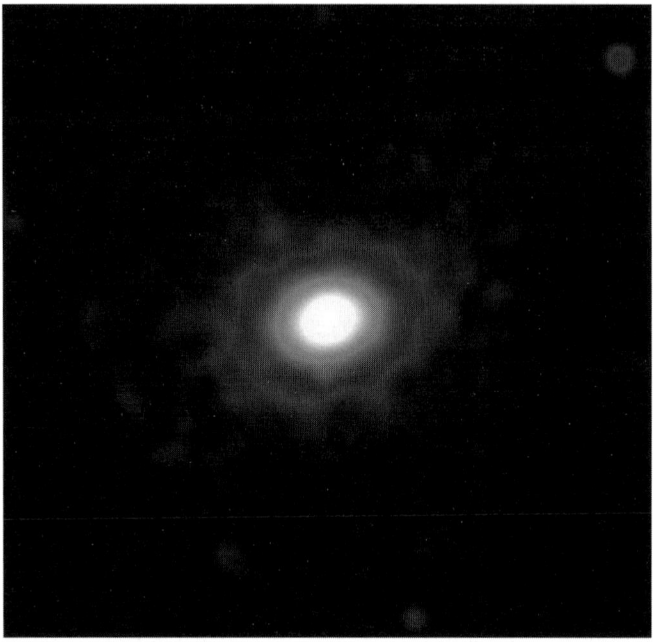

Figure 1. CHANDRA image of the cluster RBS797 (from Schindler et al., 2001). The cluster looks well virialised with a very regular surface brightness distribution.

Gas with such properties emits thermal bremsstrahlung in the X-ray range (see Table 1). Therefore groups and clusters are currently studied extensively by the X-ray observatories XMM and CHANDRA. While in almost all of the clusters X-ray emission has been found, the X-ray emission from groups is somewhat harder to detect due to their lower luminosity. So far in about 50% of the nearby groups X-ray emission has been found.

2. Gas Distribution, Dynamical State and Mass Determination

Two examples of X-ray emission from clusters of galaxies are shown in Figures 1 and 2. The cluster RBS797 (Figure 1) has a very regular shape on this scale with a somewhat elongated structure. This is a typical relaxed cluster. The cluster CL0939 +4713 (Figure 2) has a very different, irregular morphology. In the XMM image one can distinguish clearly two main subclusters. These subclusters show even some internal structure. Obviously the cluster is not in an equilibrium state, but there are still some structures visible that are in the process of falling into the cluster. So already the X-ray morphology contains a lot of information about the dynamical state of the cluster. But X-ray spectroscopy tells even more. In Figure 3 the temperature distribution of the gas in CL0939+4713 is shown. There is a hot region between the two subclusters which is an indication that the subclusters

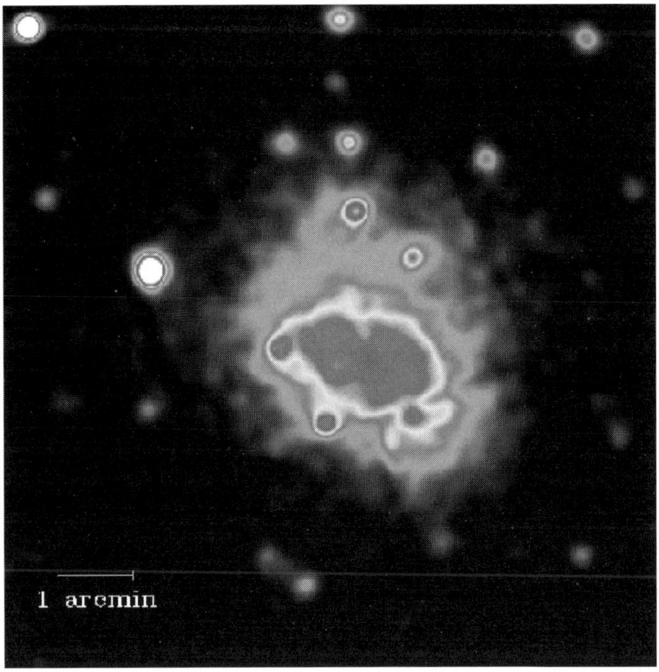

Figure 2. XMM image of the cluster CL0939+4713 (from De Filippis et al., 2002). Two subclusters are visible with even some internal structure indicating that the cluster is in a merger process.

are approaching each other – just as it is expected from hydrodynamic simulations (Schindler and Müller, 1993; Ricker and Sarazin, 2001; Ritchie and Thomas, 2002): when two subclusters start to merge the gas between them is compressed and heated, so that a hot region is produced. Therefore we conclude that this cluster in the beginning of a major merger process (De Filippis et al., 2002).

Also groups show extended emission coming from the IGM as it can be seen e.g. in HCG16 (Belsole et al., 2002) or in HCG90 (Longo et al., 2002). In groups the extended emission from the IGM is not so dominant as in clusters. One can usually see different components in the X-ray emission. Apart from the extended IGM also more compact components caused by galaxies and other sources can be seen.

The mass of the ICM and IGM is not negligible compared to the other components. Table II gives an overview of the mass fraction of gas, galaxies and dark matter in groups and clusters, respectively. In clusters there is always considerable more mass in the gas than in the galaxies, while in groups the mass in the galaxies usually exceeds the mass in the gas. Both systems have in common that most of the mass is in form of dark matter. This shows how important mass determination is. When the total mass of the system can be measured and the visible components are subtracted, the amount and distribution of the dark matter can be inferred.

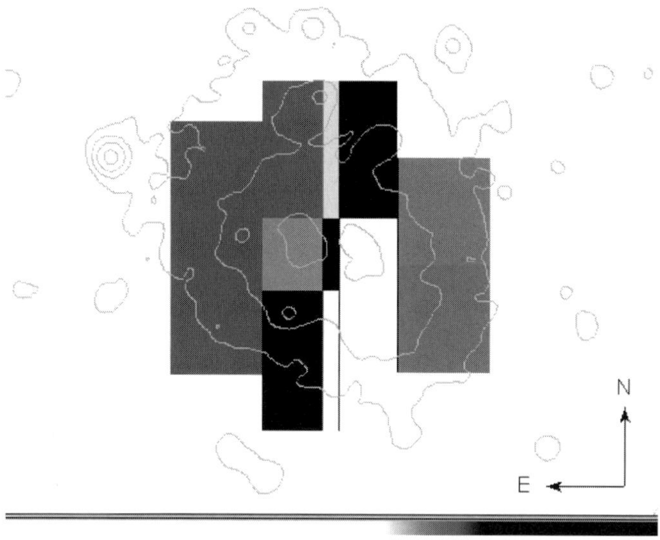

Figure 3. Temperature map of the cluster CL0939+4713 (from De Filippis et al., 2002). The temperature is increasing with increasing darkness of the grey scale. The contours show the same surface brightness distribution as in Figure 2. Between the two subclusters a region of hot gas is visible indicating that the subclusters are approaching each other.

TABLE II

Mass fraction of the different components in clusters and groups of galaxies

	Clusters	Groups
Galaxies	3–5%	3–20%
Gas	15–20%	2–10%
Dark matter	rest	rest

There are different ways to determine masses, one of them is the X-ray method. With two assumptions – spherical symmetry and hydrostatic equilibrium – the X-ray emitting gas can be used as a tracer for the total potential. Only the gas density and the gas temperature are required for this method, which can both be measured from X-ray observations.

With the gas density also the gas mass is known. The ratio between gas mass and total mass is the gas mass fraction. In Figure 4 the gas mass fraction is plotted versus the radius. We find that the gas mass fraction is not constant but is increasing with radius (Schindler, 1999; Castillo-Morales and Schindler, 2002). That means that the gas is more extended than the dark matter. Obviously, there is not only

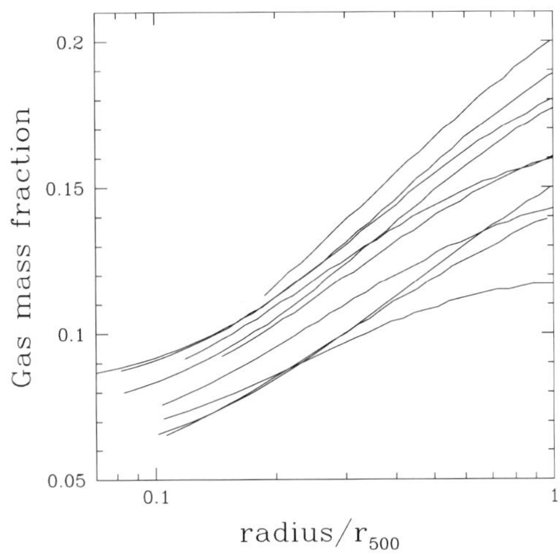

Figure 4. Gas mass fraction versus radius. For all the clusters the gas mass fraction increases with radius, i.e. the gas distribution is more extended than the dark matter distribution (from Castillo-Morales and Schindler, 2002).

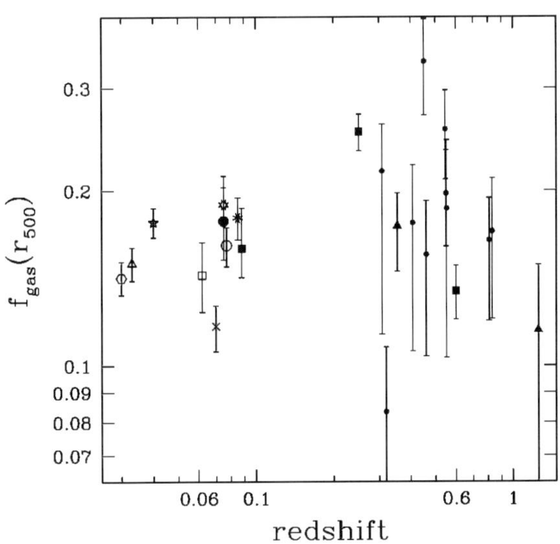

Figure 5. Gas mass fraction versus redshift. With the current data not significant evolution of the gas mass fraction in clusters is visible (from Castillo-Morales and Schindler, 2002).

gravitational energy from the collapse present but there must be additional heating processes at work like e.g. supernova-driven winds. Also interesting is the question whether the gas mass fraction is changing with time. In Figure 5 the gas mass fraction is shown for nearby and distant cluster samples. So far no significant evolution is detectable.

The gas mass fraction can also be used to derive an upper limit on the matter density of the universe Ω_m. Assuming that the matter accumulated in a cluster is representative for the universe as a whole, the ratio between the mass in baryons and the total mass $M_{gas}/M_{tot} \approx f_{gas}$ should be the same as Ω_{baryon}/Ω_m. As an upper limit for Ω_{baryon} can be determined from primordial nucleosynthesis to be about 0.06 (for $H_0 = 50$ km/s/Mpc) one finds $\Omega_m \approx 0.3 - 0.4$ in units of the critical density.

3. Interaction between ISM and ICM/IGM

The ICM and the IGM show metal lines in the X-ray spectra. These metals cannot have been produced in the gas, but they must have been produced in the galaxies and subsequently transported from the galaxies into the ICM/IGM by certain processes like e.g. ram-pressure stripping, galactic winds, galaxy-galaxy interaction or jets from active galaxies. The metallicity is the best indicator for finding out which of these processes are most important. Of special interest is the distribution of metals. So far there are only few examples of measured metallicity variations in real 2D maps and not only profiles. In CL0939+4713 we find different metallicity in the different subclusters (De Filippis et al., 2002). In the Perseus cluster also clear metallicity variations were found (Schmidt et al., 2002). 1D profiles are not very useful in this context because photons from regions in the cluster which are very far apart are accumulated in the same spectrum.

Apart from the metallicity distribution also the evolution of the metallicity is interesting. As soon as enough XMM and CHANDRA observations of distant clusters are available we can compare the metallicities in these clusters with those of nearby clusters. This is another way of distinguishing between the enrichment processes as different processes have different time dependence. In addition element ratios can be derived, e.g of Fe and α-elements to get information on the different types of supernovae that have contributed to the metal enrichment.

Various processes have been suggested for the transport of gas from the galaxies to the ICM/IGM. 30 years ago Gunn and Gott (1972) suggested ram-pressure stripping: as the galaxy moves through the cluster and approaches the cluster centre it feels the increasing pressure of the intra-cluster gas. At some point the galaxy is not able anymore to retain its ISM. The ISM is stripped off and lost to the ICM and with it all its metals. Many numerical simulations have been performed to investigate this process, first 2D models (Takeda et al., 1984; Gaetz et al., 1987; Portnoy et al., 1993; Balsara et al., 1994). With increasing computing power also more detailed 3D models could be calculated (Abadi et al., 1999; Quilis et al., 2000; Vollmer et al., 2001; Schulz and Struck, 2001; Toniazzo and Schindler, 2001). In Figure 6 such a simulated stripping process is shown for an elliptical galaxy.

Another possible process is galactic winds e.g. driven by supernovae (De Young, 1978). Also for this process simulations have been performed on order to see

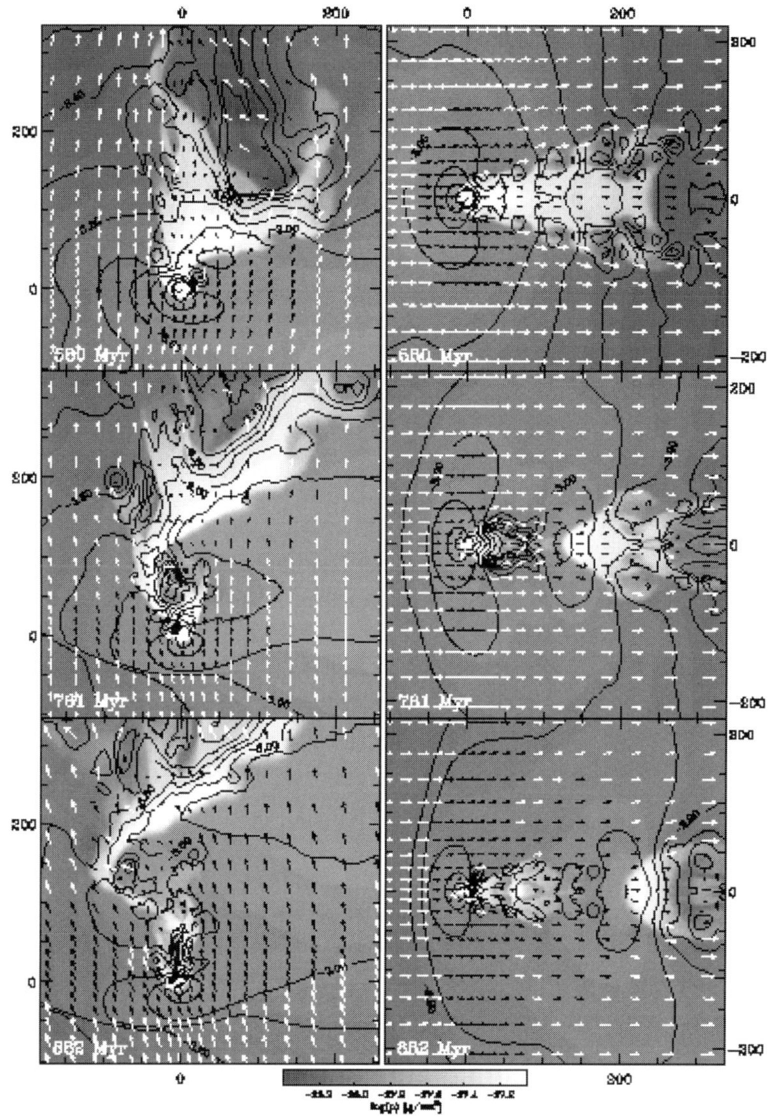

Figure 6. Gas density (grey scale) and pressure (contours) of a galaxy moving downwards towards the cluster centre. The arrows show the Mach vectors (white when $M > 1$, black otherwise). The gas of the galaxy is stripped due to ram pressure (from Toniazzo and Schindler, 2001).

whether only winds can account for the observed metallicities. The results were quite discordant as the following two examples show. Metzler and Evrard (1994, 1997) found that winds can account for the metals, while Murakami and Babul (1999) concluded that winds are not very efficient for the metal enrichment. In the simulations of Metzler and Evrard quite steep metallicity gradients showed up which are not in agreement with observations.

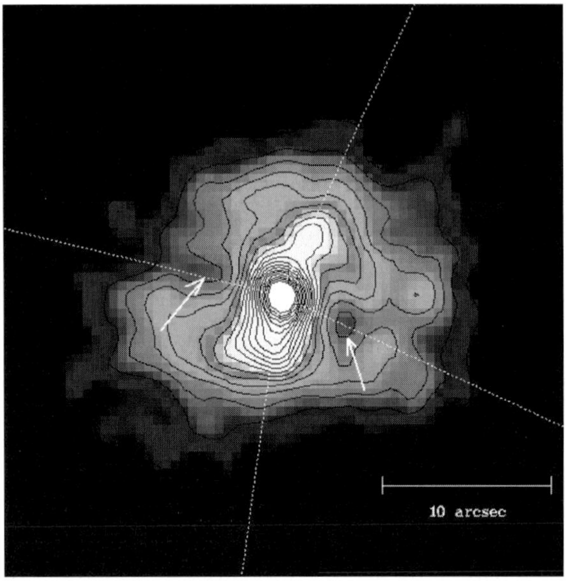

Figure 7. CHANDRA image of the central part of the cluster RBS797 (from Schindler et al., 2001). There are depressions in the X-ray emission which are located opposite to each other with respect to the cluster centre (see arrows). These depressions can be explained by an active galaxy in the centre of the cluster, which has two jets. The pressure of the relativistic particles in the jets push away the X-ray gas resulting in minima in the X-ray emission.

A third possible process is galaxy-galaxy interactions, like tidal stripping or galaxy harassment. Also during these interactions a lot of ISM can be lost to the ICM and IGM. This process is very likely more efficient in groups of galaxies, because in these systems the relative velocities are smaller and therefore the interaction timescales are longer. The ram-pressure stripping on the other hand is probably less efficient in groups because not only the pressure of the IGM is lower than that of the ICM, but also the velocities are smaller. This is also very important as the stripping is about proportional to $\rho_{gas} v^2$.

A forth possible mechanism is jets emitted by active galaxies. These jets can also carry metals. Figure 7 shows the interaction of jets with the ICM as it was discovered by X-ray observations. In the cluster RBS797 minima in the X-ray emission have been detected in a CHANDRA observation (Schindler et al., 2001). The X-ray depressions are arranged opposite with respect to the cluster centre. It is very likely that the pressure of the relativistic particles in the jets push away the X-ray gas. Preliminary radio observations with the VLA confirm this hypothesis.

Simulations with different enrichment processes were also performed on cosmological scales. Also here quite discordant results have been found as the two following examples show. Gnedin (1998) found that galactic winds play only a minor role, while galaxy mergers eject most of the gas. In contrary to these results Aguirre et al. (2001) concluded that winds are most important and ram-pressure stripping is not very efficient. The reason for these differences are probably the large ranges in scale that are covered by these simulations, from cosmological scale down to galaxy scales. Therefore only a small number of particles are left for each single galaxy and hence galaxies are not well resolved. This can be the reason for the discordant results.

In order to clarify this we are currently performing comprehensive simulations, which include the different enrichment processes.

4. Summary

So far the picture of the interaction between ICM/IGM and the ISM is not very clear. There are many different processes possible that can transport gas from the galaxies to the ICM/IGM: ram-pressure stripping, galactic winds, galaxy-galaxy interaction, jets from active galaxies and maybe even more. It is not known yet, which of the processes is most efficient, what the time variations are and how much it depends on other parameters like e.g. the mass of the cluster/group. The best property to test these different processes is the metallicity. The distribution and the time evolution of the metals are now finally measurable with high accuracy with the X-ray satellites XMM and CHANDRA, so that we will soon get answers on the importance of the different enrichment processes.

Acknowledgements

We are very grateful to Bernd Aschenbach for making the XMM data of CL0939+4713 available to us. They are part of the TS/MPE guaranteed time.

References

Abadi, M.G., Moore, B. and Bower, R.G.: 1999, *MNRAS* **308**, 947.
Aguirre, A., Hernquist, L., Schaye, J., Katz, N., Weinberg, D.H. et al: 2001, *ApJ* **561**, 521.
Balsara, D., Livio, M. and O'Dea, C.P.: 1994, *ApJ* **437**, 83.
Belsole E., Sauvageot J.-L., Ponman T.J. and Boudin H.: 2002, *A&A*, in press, astro-ph/0211350.
Castillo-Morales, A. and Schindler, S.: 2002, *A&A*, submitted.
De Filippis, E., Schindler, S. and Castillo-Morales, A.: 2002, *A&A*, submitted.
De Young, D.S.: 1978, *ApJ* **223**, 47.
Gaetz, T.J., Salpeter, E.E. and Shaviv, G.: 1987, *ApJ* **316**, 530.
Gunn, J.E. and Gott, J.R.III: 1972, *ApJ* **176**, 1.
Gnedin, N.Y.: 1998, *MNRAS* **294**, 407.
Longo G., Paolillo M. and De Filippis E.: 2002, in prep.
Metzler, C.A. and Evrard, A.E.: 1994, *ApJ* **437**, 564.
Metzler, C.A. and Evrard, A.E.: 1997, astro-ph/9710324.
Murakami, I. and Babul, A.: 1999, *MNRAS* **309**, 161.

Portnoy, D., Pistinner, S. amd Shaviv, G.: 1993, *ApJS* **86**, 95.
Quilis, V., Moore, B. and Bower, R.: 2000, *Science* **288**, 1617.
Ricker P.M. and Sarazin C.L.: 2001, *JOURNAL?* **561**, 621.
Ritchie B.W. and Thomas P.A.: 2002, *MNRAS* **329**, 675.
Schindler, S. and Müller E.: 1993, *A&A* **272**, 137.
Schindler, S.: 1999, *A&A* **349**, 435.
Schindler, S., Castillo-Morales, A., De Filippis, E., Schwope, A. and Wambsganss, J.: 2001, *A&A Lett.* **376**, L27.
Schmidt R.W., Fabian A.C. and Sanders J.S.: 2002, *MNRAS* **337**, 71.
Schulz S. and Struck C.: 2001, *MNRAS* **328**, 185.
Vollmer B., Cayatte V., Balkowski C. and Duschl W.J.: 2001, *ApJ* **561**, 708.
Takeda, H., Nulsen, P.E.J. and Fabian, A.C.: 1984, *MNRAS* **208**, 261.
Toniazzo T. and Schindler S.: 2001, *MNRAS* **325**, 509.

SELF-CONSISTENT MODELLING OF THE ISM

CHEMO-DYNAMICAL EVOLUTION OF THE ISM IN GALAXIES

STEFAN HARFST, GERHARD HENSLER and CHRISTIAN THEIS
Institute for Theoretical Physics and Astrophysics, University Kiel, Germany
E-mail: harfst@astrophysik.uni-kiel.de

Abstract. Chemo-dynamical models have been introduced in the late eighties and are a generally accepted tool for understanding galaxy evolution. They have been successfully applied to one-dimensional problems, e.g. the evolution of non-rotating galaxies, and two-dimensional problems, e.g. the evolution of disk galaxies. Recently, also three-dimensional chemo-dynamical models have become available. In these models the dynamics of different components, i.e. dark matter, stars and a multi-phase interstellar medium, are treated in a self-consistent way and several processes allow for an exchange of matter, energy and momentum between the components or different gas phases. Some results of chemo-dynamical models and their comparison with observations of chemical abundances or star formation histories will be reviewed.

Keywords: Galaxies: evolution, Galaxies: ISM

1. Introduction

Disk galaxies consist of complex structures with at least three main components: bulge, halo and disk. These differ in their fraction of ionised, atomic and molecular gas, their dust content and their stellar populations. Since the Milky Way Galaxy (MWG) is the best studied disk galaxy, the observations of its stars and its interstellar medium (ISM) can be used to test galactic evolutionary models in order to answer the addressed fundamental questions concerning the initial conditions, the formation and the evolution of the MWG and of disk galaxies in general, e.g.: When, how and on what time-scales did the galactic components form? Was there any connection between them? Which external influences have affected the structure? Is the solar neighbourhood representative for an ‘average disk’? These questions are directly related to observational facts, such as the lack of metal-poor G-dwarfs (the well-known G-dwarf problem) or the different observed effective yields in bulge, halo and disk of the MWG (Pagel, 1987). In general, these and other observational indicators, like star formation (SF) and supernova (SN) rates or colour measurements place constraints on galactic models and should be explained in a global scenario.

Another class of objects are dwarf galaxies (DGs) which differ in their structural and chemical properties from those of giant galaxies. Because of their low gravitational energies DGs are greatly exposed to energetic influences from processes like stellar winds, SN or even stellar radiation. In addition, low-mass galaxies seem to form at all cosmological epochs and by different processes, which make them ideal

Astrophysics and Space Science **289:** 431–439, 2004.

laboratories to investigate many of the astrophysical processes relevant for galaxy evolution. Again, the observations of DGs, like SF rates or metal abundances and abundance ratios, should be explained in a global model of galactic evolution.

At present, two major and basically different strategies for modelling galaxy evolution are followed: firstly, dynamical investigations which include hydro-dynamical simulations of isolated galaxy evolution and of proto-galactic interactions reaching from cosmological perturbation scales to direct mergers; and, on the other hand, studies which neglect any dynamical effects but consider either the whole galaxy or particular regions and describe the temporal evolution of mass fractions and element abundances in detail. Simulating different galactic regions, e.g. by a closed-box model, however, presupposes that these regions are neither energetically nor dynamically coupled. Thus, in this picture a galaxy is only the sum of different isolated subsystems without any connection, despite the initial conditions. Even multi-zone models which include gas exchange between different galactic regions (e.g. Ferrini et al., 1994) are devoid of self-consistent gas dynamics.

Chemo-dynamical models are a different approach to modelling galactic evolution treating self-consistently all important dynamical and energetic processes as well as their dependence on the metallicity. For this, a multi-phase description of the ISM with the inclusion of star-gas interactions is essential. Such models have been applied successfully to spherical symmetric systems (e.g. Theis et al., 1992) and to axisymmetric systems (e.g. Samland et al., 1997, hereafter SHT). More recently with increasing computational power, also 3d chemo-dynamical models have become available (e.g. Samland and Gerhard, 2002). The chemo-dynamical treatment is described in more detail in Sec. 2 and in Sec. 3 some results of two-dimensional models are reviewed.

2. Chemo-Dynamical Treatment

To approach global models of galaxy evolution which can achieve the structural differences and details, an appropriate treatment of the dynamics of stellar and gaseous components is essential. In addition, at least the following processes should be taken into account: SF, SNe, heating, cooling, stellar mass loss, condensation and evaporation. This includes the treatment of the multi-phase character of the ISM as well as the star-gas interactions and phase transitions. Since gas and stars evolve dynamically, and because several processes both depend on their metallicities but also influence the element abundances in each component, these models are called chemo-dynamical. The network of chemo-dynamical processes is sketched in Figure 1 where one can see how the different gaseous and stellar components are connected via mass, momentum and energy exchange. The important processes are: mass from the cloudy medium is transformed into stars by SF; the high and intermediate mass stars return a fraction of this mass, enriched with metals, by SNe and PNe to the ISM; the ISM is also heated by stellar radiation and feedback

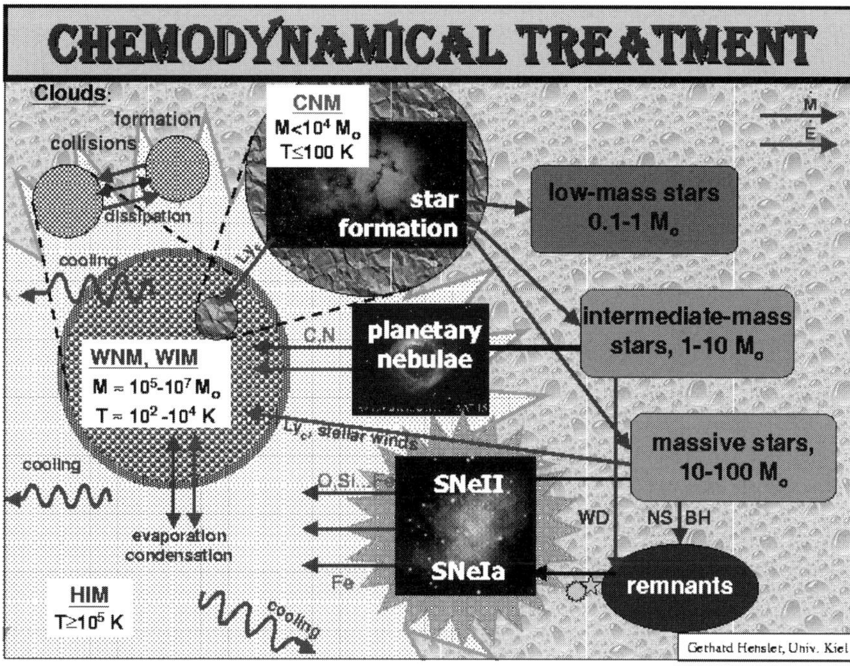

Figure 1. A schematic sketch of the chemo-dynamical treatment. The different gaseous and stellar components are connected via mass, momentum and energy exchange. The important interaction processes are shown in this diagram.

processes; remnants and low mass stars build the lock-up mass that is no longer available for the galactic cycle of matter; the cloudy medium (CM) and the hot intercloud medium (ICM) are mixed by condensation and evaporation; energy is dissipated by radiative cooling and by cloud-cloud collisions.

As can be shown and must be emphasized, however, the number of free parameters in the chemo-dynamical scheme is small, because they are either theoretically evaluated (like e.g. condensation and evaporation) or empirically determined (e.g. like stellar winds), or because they force self-regulation in a way that is independent of the parameterisation. The latter has been shown for SF by Köppen et al. (1995, 1998). The only free parameters which cannot approach particular 'equilibrium values' due to the absence of feedback are the stellar initial mass function (IMF), the momentum transfer by drag and the initial conditions, although the initial density distribution and gas-phase fraction are basically not affecting the model evolution. For a more comprehensive description of the chemo-dynamical treatment the interested reader is referred to Theis et al. (1992) and SHT.

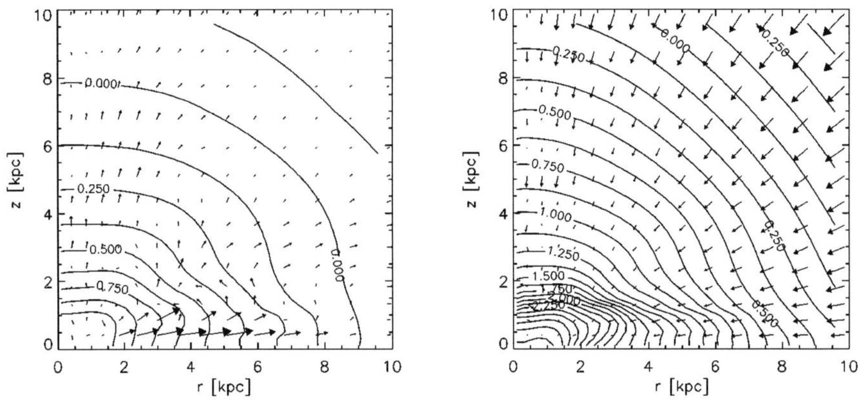

Figure 2. Surface densities and velocities of ICM (left) and CM (right) 7 Gyr after the onset of the galactic collapse. The ICM in the galactic plane has velocities up to 230 km s^{-1}, while the velocities of the infalling CM are less than 60 km s^{-1}. (taken from SHT)

3. Results from Chemo-Dynamical Models

3.1. The Milky Way's chemo-dynamical evolution

The model starts from an isolated spheroidal, rotating but purely gaseous cloud with a mass of $3.7 \cdot 10^{11}\,M_\odot$, a radius of 50 kpc, and an angular momentum of about $2 \cdot 10^7\,M_\odot\,pc^2\,Myr^{-1}$, corresponding to a spin parameter $\lambda = 0.05$. It is assumed that the protogalaxy consists initially of CM and ICM with a density distribution of Plummer-Kuzmin-type (Satoh, 1980) with 10 kpc scale length. The initial CM/ICM mass division (99%/1%) does not affect the later collapse, because the onset of SF determines the physical state within less than 10^7 years. Since almost excellent agreement of the chemo-dynamical model after 15 Gyr is found with the presently observed structure of the MWG, it may be safe to assume that also the evolutionary behaviour of different properties under consideration can be reliable deduced. The 15 Gyr old model model is e.g. able to reproduce the different metallicity distributions and effective yields of the halo, the bulge and in the solar vicinity (see Figure 7 in SHT), respectively, and also to solve the G-dwarf problem as an effect of large-scale dynamics of the metal-enriched ICM and its condensation and metal pollution of the CM (Figure 2). Furthermore, the radial abundance gradients (see Figure 6 in SHT), abundance ratios like O/Fe versus Fe/H (see Figure 9 in SHT), mass fractions of the components, SF rate in the disk, etc. fit the observations strikingly. Also interestingly, one can trace the temporal run e.g. of SN and PN rates (see Fig 8. in SHT) and of the SF rates in different MWG regions (Figure 3) which allow the dating of their formation, showing a delay of disk formation with respect to the halo extinction by almost 4 Gyr and a bulge evolution extended over 5 Gyr with a least two SF episodes.

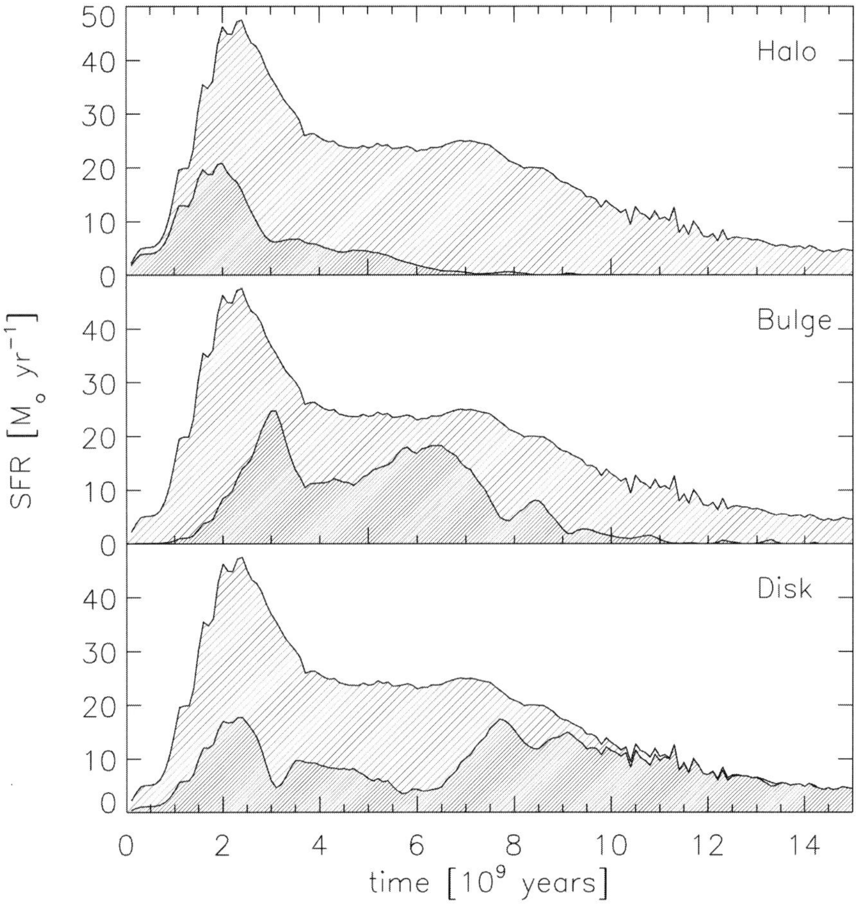

Figure 3. The evolution of SF rate in different galactic regions. The upper curve always shows the total SF rate, while the lower curve indicates the SF rate in halo, bulge and disk. (taken from SHT)

3.2. A Dwarf Galaxy model

In this model by Rieschick and Hensler (2000, hereafter RH) the initial baryonic mass is about $10^9\,M_\odot$ embedded in a static dark matter halo according to Burkert (1995) with a mass of $10^{10}\,M_\odot$. The numerical grid size is 20 kpc x 20 kpc, and the model is aimed to represent a dwarf irregular galaxy (dIrr). In Figure 4 the SF history is plotted and five distinct dynamical phases of evolution, that can be distinguished from a kinematical analysis of the model, are indicated by vertical lines. In the collapse phase (0–0.3 Gyr) the proto-galactic gas distribution cools and collapses with a net gas infall rate of $3.2 \cdot 10^{-1}\,M_\odot\,yr^{-1}$ leading to a central density increase by a factor of 100. Thus the SF rate rises steeply according to its quadratic dependence on the CM-density to its maximum value of $1.8 \cdot 10^{-2}\,M_\odot\,yr^{-1}$.

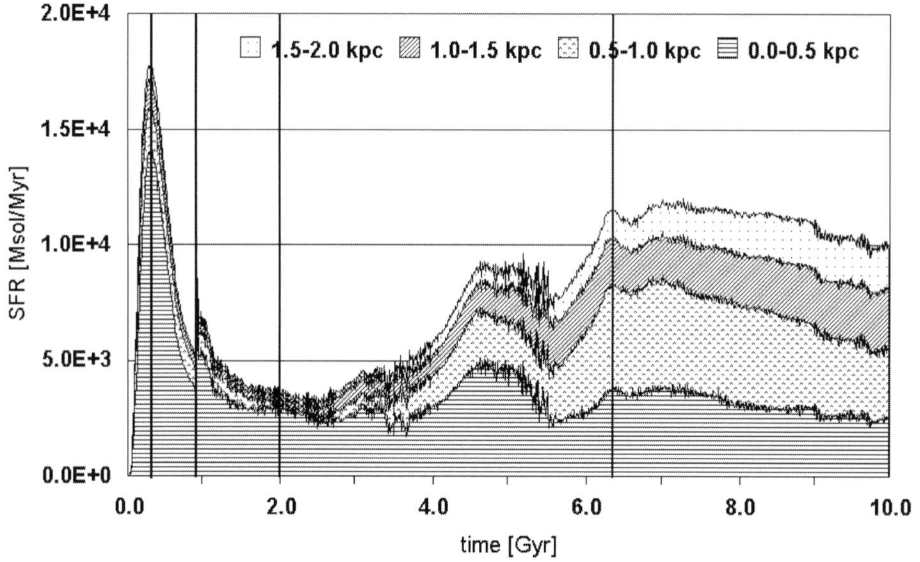

Figure 4. Star formation history of the $10^9\,M_\odot$ chemo-dynamical dIrr model in units of $M_\odot\,Myr^{-1}$ for different radial zones in the equatorial plane. The absolute values correspond to the differences between two curves. The vertical lines divide the different evolutionary phases as described in the text. (taken from RH)

After passing a post-collapse, a transitional and a turbulent phase (see RH for details) the galaxy reaches its final irregular phase after 6.2 Gyr, when a global quasi-equilibrium seems to be established. At this stage the disk is balancing itself in the sense of keeping the gas mass constant. This can be seen in Figure 5 where the mass flows between the different components averaged between 6.2–10.0 Gyr over a cylinder with $r = 2$ kpc and $z = 1$ kpc are analyzed. From the mass flow rates one can distinguish between an outer and an inner cycle. The outer one is produced by infall of CM. The CM-reservoir in the galaxy is consumed by means of SF. The SF rate amounts to about 24% of the infall rate and 10% of this matter is almost instantaneously reejected by high mass stars (HMS). The inner cycle represents SF and stellar evolution and, therefore, contains different time-scales.

The production of a minor fraction of hot ICM by SNe II leads to evaporation of remaining CM and escapes from the galactic body as outflow as a consequence of its high energy content. Significantly, almost 80% of the infall is immediately converted into outflow by only 3% of gas that has gone through the stellar cycle and is puffed up by stellar energy release. The full evaporation rate can exceed the infall rate because shell sweep-up leads to fragmentation and cloud formation in addition to a small amount of condensation. It should be emphasized that even though a large amount of the outflowing gas is gravitationally unbound and leaves the galactic body, the metals produced in HMS are for the most part kept in the outer gas flow cycle by mixing ICM with continuously infalling clouds. As a result

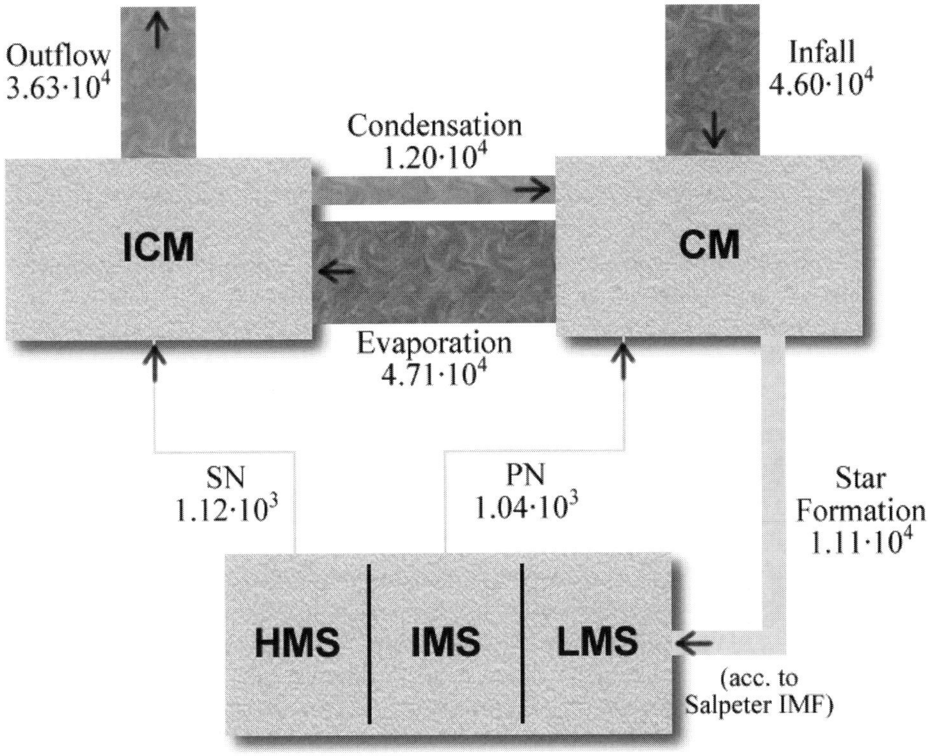

Figure 5. Flow of matter between the components in the chemo-dynamical dIrr model. The diagram shows the temporal average over the interval 6.2 to 10.0 Gyr (the irregular phase, see text) for a radius of 2.0 kpc and a z-height of 1.0 kpc, i.e. the whole visible galaxy. All numbers are in units of $M_\odot\ Myr^{-1}$. (taken from RH)

of this mechanism only a few percent of the metals leave the gravitational field of the galaxy with the outflowing ICM. While this mixing itself happens on a time-scale of about 20 Myr, the complete cycle of metal enrichment takes almost 1 Gyr because of the low infall velocity at later evolutionary stages. In contrary, the inner cycle leads to an efficient self-enrichment of SF regions within 10 Myr (see also Skillman et al., 1998). The outer enrichment time-scale would be reduced significantly, however, in a scenario of a rapidly infalling intergalactic gas cloud.

4. Summary and Outlook

Chemo-dynamical models of the MWG and dIrrs have been reviewed. Results demonstrate both the agreement of one single global model with numerous detailed and regional observations as well as with the global structures. The necessity of the full but complex chemo-dynamical treatment of the galaxy components is justified. Several observational features like abundance distributions and ratios can only be understood in a self-consistent global evolutionary scenario if one is taking the dynamics of the components and their relevant interaction processes into account. It has been shown that a large-scale coupling of different galactic regions by dynamical effects as well as the small-scale mixing between the gas phases due to condensation/evaporation affect the observational signatures. As long as the long-range streaming with inherent small-scale interactions under the inclusion of a two-phase ISM with their small-scale spatial resolution cannot be treated in other numerical simulations, e.g. in present SPH models, they miss self-consistency and cannot reproduce the observations in their global extent.

One has to emphasize that a reliable chemo-dynamical model has to reproduce the complete set of observational features on different galactic scales, i.e. for all existing components and observed variables as the whole. It is not sufficient for an understanding of the global galactic evolution to fit single particular observed properties. The time dependence of element abundances and their ratios can serve as reliable diagnostic tools of galaxy evolution and the physical state of the ISM. Their combination provides a new chance for a detailed deconvolution. While the chemo-dynamical models also provide element abundances in the hot ICM, their observational validation is still a problem.

The here applied grid code is still limited to two dimensions because of the numerically expensive treatment of complex processes and spatial resolution. It has already been mentioned, that recently also three-dimensional models were published (e.g. Samland and Gerhard, 2002; Harfst et al., 2002; Harfst et al., 2003; see also Samland, this volume). Most approaches are based on N-body simulation and a multi-phase description of the ISM is achieved by combining a SPH algorithm, to describe the hot ICM, with a sticky particle scheme, to describe the cloudy medium. With three-dimensional models infall of gas clouds or small satellites can be studied as well as the merging of giant spiral galaxies. This should help to explain e.g. the question on how and to what extend SF is triggered by such events. Nevertheless, one has to keep in mind, that even the two-dimensional chemo-dynamical investigations present a giant leap with respect to non-dynamical chemical or purely gas-dynamical studies.

Acknowledgements

S. Harfst is supported by the *Deutsche Forschungsgemeinschaft (DFG)* under the grant TH-511/2-3. Also, S. Harfst would like to thank the organisers of WS-ISM and the *Astronomische Gesellschaft* for financial support.

References

Burkert, A.: 1995, *ApJL* **447**, 25.

Ferrini, F., Molla, M., Pardi, M.C. and Diaz, A.I.: 1994, *ApJ* **427**, 745.

Harfst, S., Theis, C. and Hensler, G.: 2002, *Ap&SS* **281**, 313.

Harfst, S., Theis, C. and Hensler, G.: 2003, *A&A*, in preparation.

Köppen, J., Theis, C. and Hensler, G.: 1995, *A&A* **296**, 99.

Köppen, J., Theis, C. and Hensler, G.: 1998, *A&A* **331**, 524.

Pagel, B.E.J.: 1987, in: G. Gilmore and B. Carswell (eds.), *NATO ASIC Proc. 207: The Galaxy*, pp. 341–364.

Rieschick, A. and Hensler, G.: 2000, in: J. Franco, L. Terlevich, O. López-Cruz and I. Aretxaga (eds.), *ASP Conf. Ser. 215: Cosmic Evolution and Galaxy Formation: Structure, Interactions, and Feedback*, p. 130.

Samland, M. and Gerhard, O.E.: 2002, *A&A*, submitted.

Samland, M., Hensler, G. and Theis, C.: 1997, *ApJ* **476**, 544.

Satoh, C.: 1980, *PASJ* **32**, 41.

Skillman, E.D., Dohm-Palmer, R.C. and Kobulnicky, H.A.: 1998, in: T. Richtler and J.M. Braun (eds.), *The Magellanic Clouds and Other Dwarf Galaxies, Proceedings of the Bonn/Bochum-Graduiertenkolleg Workshop*, pp. 77–90.

Theis, C., Burkert, A. and Hensler, G.: 1992, *A&A* **265**, 465.

NON-INSTANTANEOUS GAS RECYCLING AND CHEMICAL EVOLUTION IN N-BODY DISK GALAXIES

BRUNO JUNGWIERT
Astronomical Institute, Academy of Sciences, Boční II 1401, 141 31 Prague 4, Czech Republic
E-mail: bruno@ig.cas.cz

GIOVANNI CARRARO
Dipartimento di Astronomia, Università di Padova, vicolo dell'Osservatorio 2, 35122 Padua, Italy

CLAUDIO DALLA VECCHIA
Physics Department, University of Durham, Ogden Center, South Road, Durham DH1 3LE, United Kingdom

Abstract. After briefly recalling basic facts about the metallicity gradients in spiral galaxies, we summarize two recent N-body gas-dynamical models that implement a non-instantaneous gas recycling and point our their potential for an improved treatment of the chemical evolution in galactic disks.

Keywords: N-body simulations, galaxy evolution, gas recycling, metallicity gradients

1. Introduction

Spatial and temporal variations of abundance distributions in galaxies put important constraints on galactic chemical evolution (e.g. Hou et al., 2000). For disks of spirals, the existence of radial abundance gradients is observationally well established (e.g. Vila-Costas and Edmunds 1992; Zaritsky et al., 1994) as well as the correlation of their slope with the presence and properties of a stellar bar: galaxies with strong bars tend to have shallower metallicity gradients than those with weak or no bars (Martin and Roy, 1994). Several galaxies were claimed to have two-slopes gradients within their disks (Martin and Roy, 1995; Roy and Walsh, 1997).

Numerical models, either multi-zone (e.g. Ferrini et al., 1994) or N-body (e.g. Martinet and Friedli, 1997), well explain some of the observed trends. In particular, N-body models account for the correlation of metallicity gradients slopes with bar strength: it is a result of a large-scale mixing induced by the bar dynamics. However, in many cases the existing models fail to reproduce at the same time the radial profiles of chemical abundances, star formation rate (SFR) and gas surface density (e.g. Mollá and Roy, 1999); this is true also for the Milky Way (Portinari and Chiosi, 2000). At the same time, some of the new observational abundance determinations give quite different slopes than previously reported and also put in doubt the reality of the two-slopes gradients (Pilyugin, 2002).

Astrophysics and Space Science **289:** 441–444, 2004.

It follows from the above that more efforts are needed to compare the observed and modeled chemical patterns in disk galaxies. Current models usually suffer from one or several drawbacks: either they are static (i.e. not allowing self-consistent radial flows due to non-axisymmetric perturbations), or neglect possible outflows carrying away the enriched interstellar medium (ISM), or use the instantaneous recycling approximation (IRA) (Tinsley, 1980). The last point is a core of our contribution. We summarize, in Section 2, N-body gas-dynamical models recently developed within our research groups: both use non-instantaneous (hereafter 'non-I') recycling schemes, while they differ by the approach. In Section 3, we show a simple application demonstrating how the combined effects of large-scale radial flows and non-I recycling can affect the mass distribution in the disk. Implications for the metallicity gradients are discussed in Section 4.

2. Two N-Body Codes with Non-Instantaneous Gas Recycling

We have at our disposal two N-body codes with star formation and non-I recycling. The first one (Jungwiert, 1998; Jungwiert et al., 2001; hereafter JCP01) is based on a Particle-Mesh algorithm to compute the gravity and on a sticky-particles scheme to simulate gas. It models the mass exchange between stellar and gaseous components by varying masses and number of individual particles in a time-dependent way. The second code (Carraro et al., 1998; Lia et al., 2002) relies on a gravity tree-code and on the Smoothed Particle Hydrodynamics while it treats the gas recycling in a statistical manner, keeping the masses of particles constant. In both models, the consumption of gas by star formation is implemented by means of the generalized Schmidt law. Stellar mass-loss via winds and supernova explosions is included through prescriptions reflecting stellar evolution.

Jungwiert et al. (2001) have shown that the non-I gas recycling has a considerable impact on the evolution of N-body disks: it significantly influences SFR, gas distribution, radial inflow of gas, gas consumption time-scale and bar strength.

2.1. A SIMPLE APPLICATION: GAS CONTENT EVOLUTION

Here we present a simple demonstration of how important can be the non-I recycling and the self-consistent treatment of radial flows in galactic disks. We assume an evolved barred galaxy that has been forming stars over 5 Gyr at a constant SFR, resulting in the gas-to-stars mass ratio of 0.1. We take such a galaxy as our initial conditions and follow its further evolution using two approaches: a multi-zone model without radial flows and the above mentioned N-body model (JCP01). For both of them we compare the non-I recycling and IRA cases. Figure 1 shows the temporal evolution of the gas mass fraction in the central part of the galaxy ($R < 2$ kpc). The left panel is for the multi-zone model, the right one for the N-body simulation; the dashed-lines correspond to IRA, the dash-dotted lines to non-I

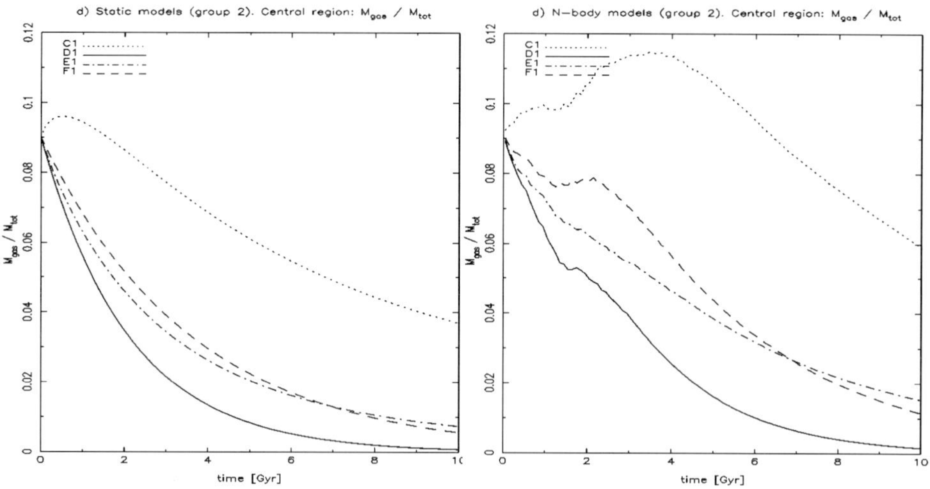

Figure 1. Gas mass fraction in the central part ($R < 2$ kpc) of the galactic disk (SFR is roughly proportional to gas mass fraction in this example). Left panel: Multi-zone model. Right panel: N-body model. Non-I recycling: dotted and dash-dotted curves (see text for details); IRA: dashed curves; no recycling: full curve.

recycling excluding the mass-loss from stars older than 5 Gyr (i.e. born prior to the start of our simulations) and the dotted lines to non-I recycling applied to stars of all ages; for reference, a model without any recycling (i.e. with star formation but no stellar mass-loss) is also given (full lines). Since we have imposed, for the purpose of this example, a linear dependence between SFR and the gas surface density, the shapes and relative vertical shifts of curves plotted in Figure 1 are approximately valid also for the temporal dependence of the SFR in the central region. It can be seen that the gas mass fraction and SFR are significantly affected by both the radial inflow of gas (present in N-body simulations) and by non-I recycling.

3. Implications for Abundance Gradients

The non-instantaneous gas recycling has naturally a direct impact on the chemical enrichment of the ISM by elements that are produced on longer time-scales, i.e. by low and intermediate-mass stars. However, here we rather want to draw attention upon an indirect effect that results from the combination of non-I recycling and radial gas flows and has influence also on the chemical enrichment by elements produced by high-mass (i.e. short lived) stars.

The reason for this indirect effect is that the non-I recycling leads locally to a different gas-to-stars ratio than IRA. This in turn globally changes the radial flows in non-axisymmetric (barred, spiral) disks since these flows are sensitive to the gas mass fraction. The radial profiles of gas surface density and of SFR are consequently affected.

This has motivated our project of studying detailed chemical evolution in galactic disks using our N-body codes with non-instantaneous gas recycling. The resulting metallicity gradients will soon be published elsewhere (Jungwiert and Carraro, 2003).

Acknowledgements

This project is carried out under the post-doc research grant of the Grant Agency of the Czech Republic (No. 202/01/D075) and under the International Relationship Program of the Padua University. The authors wish to express their gratitude to Prof. Jan Palouš for his support. G. Carraro thanks Prof. Petr Harmanec for his hospitality in Prague.

References

Carraro, G., Lia, C. and Chiosi, C.: 1998, *MNRAS* **297**, 1021.
Ferrini, F., Mollá, M., Pardi, C. and Díaz, A.I.: 1994, *ApJ* **427**, 745.
Hou, J.L., Prantzos, N. and Boissier, S.: 2000, *A&A* **362**, 921.
Jungwiert, B.: 1998, *Bars, Resonant Rings, Star Formation and Gas Recycling in Spiral Galaxies*, PhD thesis, Université Paris 7.
Jungwiert, B. and Carraro, G.: 2003, in preparation.
Jungwiert, B., Combes, F. and Palouš, J.: 2001, *A&A* **376**, 85, (JCP01).
Lia, C., Portinari, L. and Carraro, G.: 2002, *MNRAS* **330**, 821.
Martin, P. and Roy, J.-R.: 1994, *ApJ* **424**, 599.
Martin, P. and Roy, J.-R.: 1995, *ApJ* **445**, 161.
Mollá, M. and Roy, J.-R.: 1999, *ApJ* **514**, 781.
Pilyugin, L.S.: 2002, *A&A*, in press.
Portinari, L. and Chiosi, C.: 2000, *A&A* **355**, 929.
Tinsley, B.M.: 1980, *Fund. Cosmic Phys.* **5**, 287.
Vila-Costas, B. and Edmunds, M.G.: 1992, *MNRAS* **259**, 121.
Walsh, J.R. and Roy, J.-R.: 1997, *MNRAS* **288**, 726.
Zaritsky, D., Kennicutt, R.C. and Huchra, J.P.P.: 1994, *ApJ* **420**, 87.

THE INTERPLAY BETWEEN ISM, STAR FORMATION AND GALAXY EVOLUTION

MARKUS SAMLAND
Astronomisches Institut der Universität Basel, Venusstrasse 7, CH-4102 Binningen, Switzerland
E-mail: samland@astro.unibas.ch

Abstract. I present a model for the formation and evolution of a massive disk galaxy, within a growing dark halo whose mass evolves according to cosmological simulations of structure formation. The galactic evolution is simulated with a new 3D chemo-dynamical code, including dark matter, stars and a multi-phase ISM. We follow the evolution from redshift $z = 4.85$ until the present epoch. The energy release by massive stars and supernovae prevents a rapid collapse of the baryonic matter and delays the maximum star formation until redshift $z \approx 1$. The galaxy forms radially from inside-out and vertically from top-to-bottom. The feedback of stars leads to turbulent motions and large-scale flows in the ISM. As one result the galactic disk is significantly enriched by chemical elements synthesized in bulge stars.

Keywords: Galaxies: formation, evolution, stellar content, kinematics and dynamics, ISM, chemical enrichment

1. Introduction

During the last decade, significant progress has been made in understanding cosmic structure formation and galactic evolution. With high-resolution cosmological simulations, the formation of dark halos in different cosmologies has been studied in detail to determine the halo merging histories and the formation of galaxies inside these dark halos. However, it is inevitable in such simulations that structures on galactic and sub-galactic scales are not well-resolved. A way to circumvent this problem is to simulate the formation of only single galaxies or small galactic groups using a simplified, but cosmologically motivated model of the dark matter background. A brief outline of such a model and a few results are presented in the following.

2. The Galactic Model

To simulate the formation and evolution of galaxies we developed a chemo-dynamical code that models the dynamics of dark and baryonic matter (stars and a multi-phase ISM). The baryonic components are connected by an interaction network that describes the mass, momentum and energy transfer. The mass infall at the boundaries of the simulated volume is taken from cosmological simulations

Astrophysics and Space Science **289:** 445–448, 2004.

(see VIRGO-GIF-project; Kauffmann et al., 1999). The model galaxy has a mass (dark+baryonic) of $1.8 \times 10^{12}\,M_\odot$, a spin parameter $\lambda = 0.05$, and an angular momentum profile similar to the universal profile found by Bullock et al. (2001). A detailed description of the model and the initial conditions can be found in Samland and Gerhard (2003). The chemo-dynamical model provides kinematics and metallicities of individual stars. In addition, it provides gas phase metallicities and temperatures, and star formation (SF) rates, as a function of time or redshift. Using these data, Westera et al. (2002) has already calculated the colour evolution of large spiral galaxies as function of redshift.

3. Self-Regulation and Feedback

In a galactic model describing the detailed physics of stars and gas it is unavoidable to have a number of parameters. However, most of these parameters can be constrained either from theory or from observations and anyway the galactic star-gas network builds a self-regulated system in which moderate changes of the free parameters do not alter the results significantly (Samland and Gerhard, 2003). SF and phase transition between hot and warm gas are only two example for such self-regulations. There is also a self-regulation by dynamical processes. For example, the SF efficiency influences the settling of the clouds in the galactic disk, which again regulates the SF process.

4. The Galactic Formation Process

The present simulation describes the formation and subsequent evolution of a disk galaxy in a slowly growing dark halo. The halo grows by accreting dark and baryonic matter in spherical shells with an accretion rate that is derived from cosmological simulations in a ΛCDM universe. The accreted primordial matter dissipates its kinetic energy and begins to collapse inside the dark halo. The strong collapse takes place in a redshift interval between $z = 1.8$ and $z = 0.8$. This late and delayed collapse, which is in contradiction to simple collapse scenarios, is caused by stellar feedback in conjunction with the initially shallow gravitational potential of the halo.

In general terms, SF out of the dissipating cloud medium proceeds from halo to disk and from inside out. Thus the earliest stars at redshifts $z > 2$ form in the whole volume limited by the virial radius (halo formation), with a concentration to the central bulge region. At a redshift of $z = 1.3$, a (thick) disk component first appears. Later the SF is concentrated to the equatorial plane, and at late times most of the SF occurs in the outer disk. Around $z = 1.2$ the infall of gas can no longer compensate the gas consumption by SF in the inner disk. As a result, the region of highest SF density moves radially outwards from the centre. At redshift $z = 0.85$

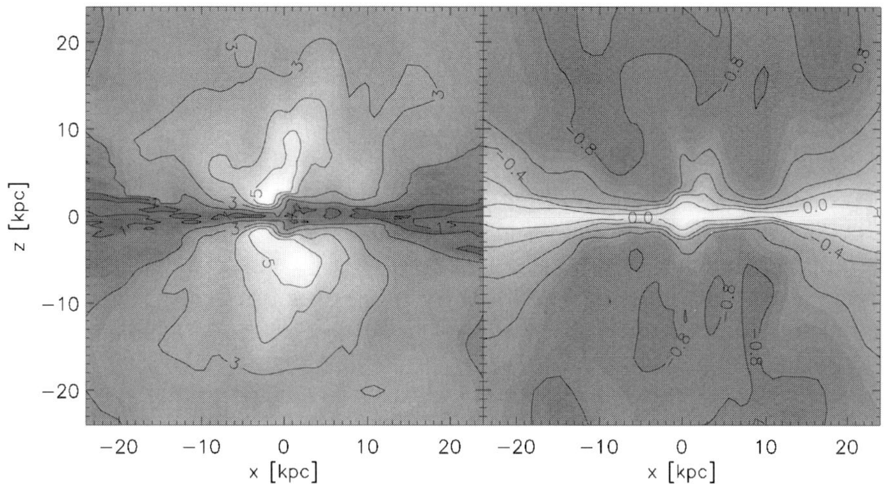

Figure 1. The left panel shows the temperature of the hot intercloud medium in the xz-plane at a redshift z=0. The contours are labeled in units of 10^6 K. The right panel shows the corresponding oxygen abundance ([O/H]).

a bar forms, which later develops into a bar-bulge. After the bar-bulge has formed, two trailing spiral arms appear in the disk which are connected to the bar-bulge. The bar induces large-scale gas motions in the ISM which mix gas from the inner galaxy with the disk ISM. Already during the bar formation process the galaxy starts to build up the outer disk. This disk is the youngest component in the model galaxy, even though the oldest disk stars are as old as the halo stars. In parallel, the vertical scale-height of the disk in a fixed radial range decreases. We expect this evolutionary sequence to be more or less typical of any dissipative collapse in a growing dark matter halo, unless it is interrupted by substantial mergers.

5. The Chemical Evolution

The chemical enrichment histories of stars, clouds, and intercloud medium are very different from the predictions of multi-phase closed box models, because stars, clouds and gas have different spatial distributions. Most of the ionized hot gas is located in the halo, while the molecular clouds and stars are preferentially found in the disk and bulge, where the gas densities and metallicities are high. After a brief initial period the average metallicity of the stars exceeds that of the cloudy medium, because the stars form from gas that is more concentrated to the equatorial plane and thus more metal-rich than the average cloud medium.

An interesting feature is a constant metallicity of the hot halo gas from redshift $z = 1$ until the present epoch. Metal-rich outflows from the bulge are balanced by the infall of low metallicity gas. In this way the metallicity in the halo gas remains constant over a long time. On average the intercloud medium in the halo has a higher temperature than the intercloud medium in the disk (Figure 1, left panel), which is surprising because the heating rate in the disk is much higher. However, the higher metallicities (Figure 1, right panel) and densities in the disk increase the cooling efficiency so that on average the disk intercloud medium has temperatures of less than 10^6 K.

The large-scale gas flows are also important for the metallicity of the cloud medium and the stars. During the collapse the SF and the metal production is concentrated to the inner galaxy. From there, the metal-rich gas expands into the disk where it can form clouds and later stars. This disk pre-enrichment causes a lack of metal-poor stars ('G-dwarf problem'), similar to that observed in the Milky Way.

Acknowledgements

The simulations were performed at the Swiss Centre for Scientific Computing (CSCS).

References

Bullock, J.S., Dekel, A., Kolatt, T.S., Kravtsov, A.V., Klypin, A.A., Porciani, C. and Primack, J.R.: 2001, *ApJ* **555**, 240.

Kauffmann, G., Colberg, J.M., Diaferio, A. and White, S.D.M.: 1999, *MNRAS* **303**, 188.

Kennicutt, R.C.: 1998, *ApJ* **498**, 541.

Samland M. and Gerhard, O.E.: 2003, *A&A*, accepted.

Westera, P., Samland, M., Buser, R. and Gerhard, O.E.: 2002, *A&A* **389**, 761.

MODELLING THE TURBULENT MAGNETIZED ISM

MAARIT J. KORPI

Astronomy Division, University of Oulu, PO BOX 3000, 90014 University of Oulu, Finland
Laboratoire d'Astrophysique, Observatoire Midi-Pyrénées, 14, Avenue Edouard Belin, 31400 Toulouse, France; E-mail: Maarit.Korpi@oulu.fi

Abstract. Briefly stated, the interstellar matter (hereafter ISM) is a highly turbulent flow, the turbulence mainly being driven by the violent stirring by supernova explosions, penetrated by dynamically significant magnetic fields both on small and large scales. Sophisticated numerical models of such a turbulent flow can give important information on some of the complicated processes occurring in the ISM, e.g. on the regulation of the observed multiphase structure, properties of interstellar turbulence and the generation and maintenance of magnetic fields by the flow itself; in this paper such models and their most important results are reviewed.

Keywords: MHD modelling, multiphase structure, interstellar turbulence, galactic dynamos

1. Introduction

The spatial and temporal scales of the multitude of phenomena occurring simultaneously in the ISM vary enormously, which makes the modelling of it as a hugely demanding task. Nevertheless, a wide variety of (often numerical) models have been developed; for a rough sketch, see Figure 1. Local computational domains are often utilized when detailed models of interstellar turbulence are developed, for high spatial resolution is vital but computational requirements high. There is a special class of these local models, aiming at self-consistent modelling of the ISM, by including as many of the basic interstellar phenomena (some of the most important discussed in Sect. 2) as possible, and numerically investigating the structure and dynamics of the resulting flow. These models and their most important results are the topic of this review. Even though the locality reduces the computational requirements significantly compared to the modelling of a whole galaxy, to achieve high degree of accuracy together with large enough spatial extents is still computationally challenging; therefore a common name for this class of models could well be NUmerically Demanding and Extremely Slow (NUDES).

2. Basic Ingredients for Self-Consistent Models of the ISM

Stars as sources of turbulence: The ISM is observed to be in a turbulent state. There are many possible sources to drive this turbulence (see the review by Mac

Astrophysics and Space Science **289:** 449–457, 2004.

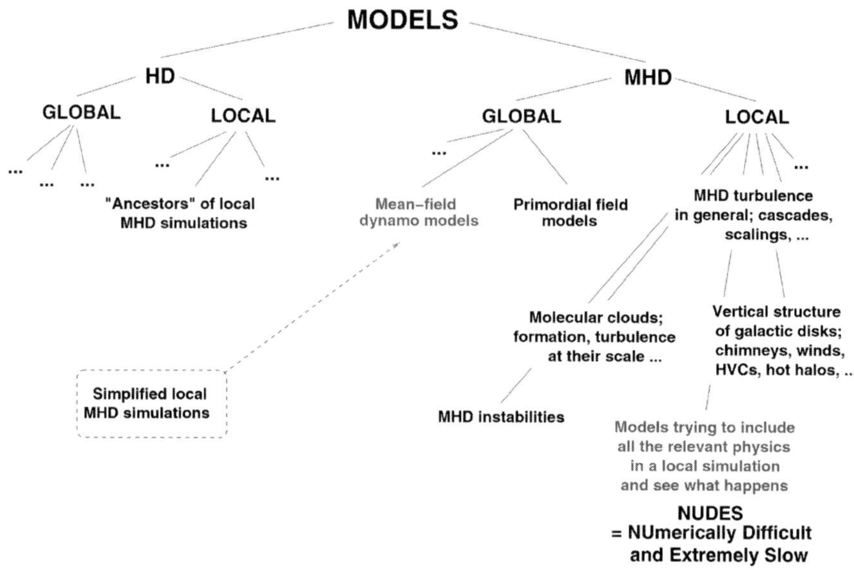

Figure 1. A rough sketch of the developed models for the ISM.

Low in these proceedings), but the dominant sources are the stars. Turbulent energy is transferred to the ISM both by supernova explosions (hereafter SNe) and stellar winds. The kinetic energy transfer rate from stellar winds can be estimated to be roughly one third of the one from SNe (e.g. Abbott, 1982), roughly one tenth when the total energy transfer rates are compared (e.g. Ferriére, 1995). It has to be noted here that in quiescent galaxies or quiescent parts of galaxies, where star-formation rate is very low, other sources may become important as well.

The SNe rapidly inject large amounts of energy (10^{51} ergs), approximately 10% of it being transformed into kinetic energy of turbulence (e.g. Lozinskaya, 1992) the rest going into heat, to the surrounding ISM. Stellar winds can be thought of as prolonged, less energetic versions of SNe, producing cavities and shells around the stars into which the SNe later on explode (taken that the star is massive enough to explode as SN). The SN forcing is not totally random in space and time, but correlated in both senses, since most of the SNe originating from the young massive stars (Type Ib and Type II) occur in OB associations (60 %, Ferrière, 2001); larger structures, called as superbubbles (hereafter SBs) will be created. Most of this SN activity occurs near the galactic midplane, where the massive molecular clouds are formed. The SNe originating from older stellar population (Type Ia) remain uncorrelated, and have larger scale heights. The impact of SNe on the ISM depends strongly on the environment where they explode: although the fraction of the kinetic energy to the total SN energy seems to stay constant (Thornton et al., 1998), the sizes of the remnants and expansion velocities are higher inside the cavities blown up by previous SN activity or stellar winds and high up from the galactic midplane. Thereby, the turbulence and its properties can be expected to be

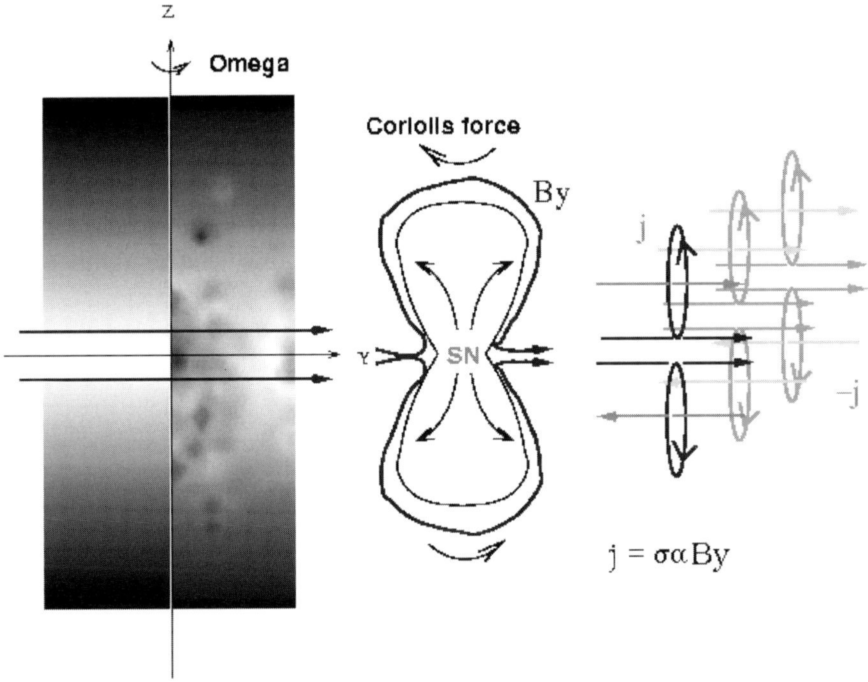

Figure 2. A simplistic illustration of the action of helical turbulence on an underlying uniform magnetic field.

intrinsically connected to the local structure of the ISM, which, in turn, is likely to be largely determined by the SN activity itself.

Heating and cooling processes: The dominating source of heating of the ISM is again provided by the stellar component, through SNe and stellar radiation, the latter mostly originating from the UV photons radiated by O and B stars. Approximately 90% of the released SN energy heats up the ISM, as the explosions produce expanding shells with hot (million K) interiors, which as an ensemble will form a complicated network of shell and bubble structures. The UV radiation from O and B stars ionizes and heats up the surrounding ISM; compared to the SNe, the temperature regime of this process is different, around 8 000 K, being an important process producing the so called warm neutral and ionized components observed in the ISM. The matter heated this way is subject to the processes of thermal conduction, radiative cooling, turbulent mixing, and in the case of the formation of dense enough structures, self-gravity; the state of the matter will dynamically change as an interplay between all these processes.

Nonuniform rotation: Another common property of galaxies is the differential rotation, i.e. the decreasing rotational velocity with increasing distance from the centre. Although it seems that the shear provided by this mechanisms is ineffective in driving turbulence in itself, with energy transfer rate of a few percent compared to SNe (e.g. Ruzmaikin et al., 1988), it can be expected to have significance in the dynamo process and possibly also igniting the magnetorotational instability. The effect of the large-scale shear on small or medium scale turbulent structures such as SN or superbubble (hereafter SB) shells is also weak. The rotation itself, however, gives rise to the Coriolis force, which in a density stratified environment, with 'bubbles' either diverging or converging in the vertical direction, will lead to the creation of small-scale twisting motions of these structures, producing helicity into the flow. From the viewpoint of the dynamo theory, helical motions are crucial to produce an ensemble of small loops of poloidal field which then, with the help of turbulent mixing, reconnect to a large poloidal loop (see Figure 2).

Magnetic fields and cosmic rays: Galaxies are observed to host dynamically significant magnetic fields both in large and small scales (see the review by Beck in these proceedings). The deduced field strengths are close to equipartition with the turbulent motions, thereby the magnetic field can be expected to play an important role in interstellar turbulence and structure formation. Globally the magnetic field pressure acts as supporting the matter in the galactic disk against its own weigth, direct local influences include the reduction of the size and morphology of expanding SN and SB structures.

Interstellar cosmic rays, being dynamically coupled to the magnetic field, also exert a pressure to the interstellar matter, and in addition contribute to the heating of the ISM. The cosmic ray pressure is of the same order as the magnetic pressure, thereby providing a significant support to the overall vertical balance of the disk. They may also have a strong impact on the dynamics and structure of the ISM via the Parker instability which they may trigger (see e.g. the contributions of Hanasch et al., in these proceedings).

3. The NUDES and Highlights of the Results

3.1. Developed models

Work in this field of research began with two dimensional (hereafter 2D) hydrodynamic (hereaftr HD) simulations. Rosen and Bregman and collaborators used a modified two-fluid version of the ZEUS code to investigate the structure and dynamics of the ISM, firstly to study the regulation of the warm and cold phases by the action of stellar winds (Rosen, Bregman and Norman, 1993), followed by a model including SN activity and thereby the hot gas created by their action (Rosen and Bregman, 1995). With the used code setup, not especially designed to make a detailed model of the turbulence with the spatial resolution of 10 pc, it was possible to adopt spatial extents of several kiloparsecs, making the model more global compared to the ones which have appeared since that time. The energy injection rate from SNe vs. stellar winds was varied, and the resulting changes in the structure and dynamics were investigated.

The group of Vázquez-Semadeni and collaborators, using a numerical method based on pseudospectral techniques, developed models with higher spatial resolution, the investigated scales ranging roughly from 1 pc to 1 kpc, therefore more suitable for studying the properties of interstellar turbulence. Their work has largely concentrated on investigating the regulation of the cold and warm phases, including the forcing by stellar winds and background radiation field due to stellar UV radiation and cosmic rays. Extensive studies on the role of thermal instability (hereafter TI) is such a flow have been undertaken. In one set of the models (Gazol-Patino and Passot, 1999) also SN forcing is taken into account. The inclusion of self-gravity makes their model especially suitable for the investigation of tmolecular cloud formation processes in the cold dense phase (see the contribution of Ballesteros-Parades in these proceedings). Their work started with 2D HD simulations (e.g. Vázquez-Semadeni et al., 1995), the effects of rotation, large-scale shear and magnetic fields have been included (e.g. Passot et al., 1995), and sets of 3D simulations produced (e.g. Vázquez-Semadeni et al., 1996).

In the series of 3D HD models, the model of de Avillez (see the contribution in this proceedings) is noteworthy: with the powerful numerical technique of adaptive mesh refinement (hereafter AMR) combined with massive parallel computing, the currently investigated spatial scales range from 0.6 pc to tens of kiloparsecs.

As evident from the discussion in the previous section, the role of magnetic fields can be expected to be important in the ISM, and therefore it would be highly desirable to be able to go from HD to MHD (and retain the resolution of the model of de Avillez). The computational requirements, however, increase significantly with the inclusion of magnetic fields, and the MHD models developed so far are suffering from poorer resolution. Nevertheless, some work has been done in this field.

In addition to the work done by the Vázquez-Semadeni group, A 3D MHD shearing-box model, based on standard finite differences, with SN forcing was developed by Korpi, Brandenburg and Shukurov; with the recently implemented parallelization and current implementation of the numerical method (Caunt and Korpi, 2001), the model is capable of investigating scales from pc to a few kiloparsecs. The model has been used to investigate the structure and dynamics of the turbulence in SN regulated flows, special attention devoted to monitor whether the generation and maintenance of magnetic field is possible in such a flow (e.g. Korpi et al., 1999).

A more recent 3D MHD model is being developed by Balsara, Kim, Mac Low and collaborators, using AMR; the first results having been published in Kim et al. (2001). Development of another MHD model is also taking place by Shukurov and collaborators (see the contribution in these proceedings).

3.2. Structure of the Simulated ISM

The regulation of phases in the ISM has been addressed by the two seminal works by Field, Goldsmith and Habing (1969; hereafter FGH) and McKee and Ostriker (1977; hereafter MO). According to the FGH model, the existence of two thermally

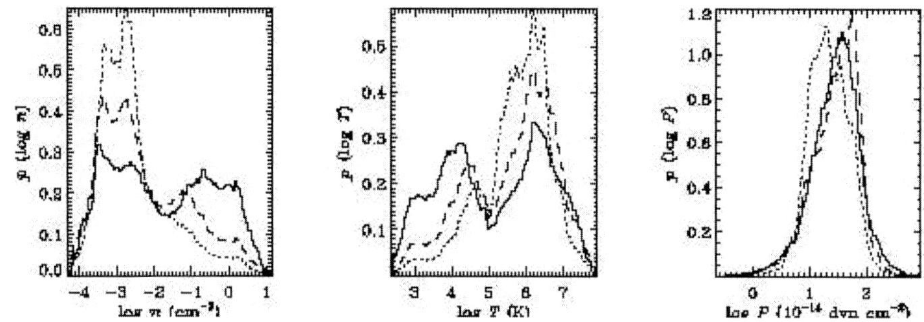

Figure 3. PDFs of density, temperature and pressure in a 3D MHD simulation similar to Korpi et al. 1999; dotted: full volume, dashed: lower 500 pc, solid: lower 250 pc.

stable phases (cold and warm) in pressure equilibrium is provided by the thermally unstable phase at intermediate temperatures. The MO model includes a hot phase generated by SNe; the condition for exact pressure equilibrium between the phases is relaxed. Recently it has become clear from observations that the pressure imbalance is larger than expected, an order of magnitude variation has been reported (e.g Jenkins and Tripp, 2001). In addition, gas has been observed to exist in the 'forbidden' thermally unstable regions of the FGH model (Heiles, 2001).

Extensive investigations applying NUDES-type models have been undertaken to investigate the regulation of phases in the ISM. The role of thermal instability (hereafter TI) has been investigated by a series of papers by Vázquez-Semadeni and collaborators. The key point in their models (e.g. Vázquez-Semadeni, Gazol and Scalo, 2000) is the background heating function, which together with the used cooling function produces a thermally unstable regime at a certain temperature range, the other regimes being thermally stable. Taken that no other types of turbulent forcing mechanisms are applied, the system developes into two phases, the matter being effectively removed from the thermally unstable region due to TI; finally the system will be characterized by bimodal density PDFs. This picture is in a good agreement with the FGH model. If, on the other hand, the action of stellar winds and rotational effects are added into the model, the system looses its character the other dynamical processes taking over; gas is observed to exist also in the forbidden thermally unstable regime (Gazol et al., 2001). This suggests that the picture of TI and discontinuous phase transitions is too simple to explain the regulation of cooler phases in the ISM.

The other temperature end of the ISM, namely the regulation of the hot gas in the ISM can be studied with models including SNe. The results of Rosen and Bregman (1995) were already suggestive of the basic property of such a flow: the heating provided by SNe, simultaneously driving the turbulence, together with the cooling function applied, determine the structure and dynamics of the system. In the model of Korpi et al. (1999), including magnetic fields, Coriolis force and large-scale shear, the simulated gas was observed to mainly prefer two physically distinct (with different temperatures, densities, vertical distributions and turbulent velocities) states, warm and hot, whilst the temperature regime around 10^5 K, with the most rapid cooling rates, was a highly improbable state of the gas (see Figure 3). The regulation of these simulated phases can be identified to be different from the FGH picture, since no local thermal equilibrium can be established due to the nature of SN heating and in the absence of background radiation field, due to which TI must be ineffective. Therefore, significant amounts of gas can exist in the thermally unstable regime $T > 10^5$K (roughly 40% of the gas in the simulated volume). This shows that phases can be regulated solely by the dynamic balance between the heating and cooling processes, which in turn depends on the properties of the turbulent flow itself; the establishment of local thermal equilibrium and TI are not necessarily needed. The simulated ISM is by no means isobaric, but large variations of pressure are seen in the pressure PDFs (right panel in Figure 3).

Note that throughout this paper, the term phase is used when discussing the preferred states, i.e. the most probable temperatures and densities in which the interstellar gas is observed/modelled to exist, regardless of the process which is regulating them. This is not in direct agreement with the 'original' definition by FGH, since many requirements (local thermodynamic equilibrium, isobaric medium) are relaxed. Some authors (e.g. Vazquez-Semadeni, Gazol and Scalo, 2001) still wish to use the original one.

3.3. Interstellar Turbulence

The NUDES models have revealed that interstellar turbulence forced by stellar injections is a very complicated phenomenon, for which no universal Kolmogorov-like theory with well-defined cascades seems to exist, having at least two principal differences. Firstly, the forcing scale of turbulence is not well-defined in the ISM, since the flow is characterized by the expansion and interactions of structures of different sizes, transporting energy both towards larger and smaller scales. This has been observed to result in ill-defined cascades and large dynamical variations (e.g. Vázquez-Semadeni et al., 1995). Secondly, the forcing does not occur isotropically, since the winds and SNe tend to be correlated in space and time; therefore the flow can be expected to develope dependencies on the local energy injection rate and environment. For example, in the model of Korpi et al. (1999), the turbulent rms velocity and amount of vorticity were observed to vary significantly between the different phases of the flow.

Whilst in incompressible flows rotational modes govern the system, the energy being carried by vortexes of different sizes, the interstellar flow could be anticipated to be dominated by compressible modes, since both winds and SNe will drive strong shocks in the ISM. This aspect has also been investigated with the NUDES models. In the HD model of Vazquez-Semadeni et al. (1995), during epochs of high star formation occurring in the flow, rotational and compressible modes were found to contain roughly similar amounts of energy, whilst ineffective star formation lead to the domination of rotational modes, when the flow behaved as the incompressible counterpart. Similar behavior of the flow was found in the SN forced MHD model of Korpi et al. (1999); approximately 60% of the energy was in the rotational modes, most of the vorticity related to the interactions of SNe in the warm phase with stronger magnetic fields.

The role of magnetic fields in interstellar turbulence has been investigated by two setups of models so far: some impose a strong large-scale magnetic field on the flow pattern (e.g. Passot et al., 1995; Kim et al., 2001), others let the magnetic field develope from an initially weak seed field (e.g. Korpi et al., 1999). In both cases, a generation of small scale magnetic field is observed by the turbulent flow, although in 2D its generation requires the presence of the large-scale field (Passot et al., 1995). The presence of the magnetic field has an immediate influence on the structures of the flow – the shells driven by explosive events get thicker, the overall structures smaller due to the magnetic pressure resisting their expansion, and anisotropic, since an imposed or generated large-scale field has a preferred direction. When rotational effects (shear and Coriolis force) were included into the 3D calculations (e.g. Korpi et al., 1999), some of the magnetic energy was observed to be transformed from the smaller to the larger scales, and a weaker mean component was observed in the azimuthal direction due the action of the shear. This dynamo-like process was observed to be capable of sustaining the magnetic field over ten turnover times.

3.4. Some aspects concerning the galactic dynamo theory

Galactic dynamo theory explains the evolution of large-scale magnetic fields in an extremely turbulent flow by the combined inductive action of helical turbulence and large-scale shear. The role of turbulence is twofold: firstly Coriolis force, density stratification and stellar explosive events together provide the inductive action, secondly the flow has to provide such a high rate of diffusion for the magnetic field that it can diffuse to small enough scales and then reconnect to form large-scale field from the ensemble of loops created by the helical flow (illustrated in Figure 2).

A considerable amount of criticism has been directed towards this picture during the recent years (e.g. Kulsrud, 1999). Although extensive semi-analytical work has been done to investigate the inductive and diffusive action of SN driven turbulence on the magnetic field, indicating that the galactic dynamo should work (series of papers by Ferrière, see e.g. Ferrière and Schmitt, 2000), the numerical results from more idealized models than the NUDES (e.g. Cattaneo and Hughes, 1996) have yielded seriously contradictory results. Namely, in the regime where the magnetic field is dynamically significant, the inductive effect of turbulence has been reported to become ineffective, argued to be due a drastic backreaction (called as quenching) of the strong magnetic field on the turbulence. Very recently it has been proposed (e.g. Blackman and Field, 2000) that this may actually be a consequence of the applied boundary conditions, which conserve magnetic helicity, and force the generation of the large-scale field to saturate 'too early'. This idea is now being at a test in several idealized models with the boundary conditions proposed unrealistic relaxed (e.g. Brandenburg et al., 2002), but remains to be addressed/taken into account by NUDES-type models.

There are other points in the dynamo theory which can easily be questioned, since direct numerical models of the turbulent processes are not yet available and very hard to produce. One of them surely is the turbulent diffusion process and the question how it can lead to the needed rate of reconnection (see e.g. the contribution of Lazarian in these proceedings). Another, very recently raised, issue is the need of rotational processes (helical motions and shear) to provide magnetic field amplification. This option is being discussed both in the case of stellar and galactic dynamos; the NUDES models will surely provide more information on this issue in the near future.

References

Abbott, D.C.: 1982, *ApJ* **263**, 723.
Blackman, E.G. and Field, G.B.: 2000, *ApJ* **534**, 984.
Brandenburg, A., Dobler, W. and Subramanian, K.: 2002, *AN* **323**, 99.
Caunt, S.E. and Korpi, M.J.: 2001, *A&A* **369**, 706.
Cattaneo, F. and Hughes, D.W.: 1996, *Phys. Review E* **54**, R4532.
Gazol-Patino, A. and Passot, T.: 1999, *ApJ* **518**, 748.
Gazol, A., Vázquez-Semadeni, E., Sánchez-Salcedo, J.F. and Scalo, J.: 2001, *ApJL* **557**, L121.
Field, G.B., Goldsmith, D.W. and Habing, H.J.: 1969, *ApJ* **155**, L149.
Ferrière, K.M.: 1995, *ApJ* **441**, 281.
Ferrière, K.M.: 2001, *Rev. Mod. Phys.* **73**, 1031.
Derrière, K.M. and Schmitt, D.: 2002, *A&A* **358**, 125.
Heiles, C.: 2001, *ApJL* **551**, L105.
Jenkins, E.B. and Tripp, T.M.: 2001, *ApJS* **137**, 297.
Kim, J., Balsara, D. and MacLow, M.-M.: 2001, *JKAS* **34**, 333.
Korpi, M.J., Brandenburg, A., Shukurov, A., Tuominen, I. and Nordlund, Å.: 1999, *ApJL* **514**, L99.
Kulsrud, R.M.: 1999, *ARA&A* **37**, 37.
Lozinskaya, T.A.: 1992, *Supernovae and Stellar Winds in the Interstellar Medium*, American Institute of Physics, New York.
McKee, C.F. and Ostriker, J.P.: 1977, *ApJ* **218**, 148.
Passot, T., Vázquez-Semadeni, E. and Pouquet, A.: 1995, *ApJ* **455**, 536.
Rosen, A. and Bregman, J.N.: 1995, *ApJ* **440**, 634.
Rosen, A., Bregman, J.N. and Norman, M.L.: 1993, *ApJ* **413**, 137.
Ruzmaikin, A.A., Shukurov, A.M. and Sokoloff, D.D.: 1988, *Magnetic Fields in Galaxies*, Klüwer, Dordrecht.
Thornton, K., Gaudlitz, M., Janka H.-Th. and Steinmetz, M.: 1998, *ApJ* **500**, 95.
Vázquez-Semadeni, E., Gazol, A. and Scalo, J.: 2000, **540**, 271.
Vázquez-Semadeni, E., Passot, T. and Pouquet, A.: 1995, *ApJ* **441**, 702.
Vázquez-Semadeni, E., Passot, T. and Pouquet, A.: 1996, *ApJ* **473**, 881.

THERMAL INSTABILITY IN TURBULENT MODELS OF THE INTERSTELLAR MEDIUM

ADRIANA GAZOL, ENRIQUE VÁZQUEZ-SEMADENI and
JAVIER SÁNCHEZ-SALCEDO
Instituto de Astronomía, UNAM, México

Abstract. We discuss numerical simulations (and issues) of the atomic ISM tailored to investigate the possible existence of significant amounts of gas in the temperature range between the warm and cold stable phases. We attribute the presence of this gas to two facts: the enhanced stability of moderate, adiabatic-like velocity fluctuations occurring when the ratio of the cooling to the dynamical crossing time is large, and the recycling of gas from the dense to the diffuse phase by stellar energy injection. We conclude by suggesting that the gas with unstable temperatures can be observationally distinguished through simultaneous determination of two of its thermodynamic variables.

Keywords: ISM, turbulence, thermal instability

For over three decades, the prevailing view of the ISM is that it has a multiphase nature, segregated into hot, warm and cold phases (Field et al., 1969; McKee and Ostriker, 1977). However, over the same period there have been other theoretical and observational suggestions that gas at intermediate (thermally unstable) temperatures exists, between the cold and warm phases (e.g. Gerola et al., 1974; Dickey et al., 1977; Heiles, 2001). One possibility for reconciling these two views is that the multi-phase models neglected the implications of the ISM being turbulent. In a previous paper (Gazol et al., 2001) we reported that the temperature histogram resulting from 2D MHD simulations including self-gravity, rotation, and small-scale, HII region-like, forcing, contained roughly half the mass at unstable temperatures (see also Kritsuk and Norman, 2002). However, the concern has been raised that perhaps the presence of this gas and the lack of sharp phase transitions in our simulations are numerical artifacts due to the relatively low resolution of these simulations (512^2) and the large physical size of the box ($L = 1$ kpc).

In this paper we first present results concerning the numerical issues, in order to assess the validity of our ISM simulations and interpret them. We then present new simulations that, based on the numerical considerations, provide reasonable evidence that significant amounts of gas at unstable temperatures should be expected in the atomic ISM. For a detailed discussion see Vázquez-Semadeni et al. (2002, Paper I). The numerical simulation of thermally unstable turbulent flows requires solving simultaneously the regions of sharp gradients occurring in the immediate neighborhood of clouds and the quasi-linear development of perturbations in the more distant, relatively quiescent, unstable medium mediating the clouds and the warm, stable, diffuse phase. The simulation of regions with strong shocks needs the use of artificial viscosities and diffusivities in order to spread out shocks over

Astrophysics and Space Science **289:** 459–463, 2004.

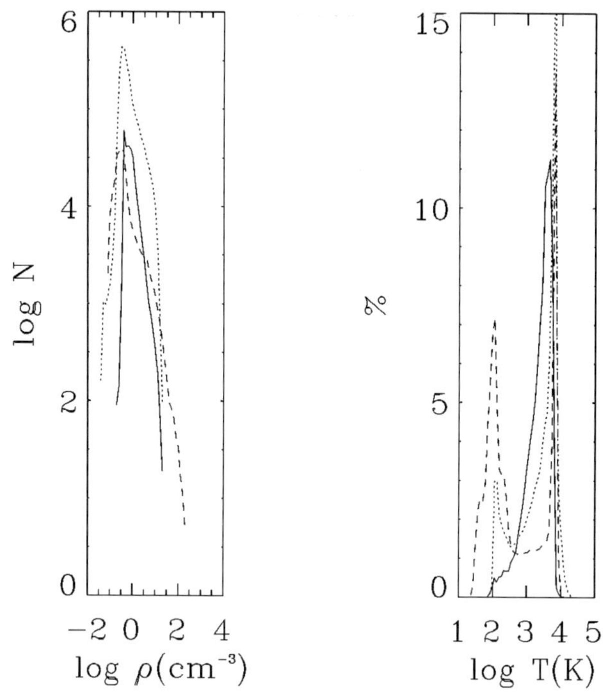

Figure 1. Top: Density histograms for the ISM simulations with $L = 10$ pc (*solid*), 150 pc (*dotted*) and 1000 pc (*dashed*). A clear trend toward higher phase segregation (more strongly bimodal shape) is seen at *large L* because the wave mode is unstable at large scales only. *Bottom:* Temperature histograms, with the same labeling.

a few grid points. Unfortunately, such diffusivities also have the effect of damping the growth of perturbations in the relatively smooth regions reducing the range of unstable scales, as well as their growth rates.

Our simulations have been performed using a pseudo-spectral scheme, described in detail by Passot et al. (1995), to solve the MHD equations in the presence of the above-mentioned physical processes, using a hyperviscosity scheme, a mass diffusion term with coefficient μ, and a thermal diffusivity term with a constant coefficient. With the adopted parameterized heating and cooling functions (Sánchez-Salcedo et al., 2002), the gas is linearly unstable under the isobaric mode (Field, 1965) of thermal instability (TI) for 313K $< T <$ 6100K and 3.16cm$^{-3} > n >$ 0.59cm^{-3}. Our numerical scheme has the advantage that it produces no numerical diffusivities at all. All artificial diffusivities are included explicitly in the equations, and can be controlled through their associated coefficients. This allowed us to now perform simulations of the non-diffusive case in the linear and weakly nonlinear regimes and of diffusive cases (Paper I), in order to establish the minimum unstable scales and the effective growth rates of small-scale perturbations as a function of the diffusive coefficients. We have found that, for the values of the diffusivities and

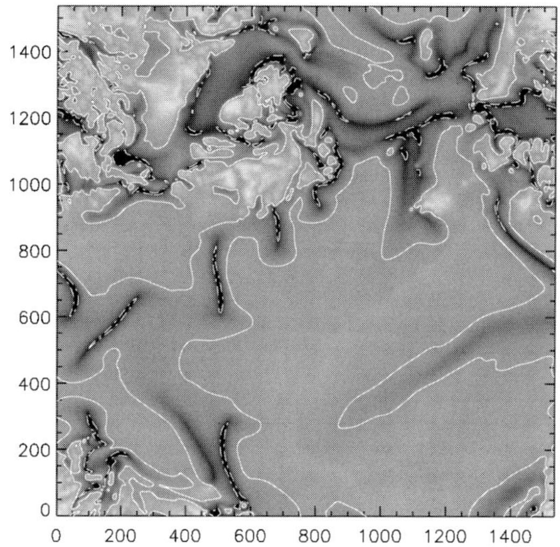

Figure 2. Temperature field of the 150-pc simulation of the ISM at the same time for which histograms are shown in Figure 1. The resolution is 1536^2 grid points.

viscosities needed for correct treatment of near discontinuities by the numerical code, only the mass diffusion has important damping effects. We have therefore performed non-diffusive and diffusive simulations with box sizes 10, 150, and 1000 pc at 128^2 resolution. Note that higher resolution is not necessary for this purpose, as all that is needed is to resolve the perturbations themselves and their initial growth. We find that for $L = 10$ pc and $\mu = 5 \times 10^{-4}$, the growth of perturbations down to $L/8 = 1.25$pc is allowed. Note that for the mean conditions of the diffuse gas, the fastest-growing scale predicted by the linear analysis, λ_{max}, is $\lesssim 10$pc. The smallest size of growing perturbations for simulations with $L = 150$pc and $\mu = 0.003$ is $L/16 = 9.4$ pc, thus barely reaching λ_{max}. Simulations with $L = 1000$pc and $\mu = 0.008$ provide the largest range of unstable scales (63-1000 pc) but it does not reach downto λ_{max}.

Having established the effect of the diffusivities, we can now discuss MHD simulations including the physical ingredients mentioned above and using previous values of L and μ. The resulting density and temperature histograms are shown in Figure 1. TI manifests itself in slight peaks at the stable densities, but a sizeable fraction of gas at unstable values of the the density and the temperature is nevertheless seen.

From the temperature histograms we can also see that as the box-size decreases, the phase segregation becomes *less* pronounced and the fraction of gas at unstable temperatures becomes larger. This is because, as smaller scales are considered, the ratio of the cooling to the turbulent crossing times, η, decreases. In this case the perturbations are adiabatic-like and are linearly stable (Paper I).

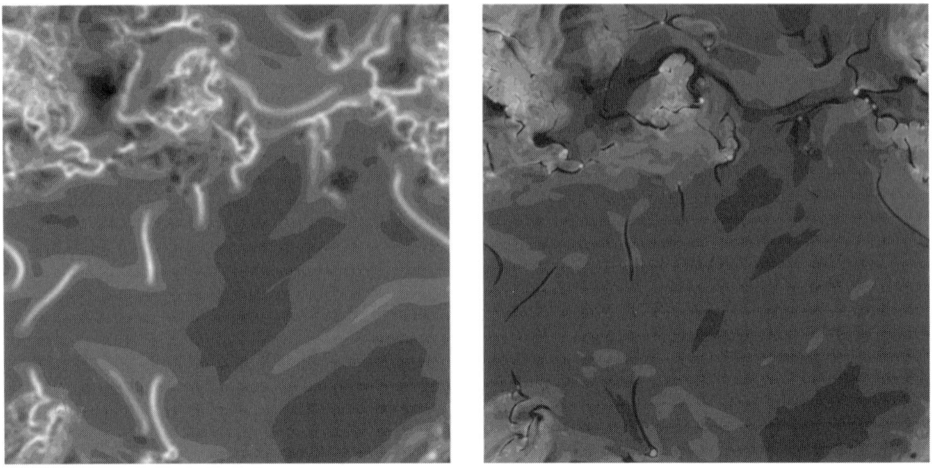

Figure 3. Density *left* and pressure *right* fields of the 150-pc simulation of the ISM at the same time for which previous figures have been made.

The temperature field for the 150 pc run ($n = 1536^2$ $\mu = 0.003$) is shown in Figure 2. Regions of gas at unstable temperatures with sizes clearly larger than 9.4 pc can be seen in this figure. This indicates that the presence of gas at unstable temperatures is not due to the artificial diffusivities. The density and pressure fields for the same run (Figure 3) show that in general, in regions with 'unstable' temperatures, the thermal pressure *increases* (although very mildly) with density, indicating stability of the flow (the isobaric mode of TI amounts to having an inverse dependence of pressure with density). This result can be understood as a consequence of the turbulent crossing time becoming shorter than the growth time of condensation, allowing turbulent fluctuations to disrupt the incipient condensations and causing the perturbations to become closer to adiabatic (and therefore linearly stable) through the heating from shocks and adiabatic compressions (Sáncez-Salcedo et al., 2002). In Figure 3 we also see that only the highest-density (small-scale) regions correspond to pressure minima, indicating that these are indeed condensing regions, within the TI paradigm (in the isobaric mode). Note that this occurs even when they are much smaller than the smallest unstable scale allowed by the mass diffusion. This occurs because the strong turbulent compressions driven by stellar-like sources, produce a local density (and cooling rate) increase, decreasing η. In turn, this causes the pressure to behave closer to the thermal equilibrium curve than the pressure of weakly compressed regions with the same sizes.

Finally, the above discussion suggests a possible observational test. Since the stabilized gas behaves out of thermal equilibrium, it may be possible to observationally determine whether the gas seen at unstable temperatures is indeed stabilized by this process by simultaneously determining two of its thermodynamic variables, in order to know whether the pressure increases or decreases with the density.

References

Dickey, J.M., Salpeter, E.E. and Terzian, Y.: 1977, *ApJ* **211**, L77.

Field, G.B.: 1965, *ApJ* **142**, 531.

Field, G.B., Goldsmith, D.W. and Habing, H.J.: 1969, *ApJ* **155**, L149.

Gazol, A., Vázquez-Semadeni, E., Sánchez-Salcedo, F.J. and Scalo, J.: 2001, *ApJ* **557**, 121.

Gerola, H., Kafatos, M. and McCray, R.: 1974, *ApJ* **189**, 55.

Heiles, C.: 2001, *ApJ* **551**, L105.

Kritsuk, A. and Norman, M.L.: 2002, *ApJ* **569L**, 127.

McKee, C.F. and Ostriker, J.P.: 1977 *ApJ* **218**, 148.

Passot, T., Vázquez-Semadeni, E. and Pouquet, A.: 1995, *ApJ* **455**, 536.

Sánchez-Salcedo, F.J., Vázquez-Semadeni, E. and Gazol, A.: 2002, *ApJ* **577**, 768.

Vázquez-Semadeni, E., Gazol, A., Sánchez-Salcedo, F.J. and Passot, T.: 2002, to appear in *Simulations of Magnetohydrodynamic Turbulence in Astrophysics: Recent Achievements and Perspectives*.

ON THE THERMAL INSTABILITY IN NUMERICAL MODELS OF THE INTERSTELLAR MEDIUM

SAMI DIB, ANDREAS BURKERT and AHMAD HUJEIRAT
Max-Planck-Institut für Astronomie, Königstuhl 17, 69117 Heidelberg, Germany
E-mail: dib@mpia.de

Abstract. We investigate with 3D hydrodynamical simulations the role played by thermal processes in the dynamical evolution of the interstellar medium (ISM). A parametric approach of the cooling process shows that the observed mass fraction of the cold (< 300 K) and unstable gas (300K < T < 6000K) can not be produced by turbulent compression or background heating of the medium alone. An analysis of the properties of the clouds that are formed by the combined effect of the thermal and gravitational instability shows that the cloud's scaling relations imprinted by the thermal instability (TI) are in good agreement with observational values.

Keywords: thermal instability, ISM, molecular clouds

1. Introduction

Recent studies confirm the important role played by thermal processes in the interstellar medium. On the observational side, there are evidences that many of the shell- and void-like structures observed in the ISM of different galaxies, especially in the outer parts of their galactic disk are unlikely to be the result of supernova explosions (Kim et al., 1999). Furthermore, a careful study by Rhode et al. (1999) of the holes of the galaxy Holmberg II shows no sign of a remnant stellar cluster inside the voids as would be expected from the supernova scenario. From a theoretical point of view, Kolesnik (1991) has shown that TI plays an important role in the formation of GMC's and Burkert and Lin (2000) showed, via 1D simulations, that TI leads to the build up of significant overdensities. Vázquez-Semadeni et al. (2000), Gazol et al. (2001) carried out a number of 2D simulations that hint to the fact that thermal processes of cooling and heating, coupled to energy feedback in the form of stellar winds can reasonably reproduce the observed mass fraction of the unstable gas. However it is still not clear from their simulations what is the time evolution of this mass fraction and what is explicitly the origin of the unstable gas, see Gazol et al. (2001) for more discussion (also Gazol et al., this volume). In §2, we use a parametric study to constrain what could be the mass fraction of the unstable gas of non-stellar origin and in §3, we present scaling relations derived from a typical model presented in §2. Calculations were performed using the ZEUS-3D code (Stone and Norman, 1992) at the Rechnenzentrum Garching of the Max-Planck-Gesellschaft.

Astrophysics and Space Science **289:** 465–468, 2004.

2. Parametric Study

We solve the equations of ideal gas dynamics for a one component fluid on a medium resolution grid (128^3 grid) which represents a 1 Kpc^3 volume. We assume a polytropic equation of state with a specific heat ratio of 1.4 and impose periodic boundary conditions in the three directions. The initial conditions of all runs are identical and taken from a fully driven isothermal (T=10^4K) simulation evolved over one dynamical timescale $\tau_{dyn} = L/V_{rms0} \approx 10^8$ years, where $L = 1$ Kpc and $V_{rms0} = 10$ km sec^{-1} is the *rms* turbulent velocity. Thus, the initial density distribution is inhomogeneous with $(\delta\rho/\rho)_{max} = 2.8$, and the velocity distribution corresponds to a Mach number $Ma_0 = 1$ turbulent medium. For a parametric study we assume the cooling and heating terms in the energy equation to be respectively $\Lambda(\rho, T) = \Lambda_0 \rho^\alpha T^\beta$ and $L(\rho) = q\Lambda_0\rho^\epsilon$. The latter is responsible for keeping a fraction of the gas in the unstable regime. Shielding of the densest condensations is taken into account, by nulling the heating inside dense clouds and by applying a shape-dependent heating at the surface cells of each condensation. Shielding applies whenever $\rho \geq \rho_{sh} = 25\bar{\rho}$ where $\bar{\rho} = 1.53 \times 10^{-27}$ g cm^{-3} is the mean density. Λ_0 can be related to the cooling timescale τ_{cool} by the relation displayed in Eq. 1.

$$\Lambda_0 = \frac{3}{2} \frac{(Ma_0 V_{rms0})^{1-\beta}}{\bar{\rho}^{\alpha-1}\tau_{cool}} \tag{1}$$

To avoid the task of fixing an arbitrary cooling timescale, we parametrize the latter as a function of the dynamical timescale by introducing the parameter $\eta = \tau_{cool}/\tau_{dyn}$. In this study α and ϵ have been fixed in all the simulations at the values of 2 and 1 respectively and minimum and maximum temperatures of 10 K and 12000 K are assumed. We varied the three other parameters η (with $\beta = 0.25$, $q = 0.1$), β (with $\eta = 0.3$, $q = 0.1$) and q (with $\eta = 0.3$, $\beta = 0.25$) and analysed the mass fractions of the cold (F_c) and unstable gas (F_u). Figure 1 shows that F_u can reach high values of the order of 0.3 close to the value of 0.4–0.5 derived from observations (Heiles et al., 2001), but at a cost of an unrealistic small value for F_c. We ran simulations with other permutations of the parameters leading to $F_u \approx$0.30–0.40, but it turned out that the outmost of what F_c can reach is a value of 0.1. We conclude by stating that the contribution of thermal processes of non-stellar origin to the mass fraction of the unstable gas is in the range of 10–20% and the location of most of the unstable gas should be directly correlated to star forming regions.

3. Scaling Relations

We use a clump-finding algorithm based on a density threshold and a friend-of-friend search method for neighbor cells to extract the scaling relations of the condensations which are formed following the TI. We use a model for which $\eta = 0.3$,

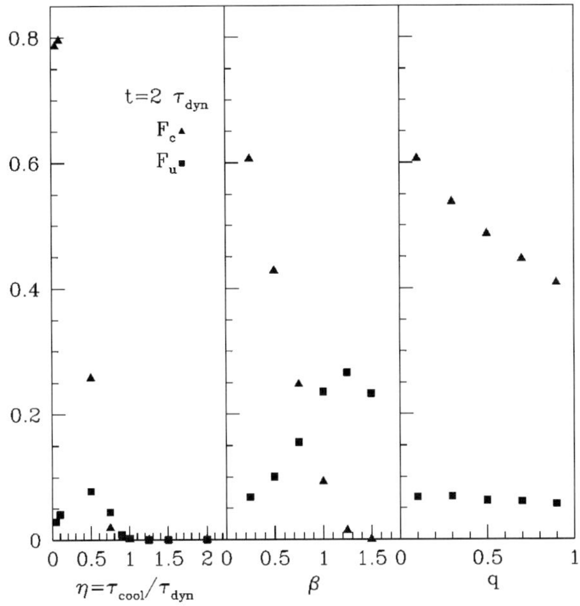

Figure 1. Mass fraction of the cold gas (F_c) and of the unstable gas (F_u) at t=2 τ_{dyn} as a function of the cooling-heating parameters.

$\beta = 0.25$ and $q = 1$, where the self-gravity of the gas has also been accounted for. Keeping in mind the conclusion of Sec. 1, we present the scaling relations at a time t$\approx 0.6\tau_{dyn} \approx 56$ Myrs where star formation is believed not to have played an important role yet and before a runaway gravitational collapse can take place (at this time the numerical density in the densest clouds has reached 300–400 cm^{-3}). The medium shows a rather filamentary structure at this stage as displayed in Figure 2a, with the densest clouds sitting at the intersection of filaments. Figure 2b displays the mean density-size relation as extracted from the 3 dimensional data cube. The coefficient m in the relation $\bar{n} \approx R^m$ is found to be equal to –0.9 and –0.7 when density thresholds of 8 and 10 times the average density are used, respectively. In comparaison, observations converge towards a value of $\approx$ –1, (Larson, 1981). We should stress that the relations we obtain are derived from the three dimensional physical data. A comparison of those relations to relations derived from the integration of the data on given line of sights as presented recently in Ballesteros-Paredes and Mac-Low (2002) (molecular cloud simulation s on a scale of 0.1 pc), will be presented in detail in a subsequent paper.

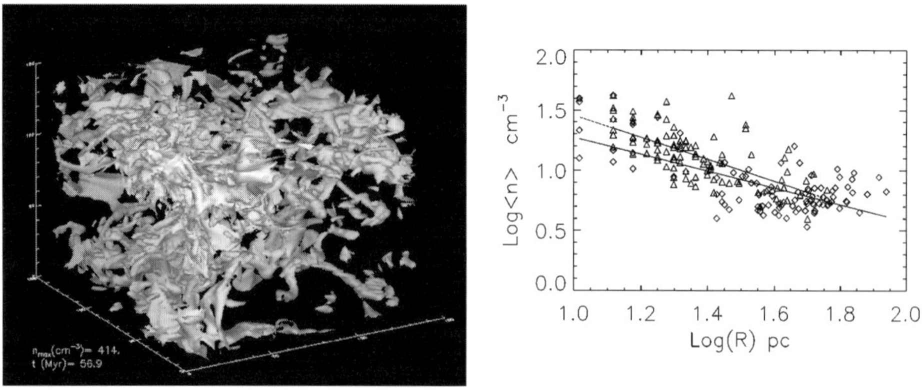

Figure 2. a) (Left) Density structure at t=0.6 τ_{dyn}. Iso-density contours correspond to n$\geq \bar{n}/2$, where $\bar{n}$=0.5 cm^{-3} is the average numerical density. b) (Right) Mean density-size relation for two density thresholds in the clump-finding algorithm. 8 $\bar{n}$ (triangles), 10 $\bar{n}$ (diamonds).

Acknowledgements

S.D. acknowleges the financial support of the EAS and would like to thank J. Ballesteros-Paredes, A. Gazol and Fabian Heitsch for stimulating discussions.

References

Ballesteros-Paredes, J. and MacLow, M.-M.: 2002, *ApJ* **570**, 734.
Burkert, A. and Lin, D.N.C.: 2000, *ApJ* **537**, 220.
Gazol, A., Vázquez-Semadeni, E. Sanchez-Salcedo, F.J. and Scalo, J.: 2001, *ApJ* **557**, L121.
Heiles, C.: 2001, *ApJ* **551**, L105.
Kim, S., Dopita, M.A., Staveley-Smith, L. and Bessell, M.S.: 1999, *ApJ* **118**, 2797.
Kolesnik, I.G.: 1991, *A&A* **243**, 239.
Larson, R.B.: 2001, *MNRAS* **194**, 809.
Rhode, K.L., Salzer, J.J., Westpfahl, D.J. and Radice, L.A.: 1999, *AJ* **118**, 323.
Stone, J.M. and Norman M.L.: 1992, *ApJS* **80**, 735.
Vázquez-Semadeni, E., Gazol, A. and Scalo, J.: 2000, *ApJ* **540**, 271.

IN ISM MODELING, THE DEVIL IS IN THE DETAILS: YOU SHOW ME YOUR OVI AND I'LL SHOW YOU MINE

DONALD P. COX
Department of Physics, University of Wisconsin-Madison, 1150 University Ave., Madison, WI 53706, USA; Email: cox@wisp.physics.wisc.edu

Abstract. The three sections of this paper illustrate the importance of rationalizing ISM theory and modeling with the observational information.

Keywords: ISM: general, ISM: local bubble, X-rays: galaxies

1. Modeling the ISM is Complicated

On this topic, see also Maarit Korpi's contribution and two other papers of mine ('Guide to Modeling the Interstellar Medium,' and 'Overall Models of the Interstellar Medium,' 1996a,b).

Imagine pouring the ISM into the Galaxy's gravitational potential, where it finds all the other constituents already in place. It is globally confined by gravity, stabilized radially by rotation, heated by starlight, disrupted by catastrophic stellar events, settles into components spanning a wide range of density and temperature, develops a magnetic field, confines cosmic rays, generates new stars, interacts with stellar density waves, and is concentrated by that into the gaseous spiral arms. It develops a vertical quasi-hydrostatics, with dynamic pressure, magnetic pressure (and tension), and cosmic ray pressure as dominant forms of support. When locally heated, it develops bubbles of hot gas that grow until their thermal pressures are in approximate equilibrium with the magnetic pressure that surrounds them. This can occur either via random heating events (accidental sequences of supernovae in proximity), or via the local production of OB associations that act collectively to build superbubbles.

Modeling this situation is approached at a variety of levels, e.g. Processes, Events, Local Region, and Global. At intermediate levels, one assumes that the processes at lower levels can be included in subgrid rules, while those at higher levels are embodied in initial or boundary conditions. One models the remnants of supernova explosions, for example, assuming an adequate understanding of the atomic and radiation processes, magnetic fields, thermal conduction, dust destruction and cosmic ray acceleration by shocks, etc., and a sufficient description of the surrounding medium. The results and their regime of applicability depend on the way in which all these lower and higher level aspects are approximated. The

Astrophysics and Space Science **289:** 469–477, 2004.

validity tends to increase with the degree to which the models are compared in intricate detail with the suite of observational data. With care, some parts may turn out to be right.

The assumptions we make are often those of convention; occasionally this may lead us astray. In general, astrophysicists are not comfortable with thermal conduction, for example. Magnetic fields interfere with it. Turbulence alters its basic nature (Lazarian and Cho, this volume). It is easy to leave it out of models because so many people believe it is quenched, while the others offer no clear prescription for how to include it. And yet it is a dangerous thing to omit in models that include hot gas. In two systems that I helped model, the Local Bubble and the W44 supernova remnant, the apparently observed characteristics appeared only when thermal conduction was included within the hot gas at roughly half the Spitzer rate. I've heard that something similar is being found in modeling cooling flows in clusters.

It is not clear how robust this hint is, but it begs for consideration. If you model the ISM, and find large volumes of hot gas, then please, to prevent our wasting too many more years in ignorance, add thermal conduction (with a variable amplitude parameter and limited by saturation) and see what its effects are. If your resolution is high enough that the turbulent and wandering field is adequately represented, let conduction cross your magnetic fields, with a reduced coefficient, perhaps of order 5% of the parallel rate (or ask Randy Jokipii what he would do). My prejudice is that thermal conduction in high temperature gas somehow manages to occur, despite magnetic fields, but that it is unable to cross boundaries to denser regions and thermally evaporate them. Maybe your models will give us a better sense of why.

Modeling superbubbles has a long history, including an early fascination with bubbles breaking out of the gaseous disk to dump their hot interiors into the Galactic halo. When it was finally realized that the gaseous disk is very thick in low density and nonthermal components, the fascination went away. Even as large as they are, the bubbles are confined within the the disk, which they help inflate, consistent with x-ray observations of other galaxies that find very few plumes of hot gas rising out of their disks. The fascination with breakout was predicated on misinformation at the larger scales.

But modeling superbubbles may have problems beyond finding a reasonable representation of the ambient medium. Have we got the process details of the interior right? Do the winds of the OB stars, augmented perhaps by entrained material from the molecular cloud birthplace, hit the shell and set up a reverse shock, applying the accumulated pressure to the bubble walls? Or is the material so dilute that there is no collisionless shock in the wind? Does the fluid description give us the wrong answer? (I ask because Sally Oey is trying to understand why so many wind bubbles appear smaller than they theoretically ought to be. See her contribution to this volume.) We have great theories, but they do not encompass the observed facts. (But, Don Osterbrock would often say something like, 'Never let questionable observations interfere with your appreciation of a good theory.' The corollary that most observations can be questioned is one of Sally's approaches.)

In short, there are areas of ISM research in which important details are too poorly known for us to be confident in the ways we include them in models at other levels. That said, I am enormously impressed by the incredibly detailed models that are starting to appear. This meeting was very good in presenting the remarkable ambition of the current enterprises.

2. There Are Many Conceptions of the ISM, All Flawed

What pictures would you draw to illustrate what you think are the dominant features of the ISM? I ask because what you draw determines what you put in your computer model, or how you describe the significance of your observations.

There are several commonly held conceptions, at different scales. Let's begin with conceptions within the disk, near us. The ones I will describe are the trivial ones. People such as Miguel de Avillez, who make models of the medium based on their best guesses of the process behaviors might draw something much more complex. One might even appear on the cover of this book, as it did on the book of abstracts.

Local conceptions tend to diverge over the component of the medium occupying the greatest volume. The three favorite possibilities appear to be:

- A warm intercloud medium, density perhaps averaging about 0.2 cm^{-3} and temperature perhaps 7000 K, in some places ionized and others not.
- A pervasive hot component with a temperature of roughly 10^6 K and pressure $p/k \sim 10^4$ cm^{-3} K.
- A tepid intercloud component with a temperature of $1 - 3 \times 10^5$ K.

The first of these views derives from the legacy of Field, Goldsmith and Habing (1969). It is somewhat supported observationally by the fact that there are warm HI and diffuse HII components that might be able to fill most of the volume. McKee and Ostriker (1977) argued that supernovae would disrupt such a medium and that the picture was therefore untenable. Slavin and Cox (1993), however, located the three fallacies leading to that conclusion, showed that supernovae might not disrupt such a medium, and found that invisible old remnants within it could provide all of the observed high stage ions (e.g. OVI) found in absorption measurements through the disk. Spitzer (private communication) however, wondered how the warm intercloud medium might go about restoring its relative uniformity. That question aside, it remains a somewhat attractive possibility, even though the detailed reasons for why it might be about right are unclear.

The second possibility follows from a suggestion by Cox and Smith (1974), and one much earlier by Spitzer (1956). The soft x-ray background appearing to derive from a large region of million Kelvin gas surrounding the Solar location led Cox and Smith to ask whether such gas might be common. I have abandoned this possibility because I think there is no observational evidence to support it, as well as no clear reason why it might be the preferred situation. I could be wrong; my distaste hinges largely on the expectation that it would have shown up more convincingly in soft x-rays. If, on the other hand, it were a few times hotter in most places, at a correspondingly lower density (to not have an unacceptably large pressure), its x-ray emission would be negligible and the scenario observationally allowed. (I like to refer to this as the Trans-Mega-Kelvin Regime.)

McKee and Ostriker (1977) suggested that there would likely be a thermal runaway in the hot component of Cox and Smith, unless one invoked a thermostatic mechanism such as evaporation of clouds to add mass. In this way, they evolved to the third scenario above. Their medium is heated to very high temperatures in local regions by supernovae, evaporates clouds, lowers its temperature as it expands and gains mass, and finally is able to radiate the supernova power when the temperature has dropped to a few hundred thousand Kelvins. The observed OVI was said to be in the thermally evaporating boundaries of the clouds, with the right average column density. The fact that far more OVI would be present in the tepid ambient medium, more than is observed by a large factor, was overlooked. For years I was convinced that the MO model was inconsistent with observation, but could not figure out how it failed. As remarked earlier, Jon Slavin and I finally found sufficient reason. It was a great example of observations insisting that the theory was lacking.

There is a fourth possibility that I try to remember to mention, thanks to a casual remark by Priscilla Frisch, in 1986. Perhaps most of the space is empty! This appears to be the impression of observers who study galactic sightlines and find components that fill only 20 or 30% of the total line of sight. Their copout, if they mention it at all, is that the rest of the sightline must be occupied by hot gas as advocated by . . . someone else.

I'll try to make empty space sound not so crazy. It isn't that there is nothing in the empty regions. The magnetic field and cosmic rays are still there; the field lines just have very little mass on them. It's like an extreme version of the Trans-Mega-Kelvin regime. There is no observational reason to reject this conception; we just have no theoretical framework in which it is plausible. How would such empty regions come into existence (McKee and Ostriker's thermal runaway perhaps)? How do they evolve? Do they exit the Galaxy because of their buoyancy, or evolve out of existence via material leaking (or being driven) into them? Sound crazy still? Well, the last I heard, the outer shock of the Crab Nebula had not been found. The explosion appears to be expanding into a vacuum.

With regard to the more complicated pictures being drawn by large scale modelers, I can only agree with Joel Bregman that until they are able to tell us how much OVI is in their descriptions, we can't look at them with a sufficiently appreciative eye. I'll wager they get at least an order of magnitude too much! It is a powerful applicability check, hence the subtitle of this paper.

Then there are the vertical and the global conceptions. If the galaxy were truly filled with hot gas (not tied to a magnetic field anchored by the weight of the cold gas), such gas would extend far above the disk of the Galaxy, filling out a scale height at least. If many superbubbles grow so large that they break out of the disk, the hot gas from their interiors might plausibly fill the halo, the density rising until the hot gas is able to radiate the supernova power. Or, if the escaping gas were fairly dense, it might generate a fountain, cooling before it gets very far, condensing into clouds that eventually rain back. So, we have two scenarios, a hot quiescent halo, and a fountain. But we know from observations that the halos of galaxies do not radiate even a small fraction of the total supernova power, not at high temperatures. Both scenarios failed badly, to the great surprise of many of us who were expecting otherwise when x-ray data first became available.

What's left then? Well, cosmic rays escape the Galaxy. The phenomenon is frequently described as a wind. Perhaps the Galaxy has a thermal wind too, arising from the disk, high enough in temperature (and therefore low enough in density) that its x-ray emission is negligible (the Trans-Mega-Kelvin regime again). I believe that such a wind cannot be carrying away thermal energy faster than the cosmic ray loss rate, which is roughly 10% of the total supernova power. The cosmic rays seem to diffuse out of the disk, forming their 'wind' rather high off the plane. If hot gas flows out rather unimpededly, it ought to take cosmic rays with it, in which case the two would leave the Galaxy in proportion to their relative energy densities in the hot component in the disk, a ratio of order unity in this scenario.

An entirely different possibility is that except for cosmic rays leaving the Galaxy, all other disturbances are trapped in the disk by the inelasticity of the great interstellar feather pillow. Superbubbles are the largest disturbances, but few reach such great size they are other than trapped volumes of hot gas that eventually dissipate their energies radiatively. This has been my favorite view for some time. It does not mean that there is no hot gas above the plane of the Galaxy; there are superbubbles that expand several hundred parsecs. We see several, and hot gas within them. There are also supernovae that occur at high $|z|$. These may be an important contributor to the thick layer in which high stage ions are found, as explored by Robin Shelton (1998). It just means that violent disturbances are generally locally confined, and dissipated.

What picture would you draw? Mine has all the usual components. There is a thick layer of low density warm intercloud gas, reaching up a kiloparsec or two, with here and there a confined supernova remnant within it. Quite a bit of it is ionized and visible in H alpha, the Reynolds Layer. It is pervaded by a largely random magnetic field and cosmic rays. Within it, dense cold clouds lie close to the midplane creating OB associations, the latter forming superbubbles that expand, but not to the extent that they break out of the disk. To be more precise, I think they help inflate the disk, rather than breaking out of it, except in the extreme case of starburst galaxies. Then there is the emptiness, emptiness that fills the ancient and dying carcasses of once grand superbubbles, à la the models of Ferrière (1998), and maybe even the emptiness of Frisch or the Trans-Mega-Kelvin regime. And finally, the whole thing is circulating around the Galaxy, now and then provoked into unusual activity by its encounter with a spiral arm. This spiral arm encounter is often described as a shock, but, Martos and Cox (1998) showed that it should have some elements of both a shock and a hydraulic jump, or bore. That work has been extended to three dimensions and a significant fraction of the Galaxy by Gómez and Cox (2002). A brief report appears in the poster contributions to this volume. It's exciting and an excellent example of a large-scale model that has to make very simple approximations to the character of the medium. There could also be a wind from the Galactic Bulge, providing a very interesting outer boundary condition on the ISM of the disk.

3. Local Matters Matter

This is about building a house of cards. It concerns the structure of the ISM in the vicinity of the Sun, starting out about 100 pc away and working in to about 5 AU. It is a progress report on years of work, and the fearful tests that lie ahead. Parts of it are discussed at length in Smith and Cox (2001) and Cox and Helenius (2003). Other authors have taken different routes with the observational data; references and discussions of their views appear in the papers above. But I am here offering an example of a modeling effort that has gone out on a limb, not surveying the whole realm of possibilities.

The Local Cavity is a large region of exceptionally low density surrounding the Solar location. Within it, there appears to be a sub-region of million Kelvin gas, the Hot Local Bubble, providing much of the soft x-ray background. To get the observed x-ray surface brightness, given the scale of order 100 pc, the required thermal pressure is $p/k \sim 2 \times 10^4$ cm^{-3} K. This is just what would be expected for such a bubble quasi-statically confined by the surrounding ISM, where the pressure of the latter equals its weight per unit area. Within that hot gas, there are wisps of higher density cooler material, the Local Fluff. The densities in the Local Fluff are of order 0.2 cm^{-3}, the temperatures about 7000 K, characteristic sizes of about 5 pc. One particular piece of this Fluff, the Local Cloud, surrounds the Solar System. The relative velocity between the Local Cloud and the Sun is about 26 km s^{-1}. Thus, in the frame of the Sun, the material of the Local Cloud (the LISM, and VLISM), constitutes an interstellar wind blowing past the Solar System.

The Solar wind interacts with this interstellar wind, its momentum creating a large cavity around which the flow of interstellar ions and magnetic field is diverted. To first order, the neutral component of the Local Cloud does not notice this disturbance. The scale is sub-fluid for the neutrals, which tend to flow undisturbed into the heliosphere. (There is some charge exchange between the neutrals and diverted interstellar ions that has important consequences for the details.) Once they are deep into the Solar System, these neutrals suffer various fates. When one becomes ionized, it is immediately picked up by the Solar Wind magnetic field and joins the outflow; these are observed as 'pickup ions' by interplanetary probes. On reaching the Solar Wind termination shock, some are accelerated to higher energies, diffuse back in toward the Earth, and are observed as the 'anomalous cosmic rays.' Radiation pressure from scattered Solar Lyman alpha approximately cancels gravity, so the hydrogen trajectories are straight. Solar radiation ionizes most of it before it reaches z AU. The distribution of interstellar hydrogen in the inner Solar System is therefore roughly uniform in the upstream direction, outward of z AU, and absent at smaller radii and in much of the downstream direction. It is observed via its backscattering of the Solar Lyman alpha. Helium suffers less radiation pressure and ionization, and is gravitationally focused in the downstream direction. This is observed via resonant backscattering of 304 Å Solar radiation.

Smith and Cox (2001) explored the possibility that the Local Bubble could have been created by an accidental sequence of supernovae in a warm intercloud medium as described in Section 2. They found that a sequence of three supernovae over the course of several million years, the second and third each occurring within the region previously evacuated, could create a bubble of radius about 100 pc, and leave it with the right density and temperature so that its radiation would resemble the soft x-ray background. Considering that the Local Cavity is much larger in some directions, and potentially much older, a generalization is that something created the large cavity, and a supernova reheated a portion of it roughly three million years ago to leave us surrounded by the Hot Local Bubble. Evidence for the nearby occurrence of a supernova about that long ago is discussed in their paper and further in Cox and Helenius(2003).

But why would the Local Fluff be sitting in the middle of such a remnant? A possible answer was provided by Cox and Helenius (2003) who examined the evolution of a magnetic flux tube pulled from the wall of the bubble by its magnetic tension, starting about three million years ago, after it had been swept to the wall by the supernova. As the flux tube springs from the wall, it is initially accelerated very rapidly; but soon drag through the hot gas becomes important, and it gradually decelerates. Material within the tube starts out very diffuse, part of the inner wall of the bubble, but the initial acceleration due to tension is largely radial. As the tube itself slows, the material within it flows toward its center, concentrating into a low-density cloud, or collection of interacting clouds, depending on the complexity of the initial conditions. By the time the tube reaches the bubble middle, the velocity distribution of the material within resembles that observed in the Local Fluff, in magnitude, and in the generally converging flow pattern. Its density and ionization state are also appropriate. The details need more study, but the scenario creates something very much like the observed Local Fluff.

Now we turn to the problems. Understanding of the soft x-ray background is not complete. High spectral resolution will eventually give us a much better idea of what we are looking at. For the present, there is at least one disturbing possibility.

The sky maps of the SXRB from ROSAT were made from data that had contamination that was not understood at the time, the 'Long Term Enhancements.' In mapping, periods of time during which this contamination was large were omitted. The maps thus present the lowest signal detected from each direction. Their spatial structure correlates very well with maps from rockets, and Galactic structure, so it is believed that the process was successful in removing the contamination. But the resulting 1/4th keV surface brightness is pretty uniform in the Galactic Plane, getting brighter at higher latitudes. The worry is that the midplane intensity could be just the DC level of the contamination. If so, we would lose the Local Bubble as we think we know it, replacing it with smaller bubbles at positive and negative higher latitudes. After the observations of x-rays from comets, and their interpretation as arising from charge exchange recombination of highly ionized elements in the Solar Wind, we finally had a mechanism for generating x-rays within the Solar System. It now seems likely that the Long Term Enhancements arise from the same mechanism, with the charge exchange occurring on the atoms of the Earth's exosphere. There should be also be a significant steady contribution from the Solar Wind ions charge exchanging on the interstellar neutrals penetrating the Solar System (e.g. Cox, 1998; Cravens et al., 2001). The optimistic view from the estimates so far is that the level of the overall contamination is modest, at 30% or so, and that our picture will remain largely undisturbed. We'll see.

The other major problem is the distance to the Solar Wind Termination Shock. I told you what the pressure in the Local Bubble had to be to produce the observed x-rays, and that it is appropriate for a bubble in equilibrium with the surrounding ISM. I also gave you data from which you can estimate the thermal pressure in the Local Fluff. The latter is less than 20% of the former. In the model of Cox and Helenius, the magnetic pressure of the flux tube is significantly higher than the thermal pressure of the gas within it. The field strength is several microGauss, to balance the thermal pressure of the surrounding hot bubble. Forget the model. In order for the Local Fluff to be in pressure equilibrium with the surroundings, such a field is required. But modelers of the interaction of the Solar Wind with the Local Cloud will tell you that if the magnetic pressure in the Local Cloud were that high, the termination shock would have been pushed inward to the point that Voyager 1 would already have encountered it, which it has not! There are only a few ways out of this difficulty. One is that making such models is difficult (as in the theme of Section 1) and details may have been overlooked. Astronomically insignificant things like a measly factor of two are critical. Another is that the Cox and Helenius model makes a specific prediction for the direction of the magnetic field, from the directionality of the local distribution of Fluff, and that direction is roughly parallel to the relative wind flow. This is a case that has not been examined and offers a radically different structure.

A third alternative is the scary one, involving both of the above problems. The field is actually much weaker. The Local Bubble does not exist. The local interstellar total pressure is very significantly lower than the average expected from the weight of the interstellar medium. The whole house of cards collapses. We'll see.

One last thing: If you model the ISM, don't forget to tell us how much OVI you get. Send me a telegram when you no longer find more than a factor of two too much!

This work was supported in part by NASA grant NAG5-8417 to the University of Wisconsin-Madison. It has benefited from critical readings by Gilberto Gómez, Bob Benjamin, Sally Oey, Maarit Korpi, and Dieter Breitschwerdt. I apologize to the many observers whose work made the above discussion possible yet who were not mentioned by name or referenced. To others, please do not suppose that I have told you the truth about observations. Go to the sources.

References

Cox, D.P.: 1996a, in: S. Bowyer and R.F. Malina (eds.), *Astrophysics in the Extreme Ultraviolet*, Kluwer, p. 247.

Cox, D.P.: 1996b, in: Kunth et al. (eds.), *Proceedings of the Institute d'Astrophysique Meeting On The Interplay Between Massive Star Formation, the ISM, and Galaxy Evolution in Paris*, Editions Frontieres, p. 233.

Cox, D.P.: 1998, in: D. Breitschwerdt, M.J. Freyberg and J. Trumper (eds.), *IAU Colloq. 166, The Local Bubble and Beyond*, Springer, Berlin, p. 5.

Cox, D.P. and Helenius, L.: 2003, *ApJ* **583**, 205.

Cox, D.P. and Smith, B.W.: 1974, *ApJLett* **189**, L105.

Cravens, T.E., Robertson, I.P. and Snowden, S.I.: 2001, *JGR* **106**, 22883.

Ferrière, K.: 1998, *ApJ* **503**, 700.

Field, G.B., Goldsmith, D.W. and Habing, H.J.: 1969, *ApJL* **155**, L149.

Gómez, G.C. and Cox, D.P.: 2002, *ApJ* **580**, 235.

Martos, M.A. and Cox, D.P.: 1998, *ApJ* **509**, 703.

McKee, C.F. and Ostriker, J.P.: 1977, *ApJ* **218**, 148.

Shelton, R.L.: 1998, *ApJ* **594**, 785.

Slavin, J.D. and Cox, D.P.: 1993, *ApJ* **417**, 187.

Smith, R.K. and Cox, D.P.: 2001, *ApJS* **134**, 283.

Spitzer, L., Jr.: 1956, *ApJ* **124**, 20.

FROM LARGE TO SMALL SCALES: GLOBAL MODELS OF THE ISM

MIGUEL DE AVILLEZ
Department of Mathematics, University of Évora, R. Romão Ramalho 59, 7000 Evora, Portugal
E-mail: mavillez@galaxy.lca.uevora.pt

DIETER BREITSCHWERDT
Max-Planck-Institut für Extraterrestrische Physik, Giessenbachstraße, Postfach 1312, 85741 Garching, Germany; E-mail: breitsch@mpe.mpg.de

Abstract. We review large scale modelling of the ISM with emphasis on the importance to include the disk-halo-disk duty cycle and to use a dynamical refinement of the grid (in regions where steep variations of density and pressure occur) for a realistic modelling of the ISM. We also discuss the necessity of convergence of the simulation results by comparing 0.625, 1.25 and 2.5 pc resolution simulations and show that a minimum grid resolution of 1.25 pc is required for quantitatively reliable results, as there is a rapid convergence for $\Delta x \leq 1.1$ pc.

Keywords: Hydrodynamics, ISM: general, ISM: structure, ISM: kinematics and dynamics, galaxies: halos, Galaxy: disk

1. Introduction

The interstellar gas exhibits time-dependent structures on all scales and is heavily non-linear and subject to numerous instabilities in certain parameter regimes, which may lead to chaotic behaviour, while in others the evolution is fairly predictable. The key to a realistic description is the highest possible spatial resolution and a realistic input of the basic physical processes with appropriate boundary conditions. The former is required due to the formation of small scale structures resulting from instabilities in the flows, in particular from thermal instabilities and condensations as a result of radiative cooling. The amount of cooling may be substantially increased if the resolution is high enough to trace regions of high compression by shocks, rather than smearing them out over a larger volume thus decreasing the average density. Such a realistic description requires therefore the appropriate tools, which are computer clusters with parallelized HD and MHD codes and a sophisticated method of tracking non-linear structures, such as shock waves on the smallest possible scales in conjunction with adaptive mesh refinement. Thus the global evolution of the ISM resulting from mass, momentum and energy input due to supernovae can be adequately described.

The structure of the paper is as follows. Section 2 deals with the importance of the disk-halo-disk cycle for global modelling of the ISM. In Section 3 the signature of the initial evolution in the temperature and density PDFs is discussed, as well

Astrophysics and Space Science **289:** 479–487, 2004.

as the resolution effects in the time histories of the volume filling factors and the maximum density and minimum temperature in the ISM. Moreover the necessity for convergence of the simulations is addressed. In section 4 a summary and final remarks are given.

2. Duty Cycle and Grid Sizes

2.1. The Duty Cycle

The evolution of the ISM in disk galaxies is intimately related to the vertical structure of the thick gas disk and to the input of energy by supernovae per unit area in the Galactic disk. The system can evolve towards some sort of dynamical equilibrium state on the global scale if the boundary conditions vary only in a secular fashion. Such an equilibrium is determined by the input of energy into the ISM by SNe, diffuse heating, the energy lost by radiative cooling and magnetic energy, and is only possible *after* the full establishment of the duty cycle of the warm and hot gas and its circulation between the disk and the halo, which takes several hundred Myr. An upper limit for this can be estimated by calculating the flow time, t_f, that the gas needs to travel to the critical point of the flow in a steady-state (see Kahn, 1981). This is the characteristic distance from which information in a thermally driven flow can be communicated back to the sources. Then $t_f \sim r_c/c_s$, where r_c and c_s are the location of the critical point and the speed of sound, respectively. For spherical geometry, the critical point can be simply obtained from the steady state fluid equations, $r_c \sim GM_{gal}/(2c_s^2)$, which yields for a Milky Way mass of $M_{gal} = 2 \times 10^{11}$ $M_\odot$ a distance $r_c \approx 21.6$ kpc, and thus $t_f \sim 1.5 \times 10^8$ years as an upper limit for the ascending flow time. This value is of the same order as the radiative cooling time, ensuring that the flow will not only cool by adiabatic expansion, but also radiatively, thus giving rise to the fountain return flow, which is the part of the outflow that loses pressure support from below and therefore cannot escape from the galaxy. Note, that r_c is the maximum extension of the fountain.

In the disk, on the other hand, the heating and cooling time scales are much shorter than in the halo, since they scale typically with the gas density there, i.e. $\propto 1/n$ and $\propto 1/n^2$, respectively. Avillez and Breitschwerdt (2004) show from large-scale simulations of the ISM that typical time scales of pressure fluctuations about a mean value of $P/k_B \sim 2600\,\mathrm{cm}^{-3}$ K are about 30 Myr, which is by the way a typical time scale for superbubble evolution.

It should be emphasized, since disk and halo are coupled dynamically, not only by the escape of hot gas, but also by the fountain return flow striking the disk, that the disk equilibrium will also suffer secular variations on time scales of the order of 100 Myr.

2.2. Vertical extension of computational domain

The need of a duty cycle and the establishment of a global dynamical equilibrium requires the use of an extended grid in the direction perpendicular to the Galactic plane. The lack of such an extended z-grid inhibits the disk-halo-disk circulation of matter, which otherwise would return gas to the disk sometime later, with noticeable effects for the dynamical evolution there. In runs in which the vertical extension is small, e.g., 1 kpc above and below the midplane[1], compared to the maximum height to which the hot plasma rises due to the injection of energy and momentum from the sources, most of the gas escapes from the disk in less than 100 Myr (without ever returning to it), unless the loss of matter is compensated by some injection of mass into the grid through the top and bottom boundaries. This can be accomplished by the use of (1) periodic boundary conditions (the gas leaving the top/bottom boundary enters in the bottom/top boundary), (2) reflection boundary conditions where the gas hitting the top/bottom boundary reflects in it returning back, or by (3) calculating the amount of mass leaving the grid on each boundary and inject that mass in the form of clouds into the top/bottom boundaries. These schemes assume that the amount of gas that leaves the grid is the same that enters it, neglecting that some of this matter may actually escape from the Galaxy altogether as a wind[2] or return to the disk at a different locations due to colliding gas streams. The wind effect may be included in scheme (3) by assuming that the gas injected into the grid is only a fraction of the escaping gas mass.

In any case these schemes neglect the effects of the duty cycle in the dynamics of the disk gas as there is no recycling of gas, and as a consequence the simulations can only run for a small evolution time thus reflecting the *initial evolution* of the system. Therefore, the volume weighted histograms of the thermodynamic properties (e.g., density, pressure and temperature) of the disk gas will retain a memory of the original set-up.

2.3. Grid extension parallel to the midplane

The extension of the grid in the direction parallel to the Galactic plane is also an important factor. Large xy-grids, parallel to the midplane are required because counter pressure effects are very important in the bubbles confinement (Avillez, 2000; Avillez and Breitschwerdt, 2003). If the grid in the plane is too small, say less than 500 pc, then large scale horizontal flows resulting from superbubbles

[1] Note, that the maximum extension of the fountain is more than an order of magnitude larger, and therefore such restricted calculations definitely miss an important component of the galactic gas dynamics.

[2] In the case in which the appropriate grid is used there is still gas leaving the upper and bottom boundaries amounting to some 10% of the total mass with which the simulations started (Rosen and Bregman, 1995; Avillez, 2000). Thus, a smaller grid with open boundaries on the top and bottom of the grid will imply the loss of a *large amount* of matter within a short time thereby evacuating the gas disk.

expanding into low density regions and eventually colliding with other hot tenuous bubbles will not be captured adequately. We therefore consider an extension of 1 kpc $\times$ 1 kpc as the minimum grid extension in the plane.

3. Global modelling of the ISM

3.1. SN-DRIVEN ISM MODEL

In order to gain insight into the nonlinear physics of a complex system like the ISM, driven by temporally and spatially variable energy input we first try to understand the interplay and effects of the most dominant processes, and therefore only the most important ingredients are included in the models, neglecting other processes that may be relevant too. We therefore, do not strive for completeness, but for a better understanding of the ISM gas dynamics. The discussion that follows is based on the results from three-dimensional kpc-scale modelling of the ISM with a grid centred on the solar circle and extending from $z = -10$ to 10 kpc with a disk area of 1 kpc^2. These simulations use adaptive mesh refinement with resolutions of 2.5, 1.25 and 0.625 pc and were obtained with a modified version of the 3D SN-driven ISM model of Avillez (2000).

The basic processes included in the model are the gravitational field provided by the stars in the disk, radiative cooling (assuming an optically thin gas in collisional ionization equilibrium) with a temperature cut off of 10 K, and uniform heating due to starlight varying with z. In the Galactic plane background heating is chosen to initially balance radiative cooling at 9000 K. With the inclusion of background heating the gas at $T < 10^4$ K becomes thermally bistable. The interstellar gas is initially set up with a density stratification distribution that includes the cold, cool, warm, ionized and hot gases according to observations. The prime sources of mass, momentum and energy are supernovae types Ia, Ib+c and II with scale heights, distribution and rates according to observations. OB associations can be setup in regions with density and temperature thresholds of 10 cm^{-3} and 100 K, respectively, and an initial mass function is applied to determine the number of OB stars and their masses, forming an OB association. The time interval between the explosions of all OB stars is determined by their main sequence life time, and the kinematics of both the association and of the low mass stars in the association can be followed in detail.

3.2. GENERAL EVOLUTION

Our simulations reproduce many of the features that have been observed in the Milky Way and other star forming galaxies, namely: (i) A thick frothy gas disk composed of a warm, neutral medium overlying a thin HI disk, with a variable thickness up to $\sim$ 80 pc; (ii) bubbles and superbubbles and their shells distributed on either side of the midplane; (iii) tunnel-like structures (chimneys) crossing the

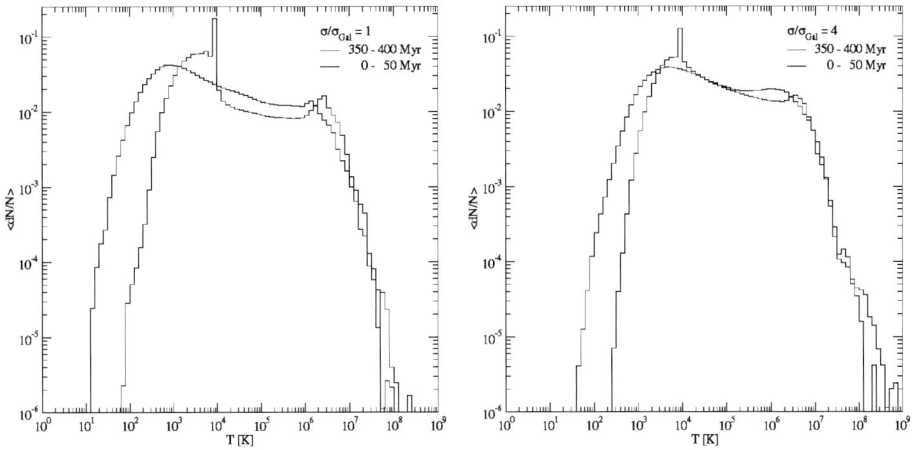

Figure 1. Averaged volume-weighted temperature PDFs over the periods of 0-50 Myr (red) and 350-400 Myr (black) calculated using 51 snapshots taken at time intervals 1 Myr. The supernova rates used in these models are: $\sigma/\sigma_{Gal} = 1$ (left panel) and 4 (right panel). The resolution of the finest AMR level is 1.25 pc.

thick gas disk and connecting superbubbles to the upper parts of the thick gas disk; (iv) thick gas disk with a distribution compatible with the presence of two phases having different scale heights: a neutral layer with $z_n \sim 500$ pc (warm HI disk) and an ionized component extending to a height $z_i \sim 1.5$ kpc above the thin HI disk; (v) cold gas mainly concentrated into filamentary structures running perpendicular to the midplane. These clouds form and dissipate within some 10-15 Myr. Compression is the dominant process for their formation, but thermal instability also plays a rôle; (vi) for supernova rates $\sigma/\sigma_{Gal} \leq 2$, σ_{Gal} is the Galactic value, the hot gas has a moderately low volume filling factor in agreement with observations ($\sim 20\%$) even in the absence of magnetic fields and is mainly distributed in an interconnected tunnel network and in some case it is confined to isolated bubbles (see Figure 2).

3.3. Effects of the Initial Evolution

A major consequence of the set-up of the duty cycle is that the system loses any memory of its initial conditions and evolution as most of the disk gas has already travelled into the halo and back to the disk. Thus, as should be in representative global simulations, memory effects are not present in the volume weighted histograms (PDFs) of the different thermodynamic properties as well as in the history of the volume filling factors of the different ISM phases. Figure 1 compares PDFs of the temperature over the periods of 0-50 Myr (red) and 350-400 Myr (black) for two AMR simulations of the ISM using supernova rates $\sigma/\sigma_{Gal} = 1$ (left panel) and 4 (right panel). The finest AMR level resolution is 1.25 pc.

These PDFs indicate that for low σ, the temperature peak is at about 2000 K, making the cold/warm HI gas the most abundant gas phase, consistent with observations. The relative importance of the hot phase increases with SN rate, and one is moving towards a bimodal distribution. It should be emphasized that a simulation time of only 50 Myrs is not sufficient to establish this result, mainly because the effect of upwards transport has as a duty cycle of the order of a few hundred Myrs. In any of the cases shown in the figures the averaged PDFs for the initial 50 Myr have two pronounced peaks, one around 8000 K and the other around 5×10^6 K. Note that with the increase of SN rate the PDFs of the first 50 Myrs suffer large variations, indicating that the loss of recollection of the initial conditions will occur earlier for the highest SN rates.

The inclusion of snapshots taken during the first tens of Myrs in the calculation of PDFs, and consequently also for the volume filling factors, has a considerable effect in the overestimation of the coolest gas in the beginning of the simulations when cooling and gravity are the main physical dominant processes as the system is only *building up* the pressure support against gravity. Thus, during the first Myrs much of the gas collapses into the plane due to gravity, and simultaneously cools towards low temperatures. As soon as the pressure has built up, there is a redistribution of matter in the grid filling it. Therefore, the PDFs of the disk gas including the first say 40-50 Myr snapshots will show the presence of a large fraction of cold gas, while the PDFs constructed at later times, say up to 90 Myr or so, will show the dominance of the hot gas. A combination of these PDFs gives a pronounced bimodal distribution in the averaged PDFs corresponding to large volume filling factors for the hot and coolest ISM phases. Moreover, the averaged PDFs of the temperature will show a bimodal structure even if there is no background heating due to starlight. However, it should be emphasized that global ISM models, which may be considered representative for star forming galaxies, should be free of initial set-up features.

3.4. Resolution and Convergence

The simulations show how crucial spatial resolution is in order to capture small scale structures and, in particular, the cold gas as can be seen by the comparison between the left and right panels of Figure 2. Both panels show a snapshot of the density distribution in the Galactic plane for $\sigma/\sigma_{Gal} = 2$, at different evolution times, for runs with AMR resolutions of 2.5 pc (left panel) and 0.625 pc (right panel). High density gas is shown in blue, while low density gas (high temperatures) is shown in red. For contrast purposes, the maximum density cutoff is set at $10^{2.2}$ cm^{-3}, although in the 0.625 pc resolution image there are regions where the density is as large as 800 cm^{-3}. The panels show a clear difference between the morphology of the ISM in the two resolutions. In the 0.625 pc resolution image the filamentary structures are well defined and thinner than in the left panel. It can be clearly seen how cooling promotes the formation of small scale structures.

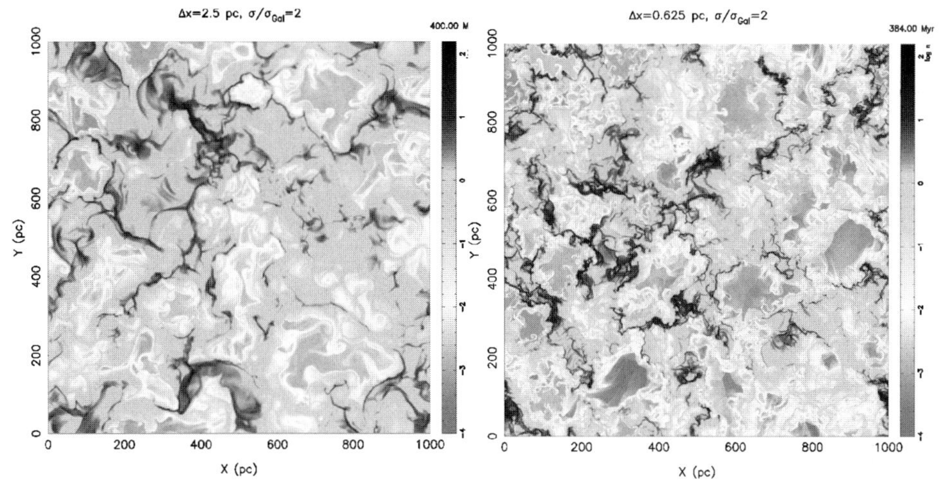

Figure 2. Density distribution in the Galactic plane for AMR resolutions of 2.5 (left panel) and 0.625 pc (right panel) at times 400 and 384 Myr of disk evolution, respectively. The SN rate is twice the Galactic value. High density gas is shown in blue, while low density (high temperatures) is shown in red. For contrast purposes, the maximum density cutoff is set at $10^{2.2}$ cm^{-3}.

A comparison between the maximum density and minimum temperature measured at the finer level resolutions of 0.625, 1.25 and 2.5 pc reveals that an increase in resolution by a factor of two from $\Delta x = 2.5$ to 1.25 pc implies an increase in the maximum density and a decrease in minimum temperature of the gas by factors greater than 4. With an increase in resolution from 1.25 to 0.625 pc the decrease in T_{min} and increase in n_{max} is smaller than a factor 2 as can be seen in Figure 3. The figure compares the average values of T_{min} and n_{max} calculated between 200 and 400 Myr of evolution for the three resolutions. The values measured for $\sigma/\sigma_{Gal} = 1$ are shown by open circles, for $\sigma/\sigma_{Gal} = 2$ by stars, and for $\sigma/\sigma_{Gal} = 4$ by triangles. The straight lines show the exponential fits associated to the variation of T_{min} (decrease with resolution) and n_{max} (increase with resolution).

Resolution also affects the occupation fraction of the gas that is observed in the different ISM phases, most importantly for the cold and hot phases. Figure 4 compares the volume filling factors of the different phases for the AMR resolution level of 0.625 pc (black), 1.25 pc (red) and 2.5 pc (green) for $\sigma/\sigma_{Gal} = 2$ and 4. In general the increase of resolution from 2.5 to 1.25 pc affects three of the ISM phases, and most importantly affects the amount of cold gas (which increases from 9% when 2.5 pc resolution is used to some 18% for 1.25 pc or 0.625 pc resolutions, respectively) for $\sigma/\sigma_{Gal} \leq 2$ after several crossing times. The relative variation of the cold gas is still very high for $\sigma/\sigma_{Gal} \geq 4$, whereas for the hot phase there is a reduction of only a few percent of the occupation fraction with resolution, since here compression and cooling are less effective.

The variations in the volume filling factors are easily understood if one takes into account that with an increase of resolution it becomes possible to resolve the

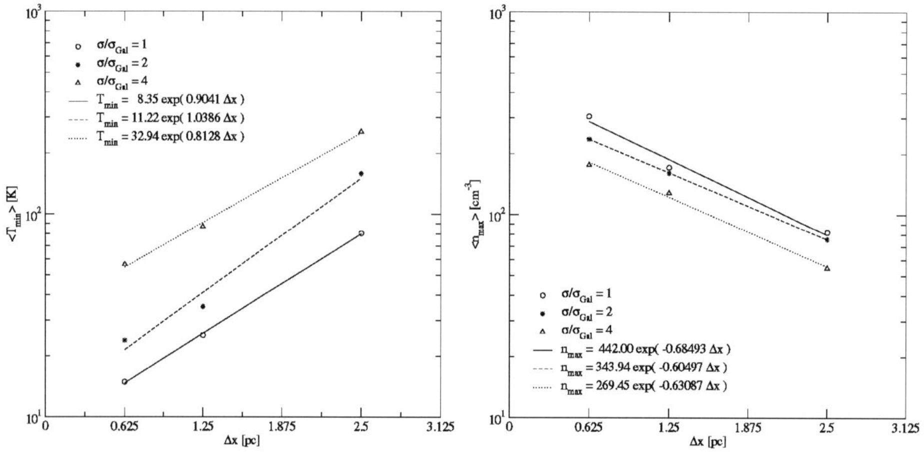

Figure 3. Comparison between the average minimum temperature (left panel) and maximum density (right panel) as function of resolution for the three SN rates: $\sigma/\sigma_{Gal} = 1$ (open circles), $\sigma/\sigma_{Gal} = 2$ (stars), and $\sigma/\sigma_{Gal} = 4$ (triangles). The plots also show exponential fits to the data points. These average values were calculated during the last 200 Myr of evolution, such that their history does not reflect any recollection of the initial conditions, and after the establishment of the duty cycle.

smallest scale structures. The rate of cooling may be substantially increased if the resolution is high enough to allow for the development of turbulent shear layers in addition to more contact surfaces. The latter is most important, because it allows for a faster mixing between parcels of gas with different temperatures (conduction or diffusion processes being of second order and hence inherently slow in nature). The mixing here is promoted by numerical diffusion rather than molecular diffusion, and therefore, the time scale for mixing to occur is somewhat smaller (because it happens on larger scales) than those predicted by molecular diffusion theory (e.g., Avillez and MacLow, 2002). However, turbulent diffusion will be most efficient.

The fact that there is only a small difference between the volume filling factors for the 1.25 and 0.625 pc resolutions in addition to the small variation of T_{min} and n_{max} is a clear indication for the convergence of the results, reproducing the physical processes involved in the dynamics and evolution of the ISM. Thus, the simulations with a resolution of 1.25 pc (or higher) can represent the real ISM. From Figure 3 and the fit for $\langle T_{min} \rangle$ and $\langle n_{max} \rangle$, e.g. $\langle T_{min} \rangle = 8.35 \exp(\frac{\Delta x}{1.106\text{pc}})$, we can see that there is rapid convergence for $\Delta x \leq 1.1$ pc.

4. Summary and Final Remarks

In this paper we discuss the effects of the inclusion of the duty cycle in global models of the ISM and show that sufficient resolution is crucial in order to obtain quantitatively reliable results that can be compared to observations, and that a min-

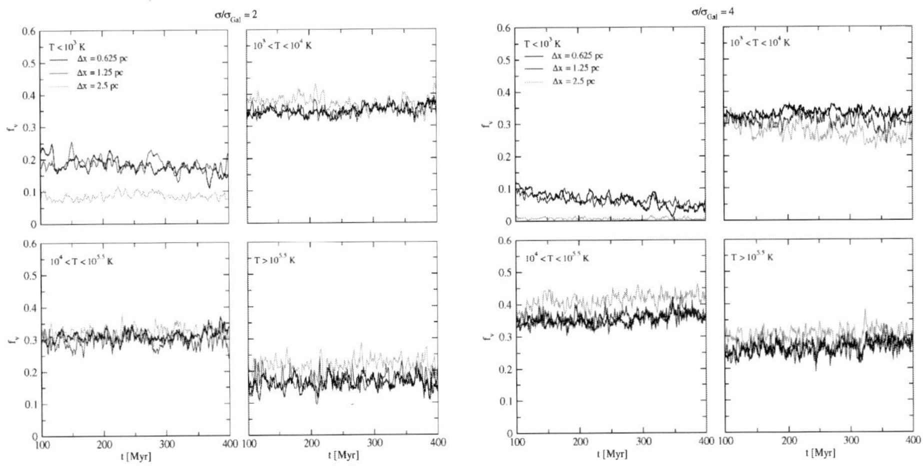

Figure 4. Comparison between volume filling factors of cold ($T < 10^3$ K), cool ($10^3 < T < 10^4$ K), warm ($10^4 < T < 10^{5.5}$ K) and hot ($T > 10^{5.5}$ K gas for the finest AMR level grid resolutions of 0.625 pc (black), 1.25 pc (red), and 2.5 pc (green) for the SN rates $\sigma/\sigma_{Gal} = 2$ (top panels), and 4 (bottom panels).

imum grid resolution of 1.25 pc is needed. This is unambiguously demonstrated by the rapid convergence towards the 0.625 pc resolution simulations. The occupation fraction of the different ISM phases depends also sensitively on the presence of a duty cycle established between the disk and halo acting as a pressure release valve for the hot phase.

The calculations presented in this paper do neither include the magnetic field nor the cosmic rays. A parameter study of their effects on the ISM is underway and will be described in forthcoming papers. If the magnetic field is present and is initially mainly orientated parallel to the disk, transport into the halo may be inhibited, although not prevented. As a consequence the hot gas in the disk should have a slightly higher volume filling factor than in the present simulations, since this is the ISM component that tries to escape first from the disk. However, on larger scales magnetic tension forces become weaker than on the smallest scales and therefore vertical expansion might still take place efficiently and therefore the occupation fraction of the hot gas could still be comparable to the values observed in the present simulations.

References

Avillez, M.A.: 2000, *MNRAS* **315**, 479.
Avillez, M.A. and MacLow, M.-M.: 2002, *ApJ* **581**, 1047.
Avillez, M.A. and Breitschwerdt, D.: 2004, *A&A*, in press.
Kahn, F.D.: 1981, in: F.D. Kahn (ed.), *Investigating the Universe*, Reidel, Dordrecht, p. 1.
Rosen, A. and Bregman, J.N.: 1995, *ApJ* **440**, 634.

SELF-CONSISTENT MODELLING OF THE INTERSTELLAR MEDIUM *

DIETER BREITSCHWERDT
Max-Planck-Institut für extraterrestrische Physik, Postfach 1312, D-85741 Garching, Germany
Email: breitsch@mpe.mpg.de

Abstract. The dynamical evolution of hot optically thin plasmas in the ISM crucially depends on the heating and cooling processes. It is essential to realize that all physical processes that contribute operate on different time scales. In particular *detailed balancing* is often violated since the statistically *inverse* process of e.g. collisional ionization is recombination of an ion with two electrons, which as a three-body collision is usually dominated by radiative recombination, causing a departure from collisional ionization equilibrium. On top of these differences in atomic time scales, hot plasmas are often in a dynamical state, thereby introducing another time scale, which can be the shortest one. The non-equilibrium effects will be illustrated and discussed in the case of galactic outflows. It will be shown, that spectral analyses of X-ray data of edge-on galaxies show a clear signature in the form of 'multi-temperature' halos, which can most naturally be explained by the 'freezing-in' of highly ionized species in the outflow, which contribute to the overall spectrum by *delayed recombination*. This naturally leads to a non-equilibrium cooling function, which modifies the dynamics, which in turn changes the plasma densities and thermal energy budget, thus feeding back on the ionization structure. Therefore *self-consistent* modelling is needed.

Keywords: ISM: general, galaxies: halos, galaxies: starburst, X-rays: galaxies

1. Introduction

Interstellar plasmas are subject to numerous dynamical and thermal processes, which are essentially driven by star formation activity in galaxies. Matter and radiation are continuously exchanged between the interstellar medium (ISM) and stars. This can loosely be described as a matter cycle. Star formation induced by gravitational instabilities is only possible if both transport of angular momentum, mass and energy work efficiently, with radiative cooling playing a key rôle both in the formation of interstellar clouds and protostars. During its main sequence lifetime a star interacts with the ISM by its constant photon output, by which especially Lyman-continuum photons can inject a substantial amount of energy creating HII regions. Moreover mass loss in the form of stellar winds gives rise to hot shock heated extended bubbles due to a mechanical wind luminosity of the

* XMM-Newton is an ESA Science Mission with instruments and contributions directly funded by ESA Member states and the USA (NASA). The XMM-Newton project is supported by the Bundesministerium für Bildung und Forschung/Deutsches Zentrum für Luft- und Raumfahrt (BMBF/DLR), the Max-Planck-Gesellschaft and the Heidenhain-Stiftung.

Astrophysics and Space Science **289:** 489–498, 2004.

order of 10^{36} erg/s, corresponding to about 30% of the energy of a type II supernova (SN) integrated over the stellar main sequence life time. However, the most efficient energy and momentum input into the ISM is by SN explosions, both randomly or in concert thus generating superbubbles. Shock velocities are initially of the order of 10^3 km/s, and therefore from simple considerations of Rankine-Hugoniot conditions for strong shocks and adiabatic coefficient $\gamma = 5/3$, the post shock temperature obeys the simple relation: $T_{sh} \approx \frac{3}{16}(\mu\bar{m}/k_B)v_8^2 \approx 2.7 \times 10^7 \mu v_8^2$ K, with μ, $\bar{m}$, k_B and v_8 being the mean molecular weight, mean particle mass, Boltzmann's constant and the upstream shock velocity (in the shock rest frame) in units of 1000 km/s, respectively. Thus radiative cooling will be by thermal bremsstrahlung, recombination lines as well as atomic line transitions, detectable mainly in soft X-rays. It should be emphasized however, that the bulk of the energy is lost by adiabatic expansion of the supernova remnant (SNR) or superbubble (SB). This can be attributed to the fact that there exists a huge overpressure with respect to the ambient medium, acting like a piston, and that the radiative cooling time of low density bubbles is usually much larger than the timescale by which they merge with the ISM. As a rough estimate, we can compare the energy loss rate per unit volume with the $p\,dV$-work done by the expanding bubble: $\Xi = n^2\Lambda(T)t/(2nk_BT) \approx 1.1 \times 10^7(n/T)t_6 < 1$, as long as $T \gtrsim 10^6$ K, using $\Lambda(T) \lesssim 10^{-22}\,\mathrm{erg\,cm^3/s}$ and t_6 being the SB lifetime in units of Myr, where T and n are the plasma temperature and density, respectively. Note that during the adiabatic stage of evolution, similarity solutions may be applied, thus $n \propto t^\alpha$, with $\alpha = -4/5$ for SBs and $\alpha = -6/5$ for SNRs. For $T \sim$ const. during the energy driven phase this implies $\Xi \propto t^{1/5}$ for SBs and $\Xi \propto t^{-1/5}$ for SNRs. Thus if $\Xi(t) \ll 1$ initially, it will be so later on due to the weak time dependence of Ξ. For $n < 10^{-3}\,\mathrm{cm^{-3}}$, $T > 10^6$ K, as it is expected for superbubbles, this will definitely hold, except, maybe, for bubbles in very dense regions.

On a larger scale, where star forming regions (SFRs) are clustered, the combined pressure of winds and SN explosions can drive a galactic fountain or wind. Such a flow is again characterized by strong adiabatic and radiative cooling. Most prominent examples are starburst galaxies, from which so-called 'superwinds' emanate (e.g. Heckman et al., 1990). Common to all examples mentioned here is a *coupling* between the *thermal* and the *dynamical* evolution, which will be described and discussed in the following sections.

2. Non-Equilibrium Cooling of Interstellar Plasmas

Interstellar cooling is on a fundamental level a complicated process. The energy loss rate per unit volume, expressed as the cooling function, $\Lambda(T, Z)$, when normalized to the electron and ion densities, not only depends on temperature T and the elemental abundances Z of the fluid, but in general also on the distribution of ionization stages. The assumption of collisional ionization equilibrium (CIE) circumvents this problem. It is assumed that the rate of ionization by inelastic collions between electrons and ions is exactly balanced by the rate of recombinations. While the former is a cooling process the latter as a statistically inverse process must consequently be a heating process. Since the energy released has to be transferred to a third particle, thereby increasing the thermal energy of the plasma, such three-body interactions are extremely improbable in a tenuous, hot optically thin plasma like in a bubble or a superwind. Therefore it is commonly taken that *radiative recombination* is dominating; however, this is also a cooling process to the plasma. Thus a hot plasma, left to itself will eventually depart from equilibrium (s. Shapiro and Moore, 1976; Schmutzler and Tscharnuter, 1993).

The situation can be more dramatic, if the plasma undergoes compression like e.g. in a shock wave, or expansion like in a SB, fountain or wind. The first case gives rise to so-called *underionized* plasmas: the electrons downstream of a collisionless shock are heated up very efficiently by randomizing their kinetic energy via turbulent electromagnetic fields, and on a somewhat larger timescale also the ions. The ionization process on the other hand depends on the cross sections and thus on atomic timescales, which are much larger. Therefore the ionization will lag behind, and when observed spectroscopically, the plasma will appear cooler than it actually is (see e.g. Cox and Anderson, 1982). The second case has the opposite characteristics, and is known as an *overionized* plasma: fast adiabatic expansion, like e.g. in an expanding wind, which occurs on much shorter timescales than radiative recombination, makes the plasma appear hotter to the observer due to *delayed recombination* (see Breitschwerdt and Schmutzler, 1994). It has been shown that in these non-equilibrium ionization (NEI) cases the dynamics of the plasma and its thermal evolution are closely intertwined. The dynamics changes the density and temperature, which in turn changes the ionization structure, which determines the cooling function, which again modifies the dynamics. Thus a self-consistent description is necessary. Under CIE assumptions, the distribution of ionization stages is a sensitive function of temperature and thus the measurement of line ratios (e.g. NV/OVI) leads to a reliable temperature determination; however, in the case of NEI this is no longer valid (s. Figure 1). In essence the plasma keeps a memory of its origin, and due to fast adiabatic cooling a particular ion can be present over a wide range of temperatures. Therefore when modelling the emissivity and the spectrum of a plasma, the time-dependent evolution of the ionization stages has to be followed closely along with the evolution of the astrophysical model (s. Figure 2). Before turning to some detailed examples, some caveats for using plasma models have to be pointed out: although there exists a large data base of atomic cross sections for the various processes included, such as collisional and auto-ionization, photo-ionization, radiative and dielectronic recombination, two-photon emission, Auger effect, charge exchange reactions, just to name the most important ones, there are still quantitative discrepancies in plasma codes on the market (cf. Masai, 1997), particularly for heavy elements like iron around 1 keV (up to a factor of 2 in emissivity). Moreover, when fitting observed X-ray spectra it is strictly speaking

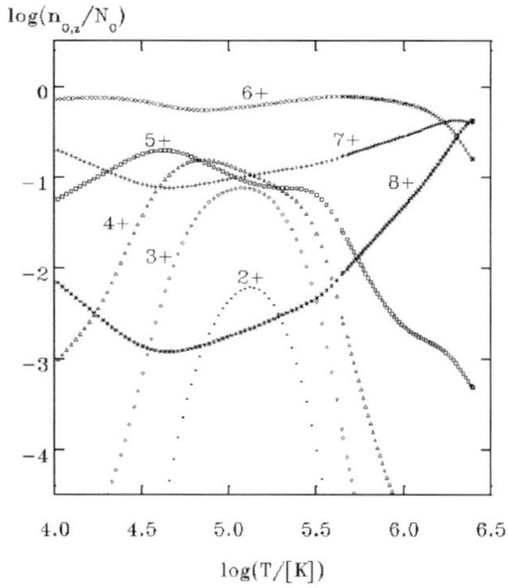

Figure 1. **Left:** Ionization fractions for oxygen as a function of temperature (taken from Böhringer and Hensler, 1989) for a plasma in collisional ionization equilibrium (CIE). **Right:** Ionization fractions for oxygen as a function of temperature for a plasma in non-equilibrium (NEI) taken from Breitschwerdt and Schmutzler (1999).

wrong to substitute a CIE fit model (e.g. RS, MEKAL etc.) by a 'non-equilibrium model' that has just one or more free parameters, like e.g. the ionization parameter $I = n_e t$, simply because it is not calculated self-consistently with the dynamical evolution. The evolution of the astrophysical model is a *necessary* although not sufficient input. It cannot even be claimed that such a 'NEI fit model is closer to the truth than the CIE model, if it cannot be physically motivated; it just fits better due to a larger number of free parameters.

3. Spectral Signatures of Galactic Outflows

In the following it will be demonstrated that NEI cooling leaves a distinct imprint on the plasma, which can already be seen by current soft X-ray observations, even at modest spectral resolutions. The reason being the temperature sensitivity of CIE spectra and the existence of diagnostic ions, such as lithium and helium like oxygen and iron L shell transitions. We have analyzed 23 ksec of ROSAT PSPC archival data in the 0.1 – 2.4 keV range of the nearby edge-on galaxy NGC 253 (Breitschwerdt and Freyberg, 2003) retaining 15.3 ksec of good background subtracted data (s. Figure 3) and obtained only a statistically acceptable fit by using *two* CIE plasma temperatures (applying the plasma code of Raymond and Smith, 1977 (RS)); alternatively decreasing the elemental abundances to about $0.05 Z_{\odot}$ was equally satisfactory. However, it seems rather odd, that a SN heated and thus chemically enriched superwind should have highly subsolar abundances (although this has been claimed in the literature). On the other hand, we have simulated the

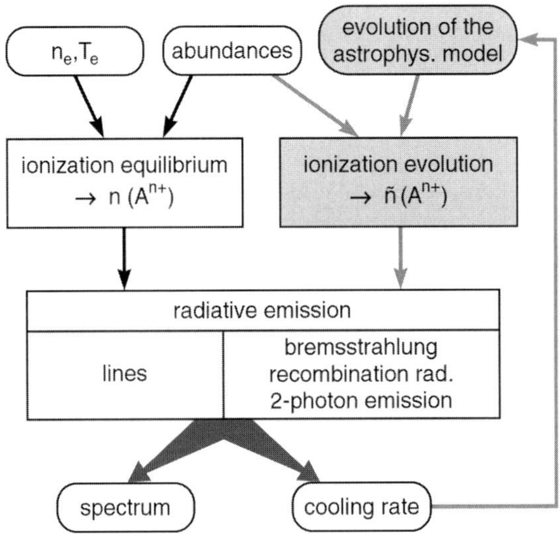

Figure 2. Flow chart of plasma emission model (s. Böhringer, 1997). White boxes refer to equilibrium (CIE), grey boxes to additions necessary for non-equilibrium (NEI) models. A^{n+} denotes atomic species A ionized n-times.

outflow by a galactic wind model (for details s. Breitschwerdt and Schmutzler, 1999) calculating self-consistently the full NEI plasma structure as discussed before. The resulting spectrum, integrated over the field of view (FOV) of the PSPC, binned into PSPC channels and folded through the instrumental response, was both quantitatively and qualitatively similar to the observed one (s. Figure 3 right). We have thus treated the *observed* and the *simulated* spectra in exactly the same manner. Moreover, spectral fitting of the *model data* was again satisfactory for 2 RS plasmas or abundances of $Z \approx 0.05 Z_\odot$. Hence it is fair to conclude that highly subsolar abundances are an artefact of the fit procedure. Clearly decreasing the abundances, washes out the prominent and characteristic lines of a spectrum and therefore gives more flexibility to accommodate all the temperature sensitive lines in a CIE spectrum! The boundary conditions of the outflow model reflect the SFR in the underlying galactic disk, but are specified at a vertical distance of $z = 1$ kpc at the base of the halo. The flow geometry is represented by a flux tube, following the transition from planar close to the disk to spherically divergent flow at large distances. The gravitational potential has been described by (oblate) spheroids, including a bulge, a disk and a dark matter halo component (to account for the flatness of the rotation curve), scaled to observations. For the successful model, hot plasma is injected at the base of the halo at a temperature $T = 2.5 \times 10^6$ K and a density $n = 7 \times 10^{-3}\,\mathrm{cm}^{-3}$. It is noteworthy that the models are fairly well constrained, even by the PSPC spectral resolution. The intrinsic outflow spectra, corresponding to disk parallel cuts through the halo at a given distance from the galactic plane, are radically different from CIE spectra (s. Figure 4 left), characterized by strong saw-

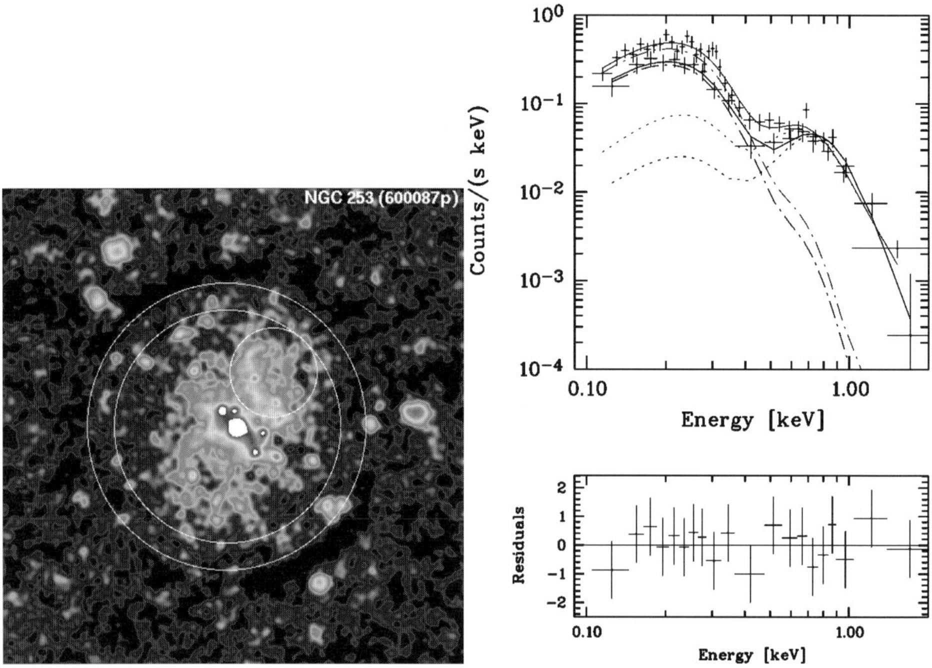

Figure 3. **Left:** ROSAT PSPC observation of the edge-on starburst galaxy NGC 253 (north is up) in soft X-rays (0.1–2.4 keV) smoothed with a Gaussian filter; colour coding is from blue or dark (low) to red or bright (high intensity). The small circle shows the halo region that has been analyzed, while the outer ring represents the region that has been used for background subtraction. **Right:** Comparison between data (red points with error bars, lower curve) and self-consistent model M4 (green points with error bars, upper curve). The solid, dashed, dash-dotted and dotted curves are plasma fit models (2 Raymond and Smith (1977) CIE models) with solar abundances, respectively). There is also good *quantitative* agreement between data and model, when taking into account missing flux in the observations due to point source masking.

tooth type recombination lines due to delayed recombination. Note that both the elements and the ionization stages of the most intense lines change as a function of time (or distance from the disk), characterized by their recombination time scale. Another marked difference between CIE and NEI spectra is the existence of high energy X-ray photons in the overionized case (at around 1 keV in Figure 4 left), although the temperature is already down to 2.5×10^5 K. This occurs, because the plasma was once hot enough to have generated these photons in the past: the plasma retains a memory of its origin. The integrated spectrum, which is folded through the detector response matrix for comparison with the observed data (s. Figure 3 right), is the sum of such intrinsic spectra over the FOV.

Another striking example of an overionized outflow is the LINER edge-on starburst galaxy NGC 3079, which we have observed with XMM-Newton for 25 ksec (Breitschwerdt et al., 2003). The image taken with the EPIC pn camera shows an extended soft X-ray halo (especially impressive in the 0.2 – 1 keV band) of

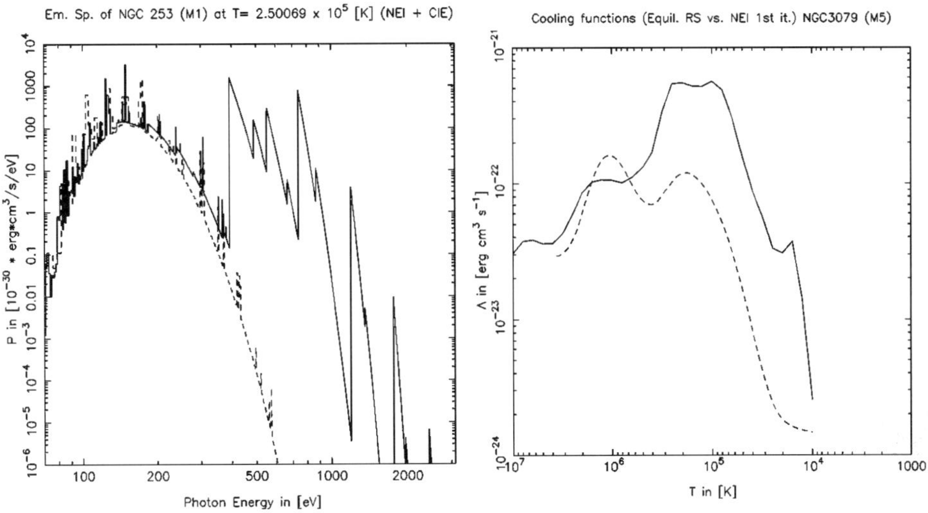

Figure 4. **Left:** Comparison of a CIE (dashed line) and a NEI (solid line) emission spectrum resulting from a simulated outflow of the starburst galaxy NGC 253 at a distance (and time) when the temperature has cooled down 2.5×10^5 K. The characteristic saw-tooth structure is due to recombination lines, here of SiXII and FeXVII (two main peaks, with energy increasing). **Right:** Comparison of CIE (solid curve) cooling function Λ and the one resulting from a self-consistent evolution (dashed curve) of ionization stages for the successful spectral fit model of the halo emission of NGC 3079.

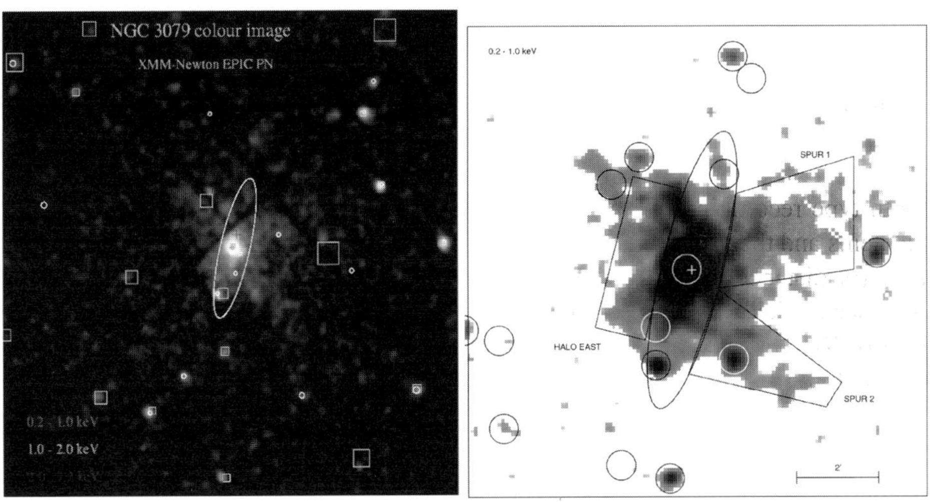

Figure 5. **Left:** RGB image of the extended X-ray halo of the nearby edge-on starburst galaxy NGC 3079, as observed by the XMM-Newton EPIC pn camera. **Right:** Smoothed image of the diffuse emission of NGC 3079 in the 0.2 - 1.0 keV energy band (Breitschwerdt et al. 2003). Point sources are shown by circles. The D_{25} ellipse is indicated to show the optical extension of the disk.

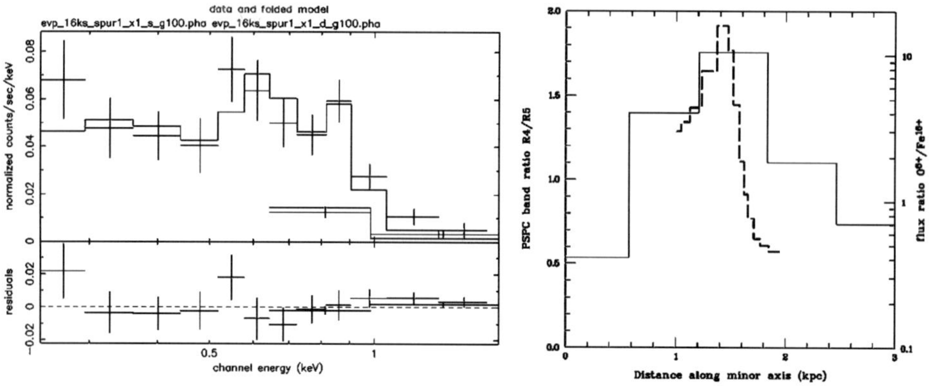

Figure 6. **Left:** Fit (solid line: $\chi^2 = 1.2$) of the XMM-Newton Epic pn observed spectrum (crosses correspond to error bars) of the emission of NGC 3079 (0.2 – 2 keV) by a self-consistent NEI outflow model (solid line; Breitschwerdt et al. 2003); the lower panel shows the fit residuals. **Right:** PSPC R4 (520 – 690 eV) and R5 (700 – 900 eV) band ratio in a disk parallel stripe ($\Delta z = 2$ kpc) along the minor axis of NGC 253 (solid histogram) compared to OVII /FeXVII flux ratio averaged over 625 – 675 eV and 805 – 865 eV, respectively (dashed histogram), derived from the best fit halo model, according to self-consistent NEI calculations (cf. Figure 3).

roughly the same size as the D_{25}-ellipse, representing the optical diameter of the disk (s. Figure 5 right). It can be clearly seen that apart from the nuclear outflow, there also exists an extended component along the major and minor axes. Its spiky morphology is probably related to the SFR in the underlying disk. We have analyzed and fitted the EPIC pn spectrum of the NE spur (Breitschwerdt et al., 2003). The data are shown as crosses (error bars) and clearly exhibit two humps, which we identify as due to oxygen (at around 0.5 keV) and iron L shell line (0.8 – 0.9 keV) complexes. This is corroborated by running again an outflow simulation, generating a self-consistent NEI spectrum and subjecting it to the same procedure as before (i.e. folding through the EPIC pn instrumental response). The best model fit (with a reduced $\chi^2 \approx 1.2$) requires a temperature of $T = 3.56 \times 10^6$ K and a density of $n = 5 \times 10^{-3}$ cm^{-3} at the base of the halo. The outflow velocity there is about 220 km/s, reaching about 450 km/s at $z = 50$ kpc. The higher initial temperature as compared to NGC 253 reflects the more violent outflow in this case and is necessary for producing the iron bump. We find again that no single RS model can fit the spectrum. The discrepancy between the cooling rates in CIE and NEI can be quite substantial at certain temperatures and thus have a pronounced effect on the outflow dynamics, as can be seen by comparing the respective cooling curves in Figure 4 right: delayed recombination can indeed reduce the cooling rate.

To go one step further, if photon statistics of the observational data permits, one can analyze the spectral behaviour of the plasma in disk parallel stripes of widths of $\Delta z \sim 2$ kpc using a ‘sliding window’ technique (s. Figure 6 right). The idea is to search for ‘diagnostic ions’ of starburst driven outflows. The double hump in Figure 6 left suggests OVII and FeXVII as spectral signatures that should vary in strength as the outflow cools down adiabatically and radiatively. We have therefore compared the PSPC R4 (520 – 690 eV) to R5 (700 – 900 eV) band ratios along the minor axis to an averaged OVII (625 – 675) eV to FeXVII (805 – 865 eV) ratio of the model calculations. Although the latter ones have arbitrarily high 'spectral resolution', we have averaged over an energy range free of other spectral features; hence the energy range is smaller than in the corresponding ROSAT PSPC bands, and the absolute fluxes do not match. However, it can be seen that the flux ratio of these characteristic ions have pronounced peak at about 1.5 kpc from the disk both for the observed and the model spectra. This gives a good hint that the time-dependent evolution of these ion species in the outflow should be basically correct. More stringent conclusions have to be deferred to an analysis of higher resolution spectra.

4. Conclusions

The discrepancy between equilibrium plasma models and observational data in the EUV and soft X-rays is commonly felt as a nuisance. There exists e.g. a plethora of high spectral resolution data for the Local Bubble. It is indeed embarrassing that presently neither SNR nor SB models, with CIE or simple NEI cooling can reproduce all these data satisfactorily. However, I would like to stress a positive aspect of this problem: the spectral fingerprints obtained contain far more information than our simple models can handle. Although being intrinsically more complicated, the advantage of NEI models is to connect the spectrum of a plasma at any given time to its history and therefore to help us to learn something about the plasma's excitation conditions and hence its origin. In case of the Local Bubble it seems that more sophisticated 3D simulations including NEI cooling, as undertaken currently (Avillez and Breitschwerdt, 2003), should lead to an improvement.

Meanwhile it is worth focusing also on somewhat simpler astrophysical problems such as galactic outflows, which are well observable in edge-on galaxies. I have shown that self-consistent modelling can explain a number of problems: (i) the extension of soft X-ray halos, (ii) their spectra (although better photon statistics and higher spectral resolution are desirable), (iii) highly subsolar abundances in starburst galaxies are artifacts of the fit procedure, (iv) depending on the origin of the plasma there exist diagnostic ions that contribute to the emission in a characteristic fashion during the evolution of the outflow (along the minor axis in edge-on galaxies). Further studies will enable us to eventually connect the spectral behaviour to the star formation activity.

References

Avillez, M.A. and Breitschwerdt, D.: 2003, *A&A*, in preparation.

Böhringer, H.: 1997, in: D. Breitschwerdt, M.J. Freyberg and J. Trümper (eds.), *Proc. of IAU Colloq. No. 166: The Local Bubble and Beyond*, Springer LNP 506, p. 341.
Böhringer, H. and Hensler, G.: 1989, *A&A* **215**, 147.
Breitschwerdt, D. and Freyberg, M.J.: 2003, *A&A*, in preparation.
Breitschwerdt, D., Pietsch, W., Vogler, A. et al.: 2003, in preparation.
Breitschwerdt, D. and Schmutzler, T.: 1994, *Nature* **371**, 774.
Breitschwerdt, D. and Schmutzler, T.: 1999, *A&A* **347**, 650.
Cox, D.P. and Anderson, P.R.: 1982, *ApJ* **253**, 268.
Heckman, T.M., Armus, L. and Miley, G.K.: 1990, *ApJS* **74**, 833.
Masai, K.: 1997, *A&A* **324**, 410.
Raymond, J.C. and Smith, B.W.: 1977, *ApJS* **35**, 419.
Shapiro, P.R. and Moore, R.T.: 1976, *ApJ* **207**, 460.
Schmutzler, T. and Tscharnuter, W.M.: 1993, *A&A* **273**, 318.